Arthropod Cell Culture Systems

Arthropod Cell Culture Systems

Edited by

Karl Maramorosch, Ph.D.
Robert L. Starkey Professor of Microbiology
Professor of Entomology
Rutgers — The State University of New Jersey
New Brunswick, New Jersey

Arthur H. McIntosh, Sc.D.
Research Scientist
Biological Control of Insects
Research Laboratory
United States Department of Agriculture
Agricultural Research Station
Columbia, Missouri

CRC Press
Taylor & Francis Group
Boca Raton London New York

CRC Press is an imprint of the
Taylor & Francis Group, an **informa** business

First published 1994 by CRC Press
Taylor & Francis Group
6000 Broken Sound Parkway NW, Suite 300
Boca Raton, FL 33487-2742

Reissued 2018 by CRC Press

Library of Congress Cataloging-in-Publication Data

Arthropod cell culture systems / edited by Karl Maramorosch, Arthur H. McIntosh.
p. cm.
Includes bibliographical references (p.) and index.
ISBN 0-8493-7642-4
1. Arthropoda—Cytology. 2. Cell culture. I. Maramorosch, Karl. II. McIntosh, Arthur H.
QL434.715.A78 1994
595'.2'0724—dc20 93-38285

A Library of Congress record exists under LC control number: 93038285

ISBN 13: 978-1-315-89080-7 (hbk)
ISBN 13: 978-1-351-06990-8 (ebk)

Visit the Taylor & Francis Web site at http://www.taylorandfrancis.com and the
CRC Press Web site at http://www.crcpress.com

PREFACE

The use of arthropod cell culture systems has increased rapidly over the past decade, assuming a prominent role in basic biological research and in biotechnology. Cell lines derived from invertebrate hosts are currently employed in studies of various biological processes in biochemistry, genetics, endocrinology, molecular biology, physiology, virology, medicine, and agriculture. Cultured arthropod cells can support the growth and replication of viruses, bacteria, and intracellular parasites, providing an important system for studying pathogens affecting plants, animals, and humans. Insect cell culture has provided a major impetus toward the commercial application of these systems in biotechnology and insect pest control. Numerous important contributions have been made employing arthropod cell culture systems but the information has been scattered in different journals. Therefore the editors felt that there remained an obvious need for a comprehensive book that would bring together the current research advances and describe the methods and diverse applications of arthropod cell culture.

This book brings together for the first time in two decades the large body of information and the significant achievements accomplished in laboratories throughout the world. The information has been presented by foremost authorities and pioneers of insect and tick cell culture, who contributed basic and applied concepts. Their international experience has been combined so that the latest developments in this fascinating and rapidly expanding field are presented in a comprehensive manner. The large body of information brings into sharp focus anticipated new directions in this field.

Specialists and pioneers from the United States, Japan, Switzerland, Slovakia and China have been invited to write chapters for this treatise based on their outstanding and authoritative knowledge of the subjects and their original contributions. This has been done with the intention to provide readers with a well balanced cross-section of current developments and to focus attention on such topics as preparation of cell culture media, cultivation of mosquito, lepidopteran, grasshopper and tick cells, the application of such cells to mammalian and plant virus research, and diverse applications in medicine, biology and agriculture. Significant strides have been made in Japan and a separate chapter has been devoted to work of Japanese cell culture pioneers.

The growing interest in insect and tick cell cultivation is closely linked with the worldwide growth of biotechnology. The editors feel that the combined efforts of eminent contributors to present new information will benefit all who are interested in arthropod cell systems. All chapters are well documented by tables, photographs, and up-to-date bibliographies. This book will be of interest to microbiologists, molecular biologists, parasitologists, virologists, entomologists, plant pathologists, geneticists and medical and agricultural researchers. It is hoped that this book will set the benchmark for future research and serve for years to come as a standard source of information for advanced research workers and students.

The editors express their sincere gratitude to the contributors for their effort and care with which they have prepared their chapters and to CRC Press for their part in indexing, proof-reading, and other aspects of production of this volume.

Karl Maramorosch
Arthur H. McIntosh

THE EDITORS

Karl Maramorosch, Ph.D., is Robert L. Starkey Professor of Microbiology and Professor of Entomology at Rutgers — The State University of New Jersey, New Brunswick, New Jersey.

Dr. Maramorosch, a native of Vienna, Austria, graduated from the Agricultural University of Warsaw, summa cum laude, and obtained his Ph.D. in 1949 from Columbia University. His scientific career began at Rockefeller University where he worked on plant viruses and insect vectors for 12 years. Later, he became Program Director of Virology and Insect Physiology at the Boyce Thompson Institute, and since 1974, he has been Professor II (Distinguished Professor) at Rutgers University, where he became the Robert L. Starkey Professor in 1983.

Dr. Maramorosch is a Fellow and former Recording Secretary and Vice-President of the New York Academy of Sciences, Fellow of the American Association for Advancement of Science, the Entomological Society of America, and the American Phytopathological Society, President of the International Association for Medicinal Forest Plants, Honorary Fellow of the Indian Virological Society, and a member of the Tissue Culture Association, Harvey Society, Society for Invertebrate Pathology, American Society for Microbiology, Electron Microscopy Society, and several other professional organizations. Dr. Maramorosch has been the recipient of grants from the National Science Foundation, National Institutes of Health, U.S. Department of Agriculture, Rockefeller Foundation, and private industry. He won the 1980 Wolf Prize in Agriculture and many other awards, including the Ciba Geigy Award in Entomology, Jurzykowski Award in Biology, AIBS Award of Distinction, Cressy Morisson Prize of the N.Y. Academy of Sciences, Waksman Award, and AAAS-Campbell Award. He was nominated by the Entomological Society of America for the National Medal of Science in 1983 and was the Founders Lecturer of the Society for Invertebrate Pathology in 1990. He is an elected member of the Leopoldina, oldest European Academy of Sciences, and a Fellow of the Indian National Academy of Sciences.

Dr. Maramorosch served twice as visiting Fulbright Professor in Yugoslavia and as visiting professor in the Peoples Republic of China, U.S.S.R., The Netherlands, Germany, Japan, India, and Poland. He was consultant to the Food and Agriculture Organization in the Philippines, India, Nigeria, Thailand, Sri Lanka, and Mauritius, and has held short-term assignments from the Rockefeller Foundation, the Agency for International Development, U.S. Department of Agriculture (FERRO), and Ford Foundation in Mexico, India, Kenya, and the Philippines. Dr. Maramorosch has edited more than 50 volumes on viruses, vectors, plant diseases, and tissue culture, and is the author or co-author of more than 600 research papers. His major research interests include comparative virology, invertebrate cell culture, parasitology, diseases caused by spirochetes, viroids, and spiroplasmas, biotechnology, and international scientific cooperation.

Arthur H. McIntosh, Sc.D., is Laboratory Director and Research Leader of the Biological Control of Insects Research Laboratory with the United States Department of Agriculture, Agricultural Research Services, in Columbia, Missouri.

Dr. McIntosh graduate in 1959 from McMaster University, Hamilton, Ontario, Canada with a B.A. He attended the University of Guelph, in Guelph, Ontario, Canada from 1960–1962 and graduated from the Department of Microbiology with an M.S. Dr. McIntosh worked on his doctorate from 1965–1969 in the Department of Microbiology, Harvard University School of Public Health and graduated in 1969. He then completed postdoctoral studies at Stanford Research Institute from 1969–1971 and accepted a position at Boyce Thompson Institute with Dr. Karl Maramorosch in Insect Virology and Cell Culture from 1971–1974. He was an Assistant Research Professor at Waksman Institute of Microbiology, Rutgers, Piscataway from 1974–1979 before joining the USDA-ARS. He holds a joint appointment in the Department of Entomology, University of Missouri as Adjunct Associate Professor and participates in the teaching of a course in Biological Control. He spent a six-month sabbatical (1992–1993) at the Commonwealth Scientific and Industrial Research Organization, in Canberra, Australia, where he initiated cell lines from *Helicoverpa-Heliothis* species for the replication of the pesticidal baculoviruses.

Dr. McIntosh conducts research in the area of Biological Control utilizing insect cell culture to study the biological characteristics of baculoviruses. His interest also includes the study of the specificity of insect viruses. He has published over 60 research papers, reviews and book chapters.

Dr. McIntosh is a member of the Tissue Culture Association, Society for Invertebrate Pathology, Entomological Society of America, Japanese Society of Applied Entomology and Zoology and Sigma Xi.

CONTRIBUTORS

Chunsheng Chen
Graduate Research Assistant
Department of Entomology
University of Minnesota
St. Paul, Minnesota

Ikuo Kimura, D.Agr.
Researcher
Department of Botany
Faculty of Agriculture
Hokkaido University
Sapporo, Japan

Hans Koblet, M.D.
Professor
Institute of Medical Microbiology
Present address:
Institut für Veterinär-Virologie
Berne, Switzerland

Yukiaki Kuroda, D.Sc.
Professor
Department of Gene and Cell Biology
Azabu University
Kanagawa, Japan

Timothy J. Kurtti, Ph.D.
Associate Professor
Department of Entomology
University of Minnesota
St. Paul, Minnesota

Jun Mitsuhashi, D.Agr.
Professor
Department of Applied Entomology
Tokyo University of
Agriculture and Technology
Tokyo, Japan

Ulrike G. Munderloh, D.Med.Vet.
Research Associate
Department of Entomology
University of Minnesota
St. Paul, Minnesota

Toshihiro Omura, Ph.D.
Senior Researcher
National Agricultural Research Center
Ibaraki, Japan

Josef Řeháček, Ph.D., D.Sc.
Senior Researcher
Department of Rickettsiae
Institute of Virology
Slovak Academy of Sciences
Bratislava, Slovakia

James L. Vaughn, Ph.D.
Research Leader
Insect Biocontrol Laboratory
United States Department of Agriculture
Beltsville, Maryland

Yan Liu
Junior Scientist
Department of Entomology
University of Minnesota
St. Paul, Minnesota

TABLE OF CONTENTS

Chapter 1

INSECT CELL CULTURE MEDIA

Jun Mitsuhashi

TABLE OF CONTENTS

0-8493-7642-4/94/$0.00+$.50

Jun Mitsuhashi, D.Agr., is Professor of Applied Entomology at Tokyo University of Agriculture and Technology. He received his Bachelor of Agriculture in applied entomology in 1955 and Doctor of Agriculture in 1965 at The University of Tokyo. 1955–1984, National Institute of Agricultural Sciences; 1984–1988. Forest Research Institute; 1988–present, Tokyo University of Agriculture and Technology; 1962–1964, Boyce Tompson Institute for Plant Research; 1968–1969, Division of Entomology, CSIRO; 1993–1994. President of Japanese Society of Applied Entomology and Zoology. Insect cell culture; insect endocrinology; artificial rearing of insects, insect endosymbiotes; insect pathology. Address: Faculty of Agriculture, Tokyo University of Agriculture and Technology, Fuchu, Tokyo 183, Japan.

I. INTRODUCTION

Since Grace[1] established the first insect cell line, very many cell lines have been established during the last 3 decades from various insects. In order to maintain these cell lines, various culture media have been used, because there is no single culture medium that could maintain the growth of every insect cell. Some cell lines can be maintained in relatively nutrient-poor media but others cannot. Formulation of widely usable media is one of the tasks in developing new culture media.

II. HISTORY OF MEDIUM DEVELOPMENT

When the first trial to culture insect tissues *in vitro* was made, insect hemolymph itself was used as the culture medium by Goldschmidt.[2] At present, we know that hemolymph itself is not suitable for culture media, because it soon turns dark due to the formation of melanin by the action of phenol oxidase, and the resulting melanin is toxic to the cultured tissues or cells. Goldschmidt did not mention how he could overcome this sort of trouble. In the early stage of insect tissue cultures, most researchers used insect hemolymph or simple saline solutions as culture media, and the results were always short survival of tissues or cells. Trager[3] formulated a culture medium by combining inorganic salts, egg albumin digest, maltose, and insect hemolymph (Table 1). This medium was the first medium formulated specifically for insect cell growth and it supported growth for a considerable time. After Trager's work, several investigators examined the ability of vertebrate cell culture media to support insect cell growth although none of them gave promising results. Wyatt[4] formulated a synthetic culture medium by combining 6 inorganic salts, 21 amino acids, 4 other organic acids, and 3 sugars (Table 2). Her formulation was based on the chemical analyses of hemolymph from several insect species.[5] Although the medium could not support cell growth by itself, it supported the cell growth well when it was fortified by the addition of heat-treated insect hemolymph. This medium actually gave a clue to the formulation of insect cell culture media, and even at present we are using various modifications of her medium. Wyatt's medium was improved by Grace,[1] and with this medium, he established the first insect continuous cell lines. His modification was the addition of ten water-soluble vitamins, change of balance of inorganic salts, and increment of sucrose (Table 3). Grace's medium is well known as the medium which can support growth of various insect tissues, when supplemented with insect hemolymph or vertebrate sera, especially fetal bovine serum (FBS), which can be purchased commercially at present. Another type of culture medium is a chemically undefined medium whose major components are natural substances. Mitsuhashi and Maramorosch[6] formulated a simple medium presently called the MM medium. This medium contained lactalbumin hydrolysate, TC-yeastolate, and FBS as natural substances (Table 4). Originally this medium was developed for culturing leafhopper cells. It turned out later that this medium was well suited for mosquito cell cultures and also for culturing cells from various other insects, belonging to many orders. Furthermore, this medium is now known to support growth of many cell lines without sera supplements.[7]

In recent years, serum-free culture media were developed by many investigators. Now many insect cell lines can be cultured in serum-free media, but usually their growth rates are inferior in serum-free media, compared to serum-containing media. Some media sold commercially are said to maintain cell growth at the same rate as serum-containing media. However, these media do not necessarily support cell growth of every cell line. Most serum-free media contain more or less chemically undefined substances. It is, therefore, required to formulate completely chemically defined media. Once such a chemically defined medium was formulated,[8] however, this medium was rather difficult to prepare and it could support only limited cell line species (Table 5). Some media sold by industries are said to be

TABLE 1
Trager's Medium

Components	Amount[a]	Components	Amount[a]
NaCl	7.6	$CaCl_2$	11.1
$NaH_2PO_4 \cdot H_2O$	20.7	Maltose	2162.3
K_2HPO_4	20.4	Egg albumin digest	0.00137
$MgCl_2 \cdot 6H_2O$	20.3		

[a] Mol/l. pH 6.7. Osmolarity: –0.290°C (as freezing point depression). Used by adding 10% *Bombyx mori* hemolymph.

From Trager, W., *J. Exp. Med.*, 61, 501–513, 1935. With permission.

TABLE 2
Wyatt's Medium

Components	mg/100 ml	Components	mg/100 ml
NaH_2PO_4	110.0	Cysteine	8.0
KCl	298.0	L-Cystine	2.5
$MgCl_2 \cdot 6H_2O$	304.0	L-Glutamic acid	60.0
$MgSO_4 \cdot 7H_2O$	370.0	L-Glutamine	60.0
$CaCl_2$	81.0	Glycine	65.0
Glucose	70.0	L-Histidine	250.0
Fructose	40.0	DL-Isoleucine	10.0
Sucrose	40.0	DL-Leucine	15.0
Malic acid	67.0	DL-Lysine	125.0
α-Ketoglutaric acid	37.0	DL-Methionine	10.0
Succinic acid	6.0	L-Phenylalanine	15.0
Fumaric acid	5.5	L-Proline	35.0
DL-α-Alanine	45.0	DL-Serine	110.0
β-Alanine	20.0	DL-Threonine	35.0
L-Arginine	70.0	L-Tryptophan	10.0
L-Asparagine	35.0	L-Tyrosine	5.0
L-Aspartic acid	35.0	DL-Valine	20.0

Note: pH: 6.35. Osmolarity: –0.53°C (as freezing point depression). Used by adding heat-treated *Bombyx mori* hemolymph.

From Wyatt, S.S., *J. Gen. Physiol.*, 39, 841–852, 1956. With permission.

almost chemically defined, but they cannot be used for biochemical studies of cultured cells because the formulation has not been disclosed.

Recent development of biotechnology using insect cell cultures requires large-scale cultures of insect cells. Developments of low cost media are desired for large-scale cultures of insect cells especially for the purpose of production of viral insecticides. By the development of serum-free media, the cost of media has decreased considerably and now we can purchase fairly low-cost media, but still further cost reduction is needed.

III. MEDIUM COMPONENTS

In general, the main ingredients of culture media consist of inorganic salts, proteins, sugars as an energy source, and vitamins.

TABLE 3
Grace's Medium

Components	mg/100 ml	Components	mg/100 ml
$NaH_2PO_4 \cdot 2H_2O$	114.0	L-Threonine	17.5
$NaHCO_3$	35.0	L-Tryptophan	10.0
KCl	224.0	L-Tyrosine	5.0
$MgCl_2 \cdot 6H_2O$	228.0	L-Valine	10.0
$MgSO_4 \cdot 7H_2O$	278.0	Sucrose	2268.0
$CaCl_2$	100.0	Fructose	40.0
L-α-Alanine	22.5	Glucose	70.0
β-Alanine	20.0	Malic acid	67.0
L-Arginine·HCl	70.0	α-Ketoglutaric acid	37.0
L-Asparagine	35.0	Succinic acid	6.0
L-Aspartic acid	35.0	Fumaric acid	5.5
L-Cystine·HCl	70.0	Thiamine·HCl	0.002
L-Glutamic acid	60.0	Riboflavin	0.002
L-Glutamine	60.0	Calcium pantothenate	0.002
Glycine	65.0	Pyridoxine·HCl	0.002
L-Histidine	250.0	*p*-Aminobenzoic acid	0.002
L-Isoleucine	5.0	Folic acid	0.002
L-Leucine	7.5	Niacin	0.002
L-Lysine·HCl	62.5	Isoinositol	0.002
L-Methionine	5.0	Biotin	0.001
L-Phenylalanine	15.0	Choline chloride	0.02
L-Proline	35.0	Penicillin G, sodium salt	3.0
DL-Serine	110.0	Streptomycin sulfate	10.0

Note: pH: 6.5. Used by adding 3% heat-treated hemolymph of *Antheraea pernyi* or 1% of the hemolymph with 1% bovine plasma albumin fraction V.

From Grace, T.D.C., *Nature,* 195, 788–789, 1962. With permission.

TABLE 4
Mitsuhashi and Maramorosch's MM Medium

Components	mg/100 ml	Components	mg/100 ml
NaCl	700.0	D-Glucose	400.0
$NaH_2PO_4 \cdot H_2O$	20.0	Lactalbumin hydrolysate	650.0
$NaHCO_3$	12.0	TC-Yeastolate	500.0
KCl	20.0	Fetal bovine serum	0–20 ml
$MgCl_2 \cdot 6H_2O$	10.0		
$CaCl_2 \cdot 2H_2O$	20.0		

Note: pH: 6.5. Osmolarity: 413 mOsm/kg.

From Mitsuhashi, J. and Maramorosch, K., *Contrib. Boyce Thompson Inst.,* 22, 435–460, 1964. With permission.

Insect cells are said to be insensitive to ion balance and change of pH. Most media contain potassium chloride, sodium phosphate, sodium bicarbonate, magnesium chloride, and calcium chloride. Some media also contain sodium chloride or magnesium sulfate. Sodium chloride is not important for insect cells probably because insect hemolymph contains only a little sodium chloride. Since most insect cells are tolerant to a wide range of pH, incorporation of

TABLE 5
Wilkie et al.'s CDM (Chemically Defined Medium)

Components	mg/100 ml	Components	mg/100 ml
$NaH_2PO_4 \cdot 2H_2O$	114.0	Arachidonic acid	0.002
$NaHCO_3$	35.0	Putrescine	0.1
KCl	287.0	Spermidine	0.1
$MgCl_2 \cdot 6H_2O$	228.0	Spermine·4HCl	0.1
$MgSO_4 \cdot 7H_2O$	278.0	Carnitine	0.1
$CaCl_2 \cdot 2H_2O$	99.7	α-Amino-*n*-butyric acid	0.1
L-α-Alanine	22.5	*o*-Phosphorylethanolamine	0.2
L-Arginine	55.0	Taurine	0.1
L-Asparagine	35.0	Riboflavin	0.02
L-Asparate (K^+ salt)	45.0	*p*-Aminobenzoic acid	0.2
L-Glutamate (K^+ salt)·H_2O	82.9	Folic acid	0.1
L-Glutamine	60.0	D-Biotin	0.005
Glycine	65.0	Calcium D-pantothenate	0.12
L-Histidine·HCl·H_2O	338.0	Isoinositol	0.2
L-Isoleucine	5.0	Ascorbic acid	0.02
L-Leucine	7.5	Cyanocobalamin	0.1
L-Lysine·HCl	62.5	Nicotinamide	0.12
L-Methionine	5.0	Thiamine·HCl	0.2
L-Phenylalanine	15.0	Pyridoxine·HCl	0.1
L-Proline	35.0	Choline chloride	2.0
L-Serine	55.0	Hypoxanthine	1.0
L-Threonine	17.5	$FeSO_4(NH_4)_2SO_4 \cdot 6H_2O$	0.5
L-Tryptophan	10.0	$ZnSO_4 \cdot 7H_2O$	0.044
L-Tyrosine	7.0	$CuSO_4 \cdot 5H_2O$	0.039
L-Valine	10.0	$MnCl_2 \cdot 4H_2O$	0.035
L-Cystine	7.5	Trilinolein	0.01
α-D-Glucose	400.0	Trilinolenin	0.01
Methylcellulose (15 cps)	200.0	Phosphatidylcholine	0.02
Stearic acid	0.01	α-Tocopherol acetate	0.001
Myristic acid	0.01	Cholesterol	0.1
Oleic acid	0.01	β-Sitosterol	0.1
Linoleic acid	0.01	Stigmasterol	0.1
Linolenic acid	0.01	Tween 80	2.0
Palmitic acid	0.01	Ethanol	0.2 ml
Palmitoleic acid	0.01		

Note: pH: 6.3. Osmolarity: 330 mOsm/kg. Used as a serum-free medium.

From Wilkie, G.E.I., Stockdale, H., and Pirt, S.V., *Dev. Biol. Standard.*, 46, 29–37, 1980. With permission.

buffer to inorganic salt composition is not common, although some media contain some buffers, such as tris buffer,[9] glycine buffer,[10] and so on. There are two types of inorganic salt compositions, one that contains about 0.8% sodium chloride and is similar to vertebrate Ringer solution, the other that does not contain sodium chloride and instead contains potassium chloride or magnesium chloride as main components. Many cell lines can be cultured in media based on either type of inorganic salt composition.

As a protein source, usually combinations of free amino acids are used. The amino acid mixture consists of 20 amino acids, which are constituents of protein, and β-alanine. In vertebrate cell culture, β-alanine has never been a component of culture media. In insect cell cultures, it has been incorporated since Wyatt's medium,[3] because insect hemolymph reportedly contains considerable amounts of β-alanine. It turned out later, however, that β-alanine was not necessary or was even detrimental for the growth of insect cells.[11] Based on this finding, some recently developed media omitted β-alanine from the amino acid mixture. Among 20 amino acids, arginine, cysteine, glutamine, histidine, isoleucine, leucine, lysine, methionine, phenylalanine, proline, serine, threonine, tryptophan, tyrosine, and valine are essential for many insect cell lines. However, nutritional requirements are different for different cell lines. For example, glutamine is essential for many insect cell lines, but some cell lines can grow without glutamine and even seem to synthesize glutamine from glutamic acid and ammonia.[12]

As for sugars, glucose is sufficient to provide the necessary energy source. Fructose, trehalose, and maltose can be utilized by insect cells. Generally sucrose is not considered to be utilizable, but some cell lines definitely can utilize it.[13,14] Moreover, many insect cell lines have α-glucosidase activity which digests sucrose.[15,16]

For vitamins, usually ten vitamins of the B group are used. They are thiamine, riboflavin, calcium pantothenate, *p*-aminobenzoic acid, biotin, inositol, pyridoxine, nicotinic acid, folic acid, and choline chloride. Some media contain vitamins A, B_{12}, C, and E. However, their roles are not clear.

Conventional media are often constructed with chemically undefined natural substances. Instead of an amino acid mixture, lactalbumin hydrolysate has been used frequently. Casein hydrolysate, egg albumin digest, or bactopeptone are also used for this purpose. As vitamin sources, yeast extracts, especially TC-yeastolate (Difco, Detroit, MI), are common. Sea water may be used in place of the inorganic salt mixture.[17]

In addition to the above basal part of culture media, the media are usually fortified by adding some cell growth-promoting substances. As such a substance, insect hemolymph has been used since the early stages of insect cell cultures. Insect hemolymph is, however, difficult to obtain in large quantities. Therefore, FBS has been widely used since it proved to promote insect cell growth.[6] After development of serum-free culture media, lipid mixture, yeast extract, cod liver oil, egg yolk, and so on have been used instead of sera.

The compositions of some representative culture media are shown in Tables 6 to 10.

IV. PREPARATION OF MEDIA

Simple media can be prepared directly according to the formulae. An example of such a medium is the MM medium (Table 4) with or without serum. Sera supplied from industries are sterile. Therefore, sera can be added under sterile conditions after media are sterilized. Even a simple medium, if prepared frequently, can easily be prepared from concentrated stock solutions of major parts of the medium. For instance, the salts in the MM medium can be stored as two separate, ten times concentrated, stock solutions. The one is the solution of $NaHCO_3$ and the other is the mixture of the remaining five inorganic salts (Table 4). When the complete medium is prepared, equal parts of both salt stock solutions are added to a 5-times volume of distilled water, and then lactalbumin hydrolysate, TC-yeastolate, and glucose are dissolved in it.

In order to prepare a complex culture medium such as Grace's medium (Table 3) and the like, it is recommended to prepare it by dividing the ingredients into several groups. Hereafter, the method for preparing MGM-450 medium (Table 10) is described as an example.

A. PREPARATION OF STOCK SOLUTIONS

1. Stock Solutions of Inorganic Salts (ISM-A and ISM-B)

These stock solutions are ten times concentrated mixtures of inorganic salts. In order to prepare 1 l stock solution, dissolve 2.92 g $NaHCO_3$ and 231.33 g $MgSO_4 \cdot 7H_2O$ in about 900 ml distilled water, and bring up the solution to 1 l by adding distilled water. This is ISM-A. Likewise, dissolve 9.58 g $NaH_2PO_4 \cdot 2H_2O$, 8.75 g KCl, 19.1 g $MgCl_2 \cdot 6H_2O$, and 8.33 g $CaCl_2$ in about 900 ml distilled water and adjust the volume to 1 l with distilled water. This is ISM-B. Both solutions can be stored in a refrigerator for several months.

TABLE 6
Weiss et al.'s IPL-41 Medium

Components	mg/100 ml	Components	mg/100 ml
$NaH_2PO_4 \cdot H_2O$	116.0	L-Valine	50.0
$NaHCO_3$	35.0	Sucrose	1650.0
KCl	120.0	Glucose	250.0
$CaCl_2$	50.0	Maltose	100.0
$MgSO_4 \cdot 7H_2O$	188.0	Malic acid	5.36
β-Alanine	30.0	α-Ketoglutaric acid	2.96
L-Arginine·HCl	80.0	Succinic acid	0.48
L-Asparagine	130.0	Fumaric acid	0.44
L-Aspartic acid	130.0	Thiamine·HCl	0.008
L-Cystine	10.0	Riboflavin	0.008
L-Glutamic acid	150.0	Calcium pantothenate	0.0008
L-Glutamine	100.0	Pyridoxine·HCl	0.04
Glycine	20.0	*p*-Aminobenzoic acid	0.032
L-Histidine	20.0	Folic acid	0.008
Hydroxy-L-proline	80.0	Niacin	0.016
L-Isoleucine	75.0	Isoinositol	0.04
L-Leucine	25.0	Biotin	0.016
L-Lysine·HCl	70.0	Cyanocobalamin	0.024
L-Methionine	100.0	Choline chloride	2.0
L-Proline	50.0	$ZnCl_2$	0.004
L-Phenylalanine	100.0	$MnCl_2 \cdot 4H_2O$	0.002
DL-Serine	40.0	$CuCl_2 \cdot 2H_2O$	0.02
L-Threonine	20.0	$(NH_4)Mo_7O_{24} \cdot 4H_2O$	0.004
L-Tryptophan	10.0	$CoCl_2 \cdot 6H_2O$	0.005
L-Tyrosine	25.0	$FeSO_4 \cdot 7H_2O$	0.055

Note: Osmolarity: 360–375 mOsm/kg. Additives: Fetal bovine serum 10%.

From Weiss, S.A. et al., *In Vitro,* 17, 495–502, 1981. With permission.

TABLE 7
Echalier and Ohanessian's D-20 Medium

Components	mg/100 ml	Components	mg/100 ml
$NaH_2PO_4 \cdot 2H_2O$	47.0	Thiamine·HCl	0.002
$MgCl_2 \cdot 6H_2O$	100.0	Riboflavin	0.002
$MgSO_4 \cdot 7H_2O$	370.0	Pyridoxine	0.002
$CaCl_2$	89.0	Niacin	0.002
Glucose	200.0	Calcium pantothenate	0.002
Glutamine[a]	1087.8	Biotin	0.001
Glycine[a]	553.8	Folic acid	0.002
Malic acid	67.0	Inositol	0.002
Succinic acid	6.0	*p*-Aminobenzoic acid	0.002
Sodium acetate·$3H_2O$	2.5	Choline chloride	0.02
Lactalbumin hydrolysate	1500.0	TC-yeastolate	150.0

Note: pH: 6.7. Osmolarity: –0.66°C (as freezing point depression). Additives: Fetal bovine serum 10–20%.

[a] Used as sodium glutamate, sodium glycinate, potassium glutamate, and potassium glycinate.

From Echalier, G. and Ohanessian, A., *In Vitro,* 6, 162–172, 1970. With permission.

TABLE 8
Shield and Sang's M3 Medium

Components	mg/100 ml	Components	mg/100 ml
$MgSO_4 \cdot 7H_2O$	440.0	Glycine	50.0
$CaCl_2 \cdot 6H_2O$	150.0	α-Alanine	150.0
Potassium glutamate·H_2O	788.0	Valine	40.0
Sodium glutamate	653.0	Methionine	25.0
$NaH_2PO_4 \cdot 2H_2O$	88.0	Isoleucine	25.0
$KHCO_3$	50.0	Leucine	40.0
Glucose	1000.0	Tyrosine	25.0
Oxaloacetic acid	25.0	Phenylalanine	25.0
BIS-TRIS	105.0	β-Alanine	25.0
TC-yeastolate	100.0	Histidine	55.0
Aspartic acid	30.0	Tryptophan	10.0
Threonine	50.0	Arginine	50.0
Serine	35.0	Lysine·HCl	85.0
Asparagine	30.0	Cysteine·HCl	20.0
Glutamine	60.0	Choline chloride	5.0
Proline	40.0		

Note: pH: 6.8. Additives: Fetal bovine serum 10%.

From Shields, G. and Sang, J.H., *Drosophila Information Service*, 52, 161, 1977. With permission.

TABLE 9
Maiorella et al.'s Medium

Components	Amount
IPL-41 medium	1000 ml
Cod liver fatty acid methyl ester	10.0 mg
Tween 80	25.0 mg
Cholesterol	4.5 mg
α-Tocopherol acetate	2.0 mg
Ethyl alcohol	1.0 ml
10% aqueous solution of pluronic polyol F-68	10.0 ml

Note: Used as serum-free medium.

From Maiorella, B. et al., *Bio/Technology*, 6, 1406–1410, 1988. With permission.

2. Stock Solution of Amino Acids, Sugars, and Organic Acids (ASO)

In order to prepare 1 l of the ASO, dissolve 105 mg L-cystine and 210 mg L-tyrosine in 20 ml of 1 *N* HCl. When both amino acids have dissolved completely, add distilled water up to 900 ml. Then dissolve the following, one by one: 1,315 mg L-α-alanine, 835 mg β-alanine, 2,915 mg L-arginine hydrochloride, 1,465 mg L-aspartic acid, 1,465 mg L-asparagine, 2,500 mg L-glutamic acid, 2,500 mg L-glutamine, 2,710 mg glycine, 10,415 mg L-histidine, 210 mg L-isoleucine, 315 mg L-leucine, 2,605 mg L-lysine hydrochloride, 210 mg L-methionine, 615 mg L-phenylalanine, 1,460 mg L-proline, 2,293 mg L-serine, 730 mg L-threonine, 415 mg L-tryptophan, 415 mg L-valine, 16,665 mg glucose, 2,085 mg fructose, 110,400 mg sucrose, 2,790 mg malic acid, 1,540 mg α-ketoglutaric acid, 250 mg succinic acid, and 230 mg fumaric acid. Bring up the solution to 1 l by the addition of distilled water. This stock solution contains

TABLE 10
Mitsuhashi's MGM-450 Medium

Components	mg/100 ml	Components	mg/100 ml
$NaH_2PO_4{\cdot}2H_2O$	95.8	L-Methionine	4.2
$NaHCO_3$	29.2	L-Phenylalanine	12.5
KCl	87.5	L-Proline	29.2
$CaCl_2$	83.3	DL-Serine	91.7
$MgCl_2{\cdot}6H_2O$	191.7	L-Threonine	14.6
$MgSO_4{\cdot}7H_2O$	233.3	L-Tryptophan	8.3
Glucose	333.3	L-Tyrosine	4.2
Fructose	41.7	L-Valine	8.3
Sucrose	2208.0	Thiamine·HCl	0.016
Malic acid	55.8	Riboflavin	0.016
α-Ketoglutaric acid	30.8	Pyridoxine·HCl	0.016
Succinic acid	5.0	Niacin	0.016
Fumaric acid	4.6	Calcium pantothenate	0.016
L-α-Alanine	26.3	Biotin	0.008
β-Alanine	16.7	Folic acid	0.016
L-Arginine·HCl	58.3	Isoinositol	0.016
L-Asparagine	29.3	*p*-Aminobenzoic acid	0.016
L-Cystine	2.1	Choline chloride	0.16
L-Glutamic acid	50.0	Inosine	20.0
L-Glutamine	50.0	Cytochrome c	10.0
Glycine	54.2	Polyvinylpyrrolidone K-90	50.0
L-Histidine	208.3	Fetuin	2.0
L-Isoleucine	4.2	Bovine plasma albumin	
L-Leucine	6.3	fraction V	1000.0
L-Lysine·HCl	52.1	Tryptose phosphate broth	300.0

Note: pH: 6.3. Osmolarity: 331 mOsm/kg. Additives: Fetal bovine serum 10% or fetal bovine serum 5% and *Antheraea pernyi* hemolymph 3%.

From Mitsuhashi, J. and Inoue, H., *Appl. Entomol. Zool.*, 23, 488–490, 1988. With permission.

five times concentrated amino acids, sugars, and organic acids (ASO), and it can be stored for several years at –20°C.

3. Stock Solutions of Vitamins (Vit-I and Vit-II)

The vitamin stock solutions are a thousand times concentrated. In order to prepare 100 ml of Vit-I and Vit-II, bring the pH of about 50 ml of distilled water to 10.0 with 0.1 *N* KOH. Dissolve in it 16 mg folic acid and 8 mg biotin. Adjust the pH to 7.0 with 1 *N* HCl. Bring up the solution to 100 ml by the addition of distilled water. This is Vit-I. Dissolve 16 mg riboflavin in 50 ml of distilled water preheated at 50°C. When the riboflavin dissolves completely, cool it to room temperature. Dissolve 16 mg calcium pantothenate, 16 mg pyridoxine hydrochloride, 16 mg *p*-aminobenzoic acid, 16 mg niacin, 16 mg isoinositol, and 160 mg choline chloride in the riboflavin solution. Add distilled water to the solution until the total volume becomes 100 ml. This is Vit-II. Both vitamin stock solutions can be stored for years at –20°C.

B. PREPARATION OF READY TO USE MEDIUM

In order to prepare 1 l of the medium, mix 100 ml ISM-A, 100 ml ISM-B, 200 ml ASO, 1.0 ml Vit-I, and 1.0 ml Vit-II. Add about 400 ml distilled water. Dissolve 10.0 g bovine plasma albumin fraction V, 20 mg fetuin, 100 mg cytochrome c (from horse heart), 200 mg inosine, 500 mg polyvinylpyrrolidone K-90, and 3.0 g tryptose phosphate broth. Adjust the pH of the solution to 6.3 with concentrated KOH. Bring up the solution to 1 l by the addition of distilled water.

Other complex culture media may be prepared in the same manner.

V. PREPARATION OF MEDIUM ADDITIVES

A. PREPARATION OF INSECT HEMOLYMPH

Insect hemolymph by itself had been used frequently as a growth medium earlier, and also as a medium additive even at present. Insect hemolymph has phenol oxidase activity and it produces melanin when added to culture media. The resulting melanin is detrimental to cultured cells. It is, therefore, necessary to inactivate the phenol oxidase in the hemolymph. The following treatment is applied in most cases. Bleed insects and collect the oozed blood in a beaker or a centrifuge tube chilled on ice. An alternative receptacle is a beaker containing a small amount of liquid paraffin. Extract sufficient blood from several insects while cooling the receptacle on ice. The vessel containing liquid paraffin need not be cooled, because the blood sinks under the oil and does not contact the air. Immediately after obtaining sufficient blood, heat the blood to 60°C for 30 min in a water bath. Freeze the heat-treated blood at −20°C. The blood can be stored at this stage for many years. Thaw the frozen blood, and centrifuge at 4000 rpm for 15 min. Separate the supernatant. This is the insect hemolymph usable as a medium additive.

B. PREPARATION OF WATER-INSOLUBLE ADDITIVES

Some media are supplemented with some water-insoluble substances such as lipids. In order to incorporate these substances into media, the following procedures are usually applied. Dissolve chemicals in an appropriate organic solvent. Lipids such as fatty acids, glycerides, sterols, and phospholipids are usually dissolved in ethyl alcohol containing about 2% Tween 80 (as surfactants, lecithin, phosphatydylethanolamine, Pluronic F-68, and so on are also used instead of Tween 80). If chemicals are not soluble in ethyl alcohol, acetone may be used. Sterilize the solution by passing through a membrane filter which is tolerant to the organic solvent used. Ethyl alcohol has a strong germicidal property. Therefore, it is not necessary to sterilize the alcoholic solution. Add the solution to the sterilized culture medium under aseptic conditions. The amount of addition should be smaller than the maximum amount of solvent which does not cause any detrimental effects on cultured cells. The sensitivity of cells to organic solvents is different for different cell lines.

VI. STERILIZATION

For the sterilization of completed culture media, usually membrane filters are used. Membrane filters with a pore size smaller than 0.2 μm are used. In order to avoid clogging a filter with tiny particles present in the media, a series of prefiltration with larger pore size filters is recommended.

Some media such as MM medium can be sterilized by autoclaving (121°C 15 min), although this results in the formation of some precipitates, and cell growth-promoting ability is impaired to some extent.

VII. TEST FOR CONTAMINATION

The sterilized culture media should be left for several days at 25°C or at room temperature. The medium contaminated with bacteria or yeasts becomes turbid, and the one contaminated with fungal spores shows growth of hyphae. Sometimes, contamination of very slowly growing microorganisms may occur. Detection of the contamination of such slowly growing organisms is very difficult.

VIII. STORAGE

The sterilized culture media can be stored for several months in a refrigerator. Media can be preserved below −20°C for several years. Some media such as MM or MGM-450 can be stored at room temperature for several months.

IX. MEDIUM IMPROVEMENT

A. SERUM-FREE MEDIA

Development of serum-free media is now a major trend of medium improvement. Serum-free media are preferable from the standpoint of medium cost as well as defining culture media. The crisis in FBS supply in the 1970s promoted development of serum-free media, and many serum-free media have been formulated. Some media were found to be adequate as serum-free media by simply eliminating the addition of sera. The MM medium was one of such media, and it supported the growth of various cell lines from different orders of insects. However, in most cases, substitutes for FBS were necessary to formulate serum-free media, because most insect cells depended on FBS for their growth. As a result of searching for such substances, the following were reported to support cell growth in lieu of FBS: bovine plasma (or serum) albumin,[22,23] pepton,[24] trypton,[25] lipid mixture,[26] liver digest,[27] cholesterol,[28] egg yolk,[29] etc.

Among these substitutes for FBS, lipids or lipid- containing substances are frequently used. Studies on lipid requirements by insect cells may give a clue to formulate chemically defined media.

Usually serum-free cultures of insect cells require some adaptive period. Most serum-free media kill insect cells when the serum-containing medium is completely replaced with serum-free medium at once. It is recommended to replace serum-containing medium with serum-free medium gradually. For this, the author has adopted the following method: when the density of cells increases enough for subculture, half of the cell suspension is discarded and an equal amount of serum-free medium is added. This procedure is repeated for more than ten passages. If the serum-free medium is not suitable for the cells, the growth rate of the cells decreases after five passages and all the cells would die before the tenth passage. If the serum-free medium is suitable for the cells, the cells grow with similar growth rate even after the tenth passage, when the concentration of substances derived from the serum is 2^{-10} of that of the original serum-containing medium. When this occurs, the medium can be replaced completely with serum-free medium. However, in the continuation of this procedure, sometimes the cells decline even after 20 passages. In this case, the reason of the decline can be hardly attributed directly to deficiency of substances carried over from serum-containing media. It appears reasonable to link the decline in activity of the cultured cells to the production of some factors, such as autocrine substances caused by the shortage of some substances in the medium. The occurrence of a very slow decline of cell multiplication points to the necessity of long-term evaluation of serum-free media.

B. CHEMICALLY DEFINED MEDIA

For chemically defined media, at present only one medium, CDM formulated by Wilkie et al. (Table 5) is available for culturing established cell lines. Notable characteristics of this medium are incorporation of trace metals, polyamines, and lipids such as fatty acids, triglycerides, sterols, and phospholipids. This medium is difficult to prepare, because some components are hardly soluble in water. Some industries once sold this medium, but currently no industries supply this medium. Chemically defined media without protein or peptides are preferable. Proteins in a medium may result in unstable amino acid composition of the medium, because of the degradation of the proteins. Protein-free media will therefore be preferable for biochemical studies on insect cells, as well as for the use of insect cell cultures for the production of foreign proteins by means of recombinant baculoviruses.

A strategy to develop such a medium is to formulate a chemically defined basic medium which can support cell growth when supplemented with FBS, and then to find chemically defined substitutes for the FBS. The development of such basic media is not difficult. However, the construction of FBS substitutes will be difficult. Many investigators have tried to substitute FBS with other substances, but unfortunately the substitutes found to be effective are all undefined (Table 11). Among them, Maiorella's supplements contain only a small amount of unknown substances (Tables 9 and 11). It may be possible to develop a chemically defined medium from this medium. Table 11 suggests the importance of lipids, because all the substitutes contain lipids.

Vaughn and Louloudes[35] have reviewed the lipid requirement of cultured insect cells. From their review, it became evident that growth of many insect cell lines is promoted by incorporation of a mixture of lipids. Among lipids, some polyunsaturated fatty acids are assumed to be growth stimulatory, based on the lipid analyses of cultured insect cells, as well as the essentiality of fatty acids for growth of intact insects. However, the effects of individual fatty acids on the growth of the cultured insect cells are still not evident.

On the other hand, it is also true that some insect cells, such as a cabbage armyworm cell line, NIAS-MB-32, or a fleshfly cell line, NIH-SaPe-4, can grow continuously in a lipid-free medium such as MTCM-1103 (Table 12). In the lipid-free medium, the cell growth was somewhat slower compared with medium containing FBS.[7] Table 13 shows fatty acid analysis of cells cultured in the lipid-free medium. Insect cells could synthesize most of saturated fatty acids, such as myristic acid, palmitic acids, stearic acid, and some unsaturated fatty acids, such as palmitoleic acid and oleic acid, even when they were cultured in the medium free of lipids.[36] They cannot, however, synthesize polyunsaturated fatty acids such as linoleic acid and linolenic acid. Therefore, addition of polyunsaturated fatty acids to the medium may improve the cell growth. Cod liver oil used in the Maiorella et al. medium has been known to contain polyunsaturated fatty acids such as octadecatetraenoic acid, eicosapentaenoic acid, docosahexaenoic acid, and so on. These fatty acids may be important in this medium. Egg yolk, known to contain various lipids, has also been used by Röder[29] and Léry and Fedière.[34] Tables 14 and 15 show lipid composition of egg yolk. Egg yolk contains various lipids such as fatty acids, phospholipids, steroids, carotenoids, and vitamins. Among the fatty acids are many polyunsaturated fatty acids.

The effect of a mixture of lipids contained in FBS should be also examined. Published data on lipid analyses of FBS are fragmental. In vertebrate cell cultures, various growth factors are used in chemically defined media. In insect cell cultures, however, no growth factors proper to insect cells nor vertebrate growth factors effective to insect cells have been known except insulin for *Drosophila* cells.[37] Insect hemolymph, FBS, and lactalbumin hydrolysate are known to contain growth-promoting substances. For a long time, several investigators have been trying to isolate growth-promoting substances from these materials, but no one has succeeded as yet. The difficulty seems to come from attenuation of the activity upon fractionation. If a growth factor effective for insect cells were found in near future, it would facilitate the development of chemically defined media.

TABLE 11
Substitutes for Fetal Bovine Serum in Serum-Free Media with Chemically Defined Base

Substances	Ref.
Lactalbumin hydrolysate, yeastolate, bactotryptose	25
Trace metal, fatty acids, sterols, polyamines, phospholipids	8
Chicken egg yolk	29
Tryptosephosphate broth	30
Lactalbumin hydrolysate, yeastolate, polyamine, sterols, trace metals, nucleic acid precursors	31
Ultroser G (LKB)	32
Nuserum (Collab. Res.)	32
Fetal bovine serum albumin (K.C. Biol.)	32
Cod liver oil, sterols, vitamin E, tryptosephosphate broth, pluronic polyol F-68, yeast ultrafiltrate	20
CPSR-1, -3 (Sigma)	33
Yeastolate, polyamines, sterols, nucleic acid precursor, liver digest, tryptose	27
Polyamines, chicken egg yolk	34

TABLE 12
Mitsuhashi's MTCM-1103 Medium

Components	mg/100 ml	Components	mg/100 ml
NaCl	700.0	Thiamine·HCl	0.016
NaH_2PO_4	20.0	Riboflavin	0.016
$NaHCO_3$	12.0	Calcium pantothenate	0.016
KCl	20.0	Pyridoxine·HCl	0.016
$MgCl_2·6H_2O$	10.0	*p*-Aminobenzoic acid	0.016
$CaCl_2·2H_2O$	20.0	Folic acid	0.016
Glucose	500.0	Niacin	0.016
Inosine	20.0	Inositol	0.016
Lactalbumin hydrolysate	1500.0	Biotin	0.008
Choline chloride	0.16		

Note: pH: 6.5. Osmolarity: 347 mOsm/kg. Used as a serum-free medium.

From Mitsuhashi, J., *Appl. Entomol. Zool.*, 17, 575–581, 1982. With permission.

C. LOW-COST MEDIA

For low-cost media, serum-free cultures are a prerequisite. Currently most insect cell lines are cultured in media supplemented with FBS. FBS was first used for insect tissue culture by Martignoni and Scallion in 1961.[38] A few years later, the growth-promoting ability of FBS for insect cells was ascertained by Mitsuhashi and Maramorosch.[6] Since then, most insect cell culturists have employed FBS for their culture media. However, FBS is very expensive, and therefore cannot be used for large-scale cultures at high concentrations. Fortunately, many serum-free culture media have been developed recently.

For large-scale cultures, media should be prepared easily and should consist of low-cost ingredients. A serum-free medium consisting of many ingredients is not practical, because the preparation would be time consuming. Another preferable property of media for large-scale culture is tolerance for autoclaving. Usually culture media are sterilized by passing through a sterilizing filter. Often, the filter gets clogged with particles in solutions. A serial prefiltration may be needed to avoid this sort of trouble. These procedures are also time consuming, and become a very big problem when large quantities of media are sterilized. If the medium is autoclavable, such troubles can be eliminated.

TABLE 13
Fatty acids Detected in Cells Cultured in a Lipid-Free Medium (MTCM-1103) for Several Generations

	Fatty acids (mg/g dry weight)						
Cell lines	Myris. C14:0	Palmi. C16:0	P.ole. C16:1	Stear. C18:0	Oleic. C18:1	Linole. C18:2	Linolen. C18:3
NIAS-MB-32	0.296	2.418	8.255	4.204	15.765	ND	ND
NIAS-AeAl-2	1.707	7.817	12.268	4.610	26.317	ND	ND
NIH-SaPe-4	0.316	4.163	7.551	0.449	8.061	ND	ND

Note: ND: not detected.

From Mitsuhashi, J., Nakasone, S., and Horie, Y., *Appl. Entomol. Zool.*, 20, 8–12, 1985. With permission.

TABLE 14
Lipid Composition of Chicken Egg Yolk (mg per 100 g Egg Yolk)

Total lipids: 31,200

Fatty acids (Total: 20,300)		β-Sitosterol	3.3
for detail see Table 15		Δ^7-Cholestenol	3.2
		Lanosterol	1.6
Phospholipids (Total: 9,400)		Δ^7-Methestenol	0.7
Phosphatidylcholine	6,900.0		
Lysophosphatidylcholine	500.0	Carotenoids (Total: 2.7)	
Phosphatidylethanolamine	1,400.0	β-Carotene	0.08
Lysophosphatidylethanolamine	200.0	Lutein	1.82
Sphingomyelin	200.0	Zeaxanthin	0.6
Phosphatidylinositol	60.0	Cryptoxanthin	0.2
Plasmalogen	80.0		
		Vitamins (Total: 5.1)	
Steroids (Total: 1,430)		α-Tocopherol	3.1
Cholesterol	1,404.0	β-Tocopherol	0.1
Desmosterol	7.6	γ-Tocopherol	1.8
Cholestanol	4.9	δ-Tocopherol	0.1
Ergosterol	3.7		

Among the serum-free media hitherto formulated, the MM-SF medium (serum-free MM medium) has the simplest composition, and is probably the most inexpensive medium (Table 4). The total cost of the ingredients is about $2.00 for 1 l of medium. It becomes about $8.00 by the addition of filtration cost without labor cost (the costs may be overestimated, because they were calculated based on the catalog price in Japan. It is needless to say the total cost depends on the prices with which we purchase the ingredients). Low-cost substitutes for the ingredients of this medium have been looked for. It became clear that 4-times diluted sea water can be used instead of an inorganic salt mixture, and table sugar can be used as the sole source of carbohydrates.[17]

These substitutes, however, reduced the cost very little. Unfortunately, no effective substitutes for lactalbumin hydrolysate and yeastolate have been found so far. The present author has tested milk, corn steep, and molasses, but none of them supported cell growth (unpublished). In order to reduce the cost sufficiently, we should search for substitutes among industrial waste products.

TABLE 15
Fatty Acid Composition of Egg Yolk
(g per 100 g Total Fatty Acids)

Myristic acid (14:0)	0.3	Octadecatetraenoic acid (18:4)	0
Myristoleic acid (14:1)	0	Arachidic acid (20:0)	0.1
Pentadecanoic acid (15:0)	0.1	Eicosenoic acid (20:1)	0.4
Palmitic acid (16:0)	25.1	Eicosadienoic acid (20:2)	0.1
Palmitoleic acid (16:1)	3.6	Eicosatrienoic acid (20:3)	0.2
Heptadecanoic acid (17:0)	0.2	Arachidonic acid (20:4)	1.7
Heptadecenoic acid (17:1)	0.3	Eicosapentaenoic acid (20:5)	0
Stearic acid (18:0)	8.6	Docosapentaenoic acid (22:5)	0.1
Oleic acid (18:1)	43.6	Docosahexaenoic acid (22:6)	1.8
Linoleic acid (18:2)	13.4	Tetracosenoic acid (24:1)	0.1
Linolenic acid (18:3)	0.3		

Autoclavable media have received limited study for insect cell cultures. The MM-SF medium (Table 4) or MTCM-1601[17] medium could be autoclaved, although they produce some precipitates. Such autoclaved MM-SF medium could support cell growth continuously, but the growth rate of cells was impaired to some extent. Koike and Sato[39] sterilized 40 l of MM-SF medium in a jar-fermenter by introducing high pressure water vapor into the fermenter, and then cultured cabbage armyworm cells in it successfully. The cultured cells are said to support replication of MB-NPV.

By adopting autoclaving for sterilization, the cost of MM-SF medium can be reduced to about $2.00 for 1 l. We may be able to further reduce the cost by reducing the quantity of lactalbumin hydrolysate and yeastolate. Also the spent medium can be recycled. The mixture of used and fresh media at the ratio of 3:7 can support the cell growth at the same rate as 100% fresh medium (T. Sato, personal communication). This may reduce the cost to some extent.

REFERENCES

1. **Grace, T.D.C.** (1962), Establishment of four strains of cells from insect tissue grown in vitro. *Nature* 195, 788-789.
2. **Goldschmidt, R.** (1915), Some experiments on spermatogenesis in vitro. *Proc. Natl. Acad. Sci. U.S.A.* 1, 220-222.
3. **Trager, W.** (1935), Cultivation of the virus of grasserie in silkworm tissue cultures. *J. Exp. Med.* 61, 501-513.
4. **Wyatt, S.S.** (1956), Culture in vitro of tissue from the silkworm *Bombyx mori* L. *J. Gen. Physiol.* 39, 841-852.
5. **Wyatt, G.R., Loughheed, T.C., and Wyatt, S.S.** (1956), The chemistry of insect hemolymphs. I. Organic components of the hemolymphs of the silkworm, *Bombyx mori* and two other species. *J. Gen Physiol.* 40, 833-847.
6. **Mitsuhashi, J. and Maramorosch, K.** (1964), Leafhopper tissue culture: embryonic, nymphal and imaginal tissues from aseptic insects. *Contrib. Boyce Thompson Inst.* 22, 435-460.
7. **Mitsuhashi, J.** (1982), Continuous cultures of insect cell lines in media free of sera. *Appl. Entomol. Zool.* 17, 575-581.
8. **Wilkie, G.E.I., Stockdale, H., and Pirt, S.V.** (1980), Chemically-defined media for production of insect cells and viruses. *Dev. Biol. Standard.* 46, 29-37.
9. **Gvozdev, V.A. and Kakpakov, V.T.** (1968), Culture of embryonic cells of *Drosophila melanogaster* in vitro. *Genetika* 4, 129-142.
10. **Echalier, G. and Ohanessian, A.** (1970), In vitro culture of *Drosophila melanogaster* embryonic cells. *In Vitro* 6, 162-172.
11. **Mitsuhashi, J.** (1976), Insect cell line: amino acid utilization and requirements. In *Invertebrate Tissue Culture, Applications in Medicine, Biology, and Agriculture,* E. Kurstak and K. Maramorosch, Eds., Academic Press, New York, 257-262.
12. **Mitsuhashi, J.** (1987), Difference in amino acid metabolism among cell lines derived from cabbage armyworm, *Mamestra brassicae* (Lepidoptera: Noctuidae). *Appl. Entomol. Zool.* 22, 533-536.
13. **Grace, T.D.C. and Brzostowski, H.W.** (1966), Analysis of the amino acids and sugars in an insect cell culture medium during cell growth. *J. Insect Physiol.* 12, 625-633.

14. **Mitsuhashi, J.** (1988), Simplification of media and utilization of sugars by insect cells in cultures. In *Invertebrate and Fish Tissue Culture,* Y. Kuroda, E. Kurstak, and K. Maramorosch, Eds., Japan Scientific Societies Press, Tokyo/Springer-Verlag, Berlin, 15-18.
15. **Kimura, S. and Mitsuhashi, J.** (1983), Preliminary notes on glycosidases in two cell lines derived from the ovary of the cabbage armyworm, *Mamestra brassicae* (Lepidoptera: Noctuidae). *Appl. Entomol. Zool.* 18, 561-563.
16. **Mitsuhashi, J.** (1990), Comparison of zymograms of some insect cell lines by means of APIZYM system. *Appl. Entomol. Zool.* 25, 535-537.
17. **Mitsuhashi, J.** (1989), Simplified medium (MTCM-1601) for insect cell lines. *J. Tissue Culture Methods* 12, 21-22.
18. **Weiss, S.A., Smith, G.C., Kalter, S.S., and Vaughn, J.L.** (1981), Improved method for the production of insect cell cultures in large volume. *In Vitro* 17, 495-502.
19. **Shields, G. and Sang, J.H.** (1977), Improved medium for culture of *Drosophila* embryonic cells. *Drosophila Information Service* 52, 161.
20. **Maiorella, B., Inlow, D., Shauger, A., and Harano, D.** (1988), Large-scale insect cell-culture for recombinant protein production. *Bio/Technology* 6, 1406-1410.
21. **Mitsuhashi, J. and Inoue, H.** (1988), Obtainment of a continuous cell line from the larval fat bodies of the mulberry tiger moth, *Spilosoma imparilis* (Lepidoptera: Arctiidae). *Appl. Entomol. Zool.* 23, 488-490.
22. **Kitamura, S., Imai, T., and Grace, T.D.C.** (1973), Adaptation of two mosquito cell lines to medium free of calf serum. *J. Med. Entomol.* 10, 488-489.
23. **Pant, U., Mascasrenhas, A.F., and Jagannathan, V.** (1977), In vitro cultivation of a cell line from embryonic tissue of potato tuber moth, *Gnoriomoschema operculata* (Zellev). *Ind. J. Exp. Biol.* 156, 303-304.
24. **Goodwin, R.H.** (1976), Insect cell growth on serum-free media. *In Vitro* 12, 303-304.
25. **Hink, W.F., Strauss, E.M., and Lynn, D.E.** (1977), Growth of TN-368 insect cells in serum-free media. *In Vitro* 13, 177 (abstract).
26. **Goodwin, R.H. and Adams, J.R.** (1980), Nutrient factors influencing viral replication in serum-free insect cell line culture. In *Invertebrate Systems in Vitro,* E. Kurstak, K. Maramorosch, and A. Dübendorfer, Eds. Elsevier, Amsterdam, 493-509.
27. **Goodwin, R.H., Adams, J.R., and Shapiro, M.** (1990), Replication of the entomopox virus from *Amsacta moorei* in serum-free cultures of a gypsy moth cell line. *J. Invert. Pathol.* 56, 190-205.
28. **Brooks, A.M., Tsang, K.R., and Freeman, F.A.** (1980), Cholesterol as a growth factor for insect cell lines. In *Invertebrate Systems in Vitro,* E. Kurstak, K. Maramorosch, and A. Dübendorfer, Eds., Elsevier, Amsterdam, 67-77.
29. **Röder, A.** (1982), Development of a serum-free medium for cultivation of insect cells. *Naturwissenschaften* 69, 92-93.
30. **Kuno, G.** (1982), Use of a polyploid cell culture adapted to a serum-free medium for the studies of dengue virus replication. *In Vitro* 18, 298 (abstract).
31. **Miltenburger, H.G.** (1983), Investigation on the cultivation of insect cell lines in serum-free media. Proc. 1st European Conf. on Serum-Free Cell Culture, 31-43.
32. **Lynn, D.E. and Hung, A.C.F.** (1986), Development of a continuous cell line from the insect egg parasitoid, *Trichogramma pretiosum* (Hymenoptera: Trichogrammatidae). *In Vitro Cell. Dev. Biol.* 22, 440-442.
33. **Vaughn, J.L. and Fan, F.** (1989), Use of commercial serum replacement for the culture of insect cells. *In Vitro Cell. Dev. Biol.* 25, 143-145.
34. **Léry, X. and Fedière, G.** (1990), A new serum-free medium for lepidopteran cell culture. *J. Invert. Pathol.* 55, 342-349.
35. **Vaughn, J.L. and Louloudes, S.J.** (1983), Lipid nutrition and metabolism of cultured insect cells. In *Metabolic Aspects of Lipid Nutrition in Insects,* T.E. Mittler and R.H. Dadd, Eds., A Westview Science Study, Boulder, CO, 223-244.
36. **Mitsuhashi, J., Nakasone, S., and Horie, Y.** (1985), Total fatty acids of some insect cell lines. *Appl. Entomol. Zool.* 20, 8-12.
37. **Seecof, R.L. and Dewhurst, S.** (1974), Insulin is a *Drosophila* hormone and acts to enhance the differentiation of embryonic *Drosophila* cells. *Cell Differ.* 3, 63-70.
38. **Martignoni, M.E. and Scallion, R.J.** (1961), Preparation and uses of insect hemocyte monolayers in vitro. *Biol. Bull., Woods Hole, Mass* 121, 507-520.
39. **Koike, M. and Sato, K.** (1989), Culture of insect cell lines originated from *Mamestra brassicae* with autoclaved serum-free medium. In *Invertebrate and Fish Tissue Culture,* Y. Kuroda, E. Kurstak, and K. Maramorosch, Eds., Japan Scientific Societies Press, Tokyo/Springer-Verlag, Berlin, 7-10.

Chapter 2

MOSQUITO CELL LINES

Jun Mitsuhashi

TABLE OF CONTENTS

0-8493-7642-4/94/$0.00+$.50

Jun Mitsuhashi, D.Agr., is Professor of Applied Entomology at Tokyo University of Agriculture and Technology. He received his Bachelor of Agriculture in applied entomology in 1955 and Doctor of Agriculture at The University of Tokyo in 1965. 1955–1984. National Institute of Agricultural Sciences, 1984–1988. Forest Research Institute, 1988–present. Tokyo University of Agriculture and Technology. 1962–1964. Boyce Tompson Institute for Plant Research, 1968–1969. Division of Entomology, CSIRO, 1993–1994. President of Japanese Society of Applied Entomology and Zoology. Insect cell culture; insect endocrinology; artificial rearing of insects, insect endosymbiotes; insect pathology. Address: Faculty of Agriculture, Tokyo University of Agriculture and Technology, Fuchu, Tokyo 183, Japan.

I. INTRODUCTION

Unlike other groups of insects, cell lines have been established from many of the major species of mosquitoes. They have been established because they were needed for arbovirus research. However, some cell lines have also been used in other fields such as biotechnology. Most mosquito species possess only six chromosomes, and their shapes are distinct. The cell lines might be, therefore, suitable as materials for cytogenetics, if their genetical background is known. In this chapter, mosquito cell lines will be described from historical, culture methodological, and practical aspects.

II. DEVELOPMENT OF MOSQUITO CELL CULTURE

Trager[1] is the first who attempted to culture mosquito tissues. He maintained pupal head tissues of *Aedes aegypti* for nearly a month in a hanging drop of a medium consisting of peptic digest of powdered egg albumen, phosphate buffer, inorganic salts, and glucose with heparinized guinea pig or chicken plasma. Moreover, the cultured mosquito tissues were inoculated with Western equine encephalomyelitis virus, and the virus was recovered after 28 days from the culture. Although Trager's cultures were organ cultures, he obtained outgrowth of cells sometimes from ovarial tissues.

After Trager's pioneering work, similar organ cultures of various mosquito tissues were attempted.[2-7] In these studies, the surviving period or actively contracting period of explants was somewhat improved. Migration of cells from the explanted tissues was also observed sporadically.[2,7] However, the migrated cells did not multiply.

Kitamura[8] initiated the culture of midgut cells from aseptically reared *Culex pipiens molestus* larvae. He obtained nice cell migration, but the cells did not multiply. He[9] also cultured ovarial tissues from the adults, and obtained active cell migration. Some of the migrated cells showed mitoses, and this led to the establishment of a continuous cell line later.

Peleg and Trager[10] cultured imaginal discs of *Aedes aegypti* larvae and obtained some cell migration. Peleg[11] later obtained multiplying cells from minced whole bodies of neonate larvae of *A. aegypti.* He also cultured embryonic cells of the same species and obtained cell migration and mitoses.

Grace[12] cultured the cells from axenically reared *Aedes aegypti* matured larvae. The heads, tails, and alimentary canals were removed. The peritrophic membranes were discarded. Then the alimentary canals and the residue of the bodies were torn apart in culture medium, and brought into culture vessels. The medium used was GMA which was used for culturing a moth cell, *Antheraea eucalypti.* Cell migration from the explants was quite active, and mitotic divisions were numerous among the migrated cells. The first subculture became possible 4 weeks after the cultures were set up. After repeating subcultures, a continuously growing cell line was obtained. However, efficiency of getting growing cells from mosquito tissues seemed to be very low, because Grace mentioned "in the past 18 months more than a hundred cultures of mosquito tissues have been made and in only two of these had growth comparable to that described for the cell line occurred." Grace's mosquito cell line was the first mosquito cell line. Unfortunately, the line was lost later.

The second and the third lines from mosquitoes were obtained by Singh.[13] He cultured minced whole tissues of the first instar larvae in Mitsuhashi and Maramorosch's medium[14] for leafhopper cell cultures, and succeeded to establish a cell line from Aedes albopictus and also a line from *Aedes aegypti.* The former line, distributed to various countries around the world, has been frequently used for the study of mosquito-borne viruses.

Since then, very many cell lines have been established from various species of mosquitoes. They are listed in Table 1.

The established cell lines from mosquitoes have been used widely in the investigations of arboviruses or other mosquito-borne pathogenic microorganisms.

TABLE 1
Reported Mosquito Cell Lines

Species	Cell line designation	Origin	Culture media	Establisher
Genus *Aedes*				
A. aegypti	—	Minced last instar larvae	GMA	Grace[12]
A. aegypti	ATC-10	Minced trypsinized neonate larvae	MM	Singh[13]
A. aegypti	59, 364	Embryos	Kitamura's medium	Peleg[15]
A. aegypti	Mos-20	Minced trypsinized larvae	MM	Varma and Pudney[16]
A. aegypti	Mos-20A	Minced trypsinized larvae	MM/VP12	Varma and Pudney[16]
A. aegypti	Mos-29	Minced trypsinized larvae	Modified Kitamura's medium	Varma and Pudney[16]
A. aegypti	Mill Hill	Larvae	Leibovitz L-15	Bailey et al.[17]
A. albopictus	ATC-15	Minced trypsinized neonate larvae	MM	Singh[13]
A. albopictus	ATC-89,-91	Minced trypsinized neonate larvae	MM	Bhat[18]
A. albopictus	—	Adult ovaries	MM	Hsu et al.[19]
A. albopictus	NIAS-AeAl-2	Minced trypsinized neonate larvae	MM	Mitsuhashi[20]
A. dorsalis	—	Neonate larval tissues	MM/VP12	Cahoon et al.[21]
A. kronbeini	NIVI-AK-453	Embryos	MM	Pant et al.[22]
A. malayensis	Mos-60	Minced trypsinized neonate larvae	MM/VP12	Varma et al.[23]
A. novalbopictus	ATC-173, -176	Minced trypsinized neonate larvae	MM	Bhat and Guru[24]
A. peudoscutellaris	Mos-61	Minced trypsinized neonate larvae	MM/VP12	Varma et al.[23]
A. triseritatus	—	Neonate larval tissues	MM/VP12	Rowley et al.[25]
A. vexans	—	Minced pupal and imaginal tissues	GMA	Sweet and McHale[26]
A. vittatus	—	Minced trypsinized naonate larvae	Modified GMA	Bhat and Singh[27]
A. w-albus	ATC-136, -137	Minced trypsinized neonate larvae	MM	Singh and Bhat[28]
Genus *Anopheles*				
A. gambiae	Mos-55	Minced trypsinized neonate larvae	Kitamura/VP12	Marhoul and Pudney[29]
A. stephensi	—	Minced larval tissues	Modified GMA	Schneider[30]
A. stephensi	Mos-43	Minced trypsinized neonate larvae	Kitamura/VP12	Pudney and Varma[31]
A. stephensi	Mos-44,-45, -46	Minced trypsinized neonate larvae	Kitamura/VP12	Pudney et al.[32]
Genus *Clista*				
Cliseta inornata	—	Minced pupal and imaginal tissues	GMA	Sweet and McHale[26]
Genus *Culex*				
C. bitaeniorhynchus	ATC-415	Embryo	MM	Pant and Dhanda[33]
C. pipiens	—	Embryos	MM	Chao and Ball[34]
C. pipiens molestus	CmOv	Adult ovaries	Kitamura's medium	Kitamura[35]
C. quinquefasciatus	—	Adult ovaries	Medium 721	Hsu et al.[36]
C. salinarius	—	Neonate larval tissues	Medium 721	Schneider[37]
C. tritaeniorhynchus	—	Neonate larval tissues	Medium 721	Schneider[37]
C. tritaeniorhynchus summorosus	—	Adult ovaries	Medium 721	Hsu et al.[38]

TABLE 1 (Continued)
Reported Mosquito Cell Lines

Species	Cell line designation	Origin	Culture media	Establisher
Genus *Haemagogus*				
H. equinus	GML-HE-12	Embryos	MM	Oro[39]
Genus *Toxorhynchites*				
T. amboinensis	TA-9,-42	Raptured eggs	MM/VP12	Tesh[40]
T. amboinensis	TRA-171	Whole larvae	MM/VP12	Kuno[41]
T. amboinensis	TRA-147, -272, 293	Neonate larval tissues	MM/VP12	Kuno[42]
T. amboinensis	RU-TAE-12V	Minced larval tissues	Leibovitz L-15	Munderloh et al.[43]

III. CULTURE METHODS

A. PRIMARY CULTURES

Since mosquitoes are very small insects, starting cultures with excised tissues is difficult. Most investigators employ the culture of fragments of neonate larvae. This method was first developed by Peleg.[11] The method does not require delicate techniques. However, it is necessary to prepare aseptic larvae. The aseptic larvae can be obtained by surface-sterilization of eggs and letting the eggs hatch under sterile conditions. For the surface-sterilization, he immersed the eggs in White's solution (0.25 g mercuric chloride, 6.5 g sodium chloride, 1.25 ml hydrochloric acid, 250 ml ethyl alcohol, and 750 ml distilled water) for 20 min, washed with sterile distilled water, and then transferred into Rinaldini's salt solution (0.8 g sodium chloride, 0.02 g potassium chloride, 0.005 g sodium phosphate, 0.1 g glucose, 0.1 g sodium bicarbonate, 0.0676 g sodium citrate, all made up to 100 ml with distilled water). Later many investigators used 70% ethyl alcohol as sterilizing agent. Usually sterilized eggs were kept in either culture media or sterile physiological saline until the larvae hatched.

The hatched larvae can survive for many days in the media. In most cases, the larvae are used within a few days after hatching. The larvae are collected and transferred into a petri dish or a Maximov slide with a small amount of culture media, and cut into several pieces with fine knives or with a pair of scissors. The still twitching pieces of mosquito larvae are then transferred into a culture vessel.

Sometimes, larvae escape from severing, or those slightly injured may get into the culture vessels with explants. Those larvae survive for a considerable period of time and swim in the medium. Such larvae often consume cells that migrated from the explants. Also by their movement, they may disturb the settling of the migrated cells on the substrate.

On the other hand, Kuno[41] reported that what he called "incapacitated larvae", which were newly hatched larvae injured mechanically at the anterior or posterior or at both ends of the body, were the best explants for *Toxorhynchites amboinensis* mosquito primary cell cultures. These larvae were incapable of feeding but remained motile more than 2 months, and during those periods many cells migrated from the cuticular wound openings. The migrated cells settled on the substrate and multiplied until the cells increased sufficiently for subculturing. Kuno's[42] possible explanation of why continuous cultivation of the attached cells in the presence of incapacitated larvae led to successful isolation of cell lines was as follows: (1) unknown substance(s) released by the incapacitated larvae stimulated replication of the glass-attached cells or (2) groups of cells that were better adapted to the *in vitro* environment were later released from the incapacitated larvae and replaced less well-adapted cells that had attached earlier.

Some investigators initiated cell cultures from embryonic tissues. According to Peleg,[44] a large amount of eggs were collected, surface-sterilized, and then crushed in culture media by means of a homogenizer. The resultant cell suspension was passed through an 80-mesh strainless mesh, and the cell suspension was washed three times and seeded in culture vessels.

The only organ which was isolated from mosquitoes and used as an explant for primary cultures was the ovary. Kitamujra[9] cultured adult ovaries from the Japanese cellar mosquito, *Culex pipiens molestus*. The surface of the adult mosquito was sterilized with 5% chloramine in 70% ethyl alcohol for 2 min. The adults were dissected in a sterile mosquito saline. The ovaries were pulled out from the body cavity, care being taken not to damage their guts. The ovaries were excised, washed in saline, then cut into two or three pieces and transferred into culture vessels with a small amount of the culture medium . Hsu and his co-workers[36,38] also adopted this method for the culture of *Culex* mosquito cells.

Notwithstanding the methods employed, cell migration occurred from the explants or tissue masses soon after the cultures were set up. In most cases, migrated cells formed patches of cells in which the cells multiplied. This resulted in formation of cell sheets. The cell sheets increased their size and finally covered the whole area of the bottom of the vessels. At this stage subculture could be made.

B. SUBCULTURES

Subcultures are made either by mechanical detaching of the cells or by chemical treatment of the cells. If the cells attached loosely onto the vessel, vigorous pipetting is sufficient to detach the cells from the substrate. However, if the cells are not detached by this treatment, chemical treatment will be necessary. Usually 0.25% trypsin or an equal parts mixture of 0.25% trypsin and 0.1% EDTA is used. Other enzymes such as pancreatine, collagenase, dispase, or pronase may be used. It is recommended to determine the appropriate enzyme by watching the effects of these enzymes in preliminary experiments. After enzyme treatment, a protein solution such as sera will be added to the enzyme solution to stop the digestion and to protect the treated cells. Then the cells will be detached by pipetting from the substrate. The resultant cell suspension has to be centrifuged and the cell pellet suspended in fresh medium, and distributed to culture vessels. After subculturing, cells attach to the substrate and continue to multiply. If the multiplication of cells continues unlimitedly, the cell population is considered a continuous cell line.

C. LARGE-SCALE CULTURES

Igarashi[45] developed a method for large-scale culture of Singh's *Aedes albopictus* cell line ATC-15(subline C6/36). The C6/36 cells do not attach to the substrate firmly. The cells, therefore, can be easily detached from the substrate by pipetting, and they can grow in suspended state in spinner bottles. In this way, the cells can multiply up to 10^6 cells per ml. If microcarrier was added to the suspending culture, the cells attached to the microcarrier sank within a few minutes when the magnetic stirrer was stopped. This property is convenient for large-scale production of viruses that multiply in this cell line.

IV. CONTINUOUS CELL LINES

The first continuous cell line of mosquitoes was reported by Grace[12] from *Aedes aegypti* in 1966. Unfortunately, this line no longer exists. There might have been some trouble with this cell line from the beginning. When Grace obtained growing cells from *A. aegypti* larval tissues, he ascertained that the cells did not react with the antiserum to the cell line of the moth, *Antheraea eucalypti*, which he had established years before. He also examined the karyotype of the new cell line, and found that most of the cells had 96 chromosomes (32n). This number was similar to that of *A. eucalypti* cell line. Later, the cell lines were distributed to several laboratories. Various works were developed with this cell line, and Greene et al.[46] found that the cell line, cultured as mosquito cells, was actually *A. eucalypti* cells. They proved this by examination of the karyotype, serological reaction, and isoenzyme patterns of the cells. They used Suitor's clone of Grace's *Aedes aegypti* cells. In their report on cloning of Grace's *A. aegypti* cell line, Suitor et al.[47] demonstrated the karyotype of their cloned mosquito cells. The karyotype was a typical lepidopteran karyotype. It is not known at which stage the real mosquito cell line was replaced with the *A. eucalypti* cells. Grace proved the new cell line he obtained was not cross-reactive to *A. eucalypti* antiserum. This means the new cell line was really mosquito cells when he examined it serologically. However, the chromosome number was very similar to that of the *A. eucalypti* cell line. After many cell lines had been established from various mosquito species, it became clear that most of the mosquito cell lines were diploid and had distinct large chromosomes. It can be said that in general dipteran cells, including *Drosophila*, flies, and other species in addition to mosquitoes, maintain their diploidy with large distinct chromosomes even when they enter the continuously growing state. On the other hand, lepidopteran cell lines have an inclination to become polyploid when they become continuously growing cell lines. Their karyotypes are characterized by a mixture of many micro- and small macrochromosomes, and by the chromosomes lacking a definite centromere. When we take these general features of established cell lines into consideration, it is most likely that Grace examined the karyotype of *A. eucalypti* cell line instead of *A. aegypti* cell line unconsciously. How this happened nobody knows. My speculation is that at least a part of the cultures of the mosquito cells were mislabeled or contaminated with *A. eucalypti* cells at some stage of subculturing. Trouble expanded when the replaced mosquito cells (actually *A. eucalypti* cells) were distributed to other laboratories. Consequently, some work done with these cell lines has to be considered as done with *A. eucalypti* cells.[47-55]

Sweet and McHale[26] reported the establishment of cell lines from *Aedes vexans* and *Cliseta inornata*. These cell lines were, however, again identified as *Antheraea eucalypti* cells by Greene et al.[46] The morphology of these cell lines and their karyotypes were similar to each other. *A. vexans* and *C. inornata* cells reacted with antisera to *A. eucalypti*, but not with antisera to mosquitoes from which they derived. Isozyme analysis tested for glucose-6- phosphate dehydrogenase, lactic dehydrogene, and malic dehydrogenase showed no difference between these cell lines.

Therefore it is natural to consider that the cell lines from *A. vexans* and *C. inornata* were replaced with *A. eucalypti* cells during the course of establishing the cell lines. In the first report on cultivation of *A. vexans* and *C. inornata* cells, Sweet and Dupree[56] mentioned that growing cells were more similar to Grace's *A. aegypti* cells than to Singh's line, because they grew as free-floating cells rather than as monolayer. This suggests that the replacement of the cell lines occurred rather early during stages of subculturing. When Sweet and McHale[26] and McHale and Sweet[57] characterized these cell lines, they showed a karyotype of *A. vexans* cells. The metaphase plate contained approximately 170 microchromosomes, and was very similar to that of the *A. eucalypti* cells. Ibrahim and Sweet[58] also showed that these cell lines reacted with antibody to *A. eucalypti* but not with that to Singh's *A. aegypti* cells. It may be considered that the mosquito cells were replaced with *A. eucalypti* cells at some early stage of the culture history by an unknown cause. Cupp[59] cultured filariae in these mosquito cell cultures, and now it should be understood that he cultured the filariae in Grace's moth cells.

Although there was some trouble with mosquito cell lines as mentioned above, many cell lines have been established from various mosquito species. They are listed in Table 1.

In general, mosquito cell lines have characteristics as follows. They are (1) substrate dependent and grow as monolayer, (2) mostly diploid (6 chromosomes) and the chromosomes are large, distinct and meta- or submetacentric, and (3) susceptible to some arboviruses. In addition to these characteristics, it may be added that some mosquito cell lines are latently infected with some unknown viruses.[3,22,60-63]

V. CULTURE MEDIA

Trager[1] used a medium based on peptic digest of egg albumen for the organ culture of pupal tissue of *Aedes aegypti* (Table 2). Gavrilov and Cowez[2] used chicken plasma and embryo extract as well as extract of mosquito embryos as a culture medium and obtained some cell migration and multiplication from the explanted salivary glands of *Anopheles maculipennis* var. *atroparvas* and *Stegomyia fasciata.* At this stage, investigators working on mosquito tissue culture mostly used Trager's medium or its modifications.

Grace[12] established the first mosquito cell line with the same medium (GMA) with which he had established the lepidopteran cell lines (Table 2).

Schneider[30] cultured *Anopheles stephensi* cells in Grace's modified medium.

Singh[13] established cell lines from *Aedes albopictus* and *Aedes aegypti* by using a leafhopper cell culture medium MM (Table 2). He visited my laboratory in Tokyo before starting mosquito cell culture in the U.S., and asked me to show him how to prepare culture media and how to start primary cultures of insects. At that time I had worked on leafhopper cell cultures and showed him how to do it. Years later, he succeeded in establishing mosquito cell lines with the MM medium. Since this MM medium was found so suitable for mosquito cell cultures, many investigators have employed this medium for mosquito cell cultures. This explains why the leafhopper cell culture medium became popular among mosquito cell culturists.

On the other hand, Kitamura[8] formulated a mosquito cell culture medium based on the contracting movement of several organs in the test medium. He further improved this medium and succeeded in obtaining a continuous cell line of *Culex pipiens molestus* later (Table 2). Hsu et al.[36,38] followed Kitamura's method and also established two cell lines from two species of *Culex* with Kitamura's modified medium (Table 2, medium 721).

Although many established mosquito cell lines have been maintained in MM medium, they may grow in other types of media. For example, *Aedes albopictus* cells can be cultured in MGM448[64] or MGM-450.[65]

Some mosquito cell lines can be cultured in serum-free media.[38,66,67] *Aedes albopictus* cells, for example, can grow in serum-free MM medium.[66]

VI. APPLICATIONS

Most mosquito cell lines have been established for studies of mosquito-borne viruses. However, some of them have been used for basic cell physiology or biotechnology studies and other purposes.

A. ARBOVIRUS STUDIES

Most mosquito cell lines are susceptible to B group arboviruses. Cell lines and their susceptibility to arboviruses are listed in Table 3.

TABLE 2
Composition of Culture Media (mg/100 ml)

Ingredients	Trager's medium[1]	GMA[12]	Kitamura's medium[35]	MM[14]	Medium 721[36]	VP-12[16]
NaH_2PO_4		114		20[a]		55
$NaHCO_3$		35	7.3	12	70	50
NaCl	40		475	700	315	390
KH_2PO_4					17.5	
KCl	3	224	36.5	20	56	55
$MgCl_2 \cdot 6H_2O$		228		10		110
$MgSO_4 \cdot 7H_2O$		278			28	120

TABLE 2 (Continued)
Composition of Culture Media (mg/100 ml)

Ingredients	Trager's medium[1]	GMA[12]	Kitamura's medium[35]	MM[14]	Medium 721[36]	VP-12[16]
$CaCl_2$	8	100	7.3[b]	20[b]	24.5	40
L-α-Alanine		22.5				
β-Alanine		20.0				
L-Arginine-HCl		70.0				
L-Asparagine		35.0				
L-Aspartic acid		35.0				
L-Cystine-HCl		70.0				
L-Glutamic acid		60.0				
L-Glutamine		60.0				30.0
Glycine		65.0				
L-Histidine		250.0				
L-Isoleucine		5.0				
L-Leucine		7.5				
L-Lysine-HCl		62.5				
L-Methionine		5.0				
L-Phenylalanine		15.0				
L-Proline		35.0				
DL-Serine		110.0				
L-Threonine		17.5				
L-Tryptophan		10.0				
L-Tyrosine		5.0				
L-Valine		10.0				
Sucrose		2668.0			420.0	
Fructose		40.0				
Glucose	900	70.0	146.0	400	112.0	200.0
Malic acid		67.0			42.0	
α-Ketoglutaric acid		37.0			28.0	
Succinic acid		6.0			4.2	
Fumaric acid		5.5			4.2	
Thiamine-HCl		0.002				
Riboflavin		0.002				
Calcium pantothenate		0.002				
Pyridoxine-HCl		0.002				
p-Aminobenzoic acid		0.002				
Folic acid		0.002				
Niacin		0.002				
Isoinositol		0.002				
Biotin		0.001			40.0	
Choline chloride		0.02			25.0	
TC-yeastolate				500	140.0	
Lactalbumin hydrolysate			730.0	650	1400.0	500.0
Bacto-peptone					350.0	
Bovine plasma albumin V						100.0
Peptic digest of egg albumen	5.0 ml					
Guinea pig or chicken plasma heparinized	33.3 ml					
Calf serum			9.0 ml			
Fetal bovine serum				0–20 ml	10.0 ml	
TC-199 medium			18.0 ml		20.0 ml	
MEM vitamin mixture (×100)						2.0 ml
pH		6.5	7.0–7.2	6.5	6.8	7.0

[a] $+H_2O$.
[b] $+2H_2O$.

TABLE 3
Arboviruses and Mosquito Cell Lines Susceptible to Them

Viruses	Susceptible cell lines
Alphaviruses	
Chikungunya	ATC-10, ATC-15, ATC-121, ATC-136, ATC-137, ATC-173, ATC-176, TA-42, TRA-171, NIVI-AK-453, Mos-20, Mos-60, Mos-61, *A. albo* (Hsu), *A. vitt, A. w-alb*
Eastern equine encephalomyelitis	ATC-10, ATC-15, TA-42, 59, 364, GML-HE-12, *A. albo* (Hsu)
Getah	TA-42, *A. aeg*
Mayaro	GML-HE-12
O'nyong-nyong	Mos-43, Mos-60, Mos-61, ATC-15
Ross river	TA-42, ATC-15
Sagiyama	*C. quin* (Hsu)
Semliki forest	ATC-10, ATC-15, 364, Mill Hill, Mos-20, Mos-60, Mos-61
Sindbis	ATC-10, ATC-15, TA-42, 59, 364, NIVI-AK-453, Mos-20, Mos-60, Mos-61, *A. albo* (Hsu), *A. aeg* (Peleg)
Venezuelan equine encephalitis	ATC-10, ATC-15
Western equine encephalomyelitis	ATC-10, ATC-15, *A. albo* (Hsu)
Bunyaviruses	
Akabane	ATC-15
Anopheles A	Mos-43, *An. st* (Varma)
Batai	ATC-15, *C. quin* (Hsu)
Bunyamwera	ATC-15
Bwamba	Mos-43, *An. st* (Varma)
Cache valley	TA-9, ATC-10, ATC-15
California encephalitis	ATC-10, ATC-15, 59, Mos-20, Mos-60, Mos-61
Calovo	ATC-15, Mos-43, *An. st* (Varma)
Ganjam	ATC-15, ATC-173
Germiston	ATC-15
Guaroa	GML-HE-12
Kairi	ATC-15
La Crosse	TA-9, *A. trir*
Maguari	ATC-15
Main Drain	ATC-15
Marituba	ATC-15
Northway	ATC-15
Sathuperi	ATC-15
Tahyna	Mos-29, Mos-43, Mos-55, ATC-15
Flaviviruses	
Banzi	ATC-15
Dengue 1	ATC-10, ATC-15, ATC-173, Mos-61, TRA-171, TRA-284, GML-HE-12, NIVI-AK-453, *A. albo* (Hsu)
Dengue 2	ATC-10, ATC-15, ATC-136, ATC-173, GML-HE-12, NIVI-AK-453, TRA-171, Mos-60, Mos-61, 59, *A. albo* (Hsu), *C. quin* (Hsu)
Dengue 3	ATC-15, ATC-173, GML-HE-12, NIVI-AK-453, TRA-171, Mos-61
Dengue 4	ATC-15, ATC-173, GML-HE-12, NIVI-AK-453, TRA-171, Mos-61, *A. albo* (Hsu), *C. quin* (Hsu)
Japanese encephalitis	ATC-15, ATC-121, ATC-136, ATC-173, TA-9, TA-42, NIVI-AK-453, 59, 364, Mos-20, Mos-60, Mos-61, CmOv, *A. albo* (Hsu), *A. vitt, An. st* (Schn), *C. quin* (Hsu), *C. trit* (Hsu)
Kunjin	59, ATC-15, *A. albo* (Hsu)
Murrey valley encephalitis	*C. quin* (Hsu)
St. Louis	ATC-10, ATC-15, TA-9, TA-42, GML-HE-12, TRA-171
West Nile	ATC-10, ATC-15, ATC-121, ATC-173, TA-9, TA-42, 59, Mos-20, Mos-29, Mos-60, Mos-61, NIVI-AK-453, *A. albo* (Hsu), *A. vitt, C. quin* (Hsu)
Yellow fever	ATC-10, ATC-15, Mill Hill, GML-HE-12, TRA-171, Mos-20, Mos-60, Mos-61

TABLE 3 (Continued)
Arboviruses and Mosquito Cell Lines Susceptible to Them

Viruses	Susceptible cell lines
Nairoviruses	
Dugbe	ATC-15
Orbiviruses	
African horse sickness	ATC-15
Blue tongue	ATC-15, Mos-61
Changuinola	TA-9, TA-42
Chenuda	ATC-10, ATC-15
Colorado tick fever	ATC-15
Corriparta	TA-9, TA-42
Epizootic hemorrhagic disease	ATC-15
Irituia	ATC-10, ATC-15
Kasba	ATC-15
Kemerovo	ATC-10, ATC-15
Lipovnik	ATC-10, ATC-15
Palyam	ATC-15
Tribec	ATC-10, ATC-15
Vellore	ATC-15
Phleboviruses	
Toscana	Mos-61
Vesiculoviruses	
Bovine ephemeral fever	ATC-15
Chandipura	ATC-121, ATC-10, ATC-15, ATC-136, ATC-173, NIVI-AK-453, Mos-20, Mos-60, Mos-61, *A. vitt, An. st* (Schn)
Vesicular stomatitis	TA-9, TA-42, TRA-171, ATC-10, ATC-15, GML-HE-12, NIVI-AK-453
Other viruses	
Arkonam	ATC-15
Carriparta	TA-9, TA-42
Equine anemia	*C. quin* (Hsu)
Minnal	ATC-15
Nodamura	ATC-10, ATC-15, Mill Hill
Rift vally fever	Mos-61

Arbovirus infection of mosquito cell culture is characterized by lack of cytopathic effects and establishment of persistent infection. Most mosquito cells are not killed by arbovirus infection and this may lead to a carrier state. However, it is not certain whether all cells are infected by arboviruses in the carrier cultures. Additional characteristics of carrier cultures are as follows:

1. The growth rates of cells in carrier cultures are similar to those of the corresponding noninfected cells
2. The strength of the virions produced in carrier cultures is weaker than that of wild type viruses
3. The persistently infected cells interfere with the superinfection of the homologous viruses
4. When a carrier culture is maintained for a long period, the viruses produced become temperature sensitive and form small-sized plaques
5. The small plaque-forming, temperature-sensitive viruses have the ability to interfere with the homologous infection, and the interference is stronger in mosquito cells than in vertebrate cells

6. In persistently infected cells RNA, containing only a part of wild virus genes, is produced, but the RNA is not taken into the virions liberated into the medium. The reported carrier culture systems are as listed in Table 4.

Mosquito cell lines can be used for field isolation of mosquito-borne viruses. Singh and Paul[92] and Pavri and Ghosh[93] applied *Aedes albopictus* cells for isolation and identification of dengue viruses from client's blood. Gubler et al.[94] has used *Toxorhynchites amboinensis* cells for detection of flavivirus by a similar method with fluorescent antibody staining. Hsu et al.[95] with *Culex tritaeniorhynchus sumorosus* cells and Igarashi et al.[96] with a C6/36 strain of *Aedes albopictus* cells developed detection methods of Japanese encephalitis virus in mosquitoes collected from fields. Similar applications have been made by several investigators.[97,98]

Only a few mosquito cell lines have been reported to produce interferon-like substances when inoculated with certain viruses,[99] whereas most other mosquito cell lines showed homologous but not heterologous interference.[68,70,78,100-102] At least a part of homologous interference seemed to be caused by defective interference particles (DI particles).[103]

Attempts have been made to produce vaccines by the use of mosquito cell cultures. However, preliminary experiments with dengue vaccine produced from infected mosquito cell cultures revealed that some recipients of the vaccine showed a positive reaction to the Prausnitz-Kustner test. It was assumed that most people had had an earlier experience of being bitten by mosquitoes, and had antibodies to proteins of mosquito origin. Later, however, it became evident that the vaccine produced from the infected nonhematophagous mosquito cell lines also caused a similar but weaker reaction in some recipients. It may be true that mosquitoes of different genera possess a common antigen.[104,105]

Besides the above cited works, there are many studies concerning cell morphology, ultrastructure, and biochemical changes of mosquito cells infected with arboviruses.

B. STUDIES OF PATHOGENS OTHER THAN ARBOVIRUSES

Next to arbovirus studies, mosquito cell lines have been used frequently to study pathogenic protozoa, especially malaria pathogens.

Since *Plasmodium* develops in the mosquito gut, often organ cultures of mosquito guts were used for infection by *Plasmodium in vitro,* Ball and Chao working with *Plasmodium relictum* developed oocysts by inoculating mosquito guts *in vitro,* and also gametocytes-forming sporozoites.[3-5,106-111] Simons et al.[112] maintained *Plasmodium* in cultured *Anopheles stephensi* guts for 30 days. Cultured mosquito cells have been used also for this purpose. Schneider[113] showed that the presence of cultured mosquito cells promoted cocultured oocysts of *Plasmodium gallinaceum.* Speer et al.[114,115] confirmed that the ookinates of *Plasmodium berghei* can grow in the culture of *Anopheles* cells, but did not invade into the mosquito cells. On the other hand, Rosales-Ronquillo and her co-workers[116,117] reported that *Plasmodium* ookinates could grow better in the nonvector cell cultures than in mosquito cell cultures. Later she succeeded in obtaining the complete whole sporogonic cycle of *P. berghei* in media conditioned with mosquito cells and non-vector cells, and she also cultured *Anopheles stephensi* guts.[118]

Chao and Ball[119,120] culturing *Hepatozoon fusifex* and *H. rarefasciens* in mosquito cell cultures succeeded in obtaining their complete sporogonous cycles within 20 days at 23 to 24°C.

Microfilaria of *Macacanema formosana, Dirofilaria immitis,* and *D. corynodesa* were reported to develop in some mosquito cell cultures such as Grace's *Aedes aegypti,* McHale and Sweet's *Cliseta inornata,* and *Aedes vexans,*[59] although these cell lines have been suspected to be identical to Grace's moth cell *Antheraea eucalypti.* Therefore, the microfilaria had been cocultured with nonvector insect cells.

Besides Protozoa, *Culex quinquefasciatus* cells have been used for assaying *Bacillus sphericus* cytotoxin.[121]

TABLE 4
Reported Carrier Culture Systems of Mosquito Cell Lines with Arboviruses

Viruses	Mosquito cell lines	Reporters
Banzi virus	ATC-15	Lee and Schloemer[68]
Bunyamwera virus	ATC-15	Newton et al.[69]
Bunyamwera virus	ATC-15	Elliot and Wilkie[70]
Chikungunya virus	ATC-10, ATC-15, *A. w-alb*	Buckley et al.[71]
Dengue virus 1	ATC-15(C6/36)	Igarashi[72]
Dengue virus 1	TRA-171	Kuno[73]
Dengue virus 2	ATC-15(C6/36)	Igarashi[72]
Dengue virus 2	TRA-171	Kuno[73]
Dengue virus 3	ATC-15(C6/36)	Igarashi[72]
Dengue virus 3	TRA-171	Kuno[73]
Dengue virus 4	ATC-15(C6/36)	Igarashi[72]
Dengue virus 4	TRA-171	Kuno[73]
Getah virus	*A. aeg*	Lisak et al.[74]
Japanese encephalitis virus	*C. tritaen* (Hsu)	Hsu et al.[75]
Japanese encephalitis virus	ATC-15	Banerjee and Singh[76]
Kemerovo virus	ATC-15	Libikova and Buckley[77]
La Crosse virus	ATC-15	Kascsak and Lyons[78]
Mariluba virus	ATC-15	Carvalho et al.[79]
Ross river virus	ATC-15	Raghow et al.[80]
Semliki forest virus	ATC-15	Davy and Dalgarno[81]
Semliki forest virus	ATC-15	Stalder et al.[82]
Sindbis virus	ATC-15	Enzman[83]
Sindbis virus	ATC-15	Stoller and Shenk[84]
Sindbis virus	ATC-15	Eaton[85]
Sindbis virus	*A. aeg* (Peleg)	Stoller et al.[86]
Tahyna virus	ATC-15	Otova and Krobova[87]
Toscana virus	Mos-61	Nicoletti and Verani[88]
Venezuelan equine encephalitis virus	ATC-10, ATC-15	Esparza and Sanchez[89]
Vesicular stomatitis virus	ATC-10, ATC-15	Artsob and Spence[90]
West Nile virus	ATC-15	Banergee and Singh[76]
Western equine encephalitis virus	ATC-15	Maeda et al.[91]

Yunker et al.[122] inoculated a rickettsia, *Coxiella burneti*, into the *Aedes albopictus* cell line, which resulted in increased infectivity to guinea pigs. The same authors[123,124] also showed multiplication of *Rickettsia tsutsugamushi, R. conori, Walbachia*-like agents, and spotted fever group rickettsia in mosquito cell cultures.

C. BIOTECHNOLOGY AND OTHER APPLICATIONS

Some experiments on cell fusion were made with mosquito cells. Zepp and Conover's group[49,50] succeeded in obtaining hybrid cells between HeLa and Grace's mosquito cells (later, this cell was identified as a moth cell, *Antheraea eucalypti*) by the use of an ultraviolet-inactivated Sendai virus. Mitsuhashi[125] attempted to fuse *Aedes albopictus* NIAS-AeA1-2 cell line by the use of polyethyleneglycol. Homologous cell fusion occurred, but heterologous fusion, for example with lepidopteran cells, did not. Synkaryon formation was not confirmed, and isolation of fused cells could not be carried out. He also tried electrofusion of the mosquito cells without success. Chiplonkar[126] reported the formation of polynucleated cells from homologous fusion of mosquito cells by means of electrofusion. Koblet's group has studied syncytium formation of *Aedes albopictus* cells by Semliki Forest virus. In a series of works, they elucidated that the development of cell fusion capacity correlated with the synthesis of the viral RNA and proteins in the cells, and especially coincided with the appearance of the viral proteins at the cell surface. The mechanism of this fusion is, therefore, a so-called fusion from within. They also proved that the extracellular viruses were not involved; in other words, a fusion from without is not the mechanism of the syncytium formation in this case.[127,128] The fusion from within required lowering the pH of the culture medium for a short time after the infection. A similar phenomenon was reported by Randolph and Stollar[129] in *Aedes albopictus* cells infected with dengue and St. Louis encephalitis viruses.

Gene manipulation of mosquito cells has been studied by Fallon and her co-workers. They developed efficient transfection of cultured *Aedes albopictus* cells by the use of polybrene, 1,5-dimethyl-1,5,-diazaundecamethylene polymethobromide.[130,131] By this method they introduced into the mosquito cells a recombinant plasmid, in which the bacterial chloramphenicol acetyltransferase (CAT) gene is under the control of *Drosophila* heat-shock protein 70 promoter. They could also induce the activity of the introduced CAT gene by heat-shock treatment. They examined the efficiency of transfection of mosquito cells by DNA-calcium phosphate coprecipitation and found that under optimal conditions (especially with appropriate growth rate of the recipient cells and the temperature), transfection frequency was up to tenfold higher than by using the polybrene procedure.[132]

Mainly, for biotechnological studies, some mutants or strains of mosquito cells resistant to certain drugs have been developed, for example, mutants resistant to 5-bromodeoxyuridine (BrdU), cycloheximide (138), and methotrexate (139).

In addition to the above cited works, there are many studies concerning basic physiology, biochemistry, morphology, chromosomes, and cell cycles of mosquito cells cultured *in vitro.*

REFERENCES

1. **Trager, W.**, Multiplication of the virus of equine encephalomyelitis in surviving mosquito tissues. *Am. J. Trop. Med.* 18, 387–393, 1938.
2. **Gavrilov, W. and Cowez, S.**, Essai de culture in vitro de tissus de moustiques et d'intestins de lapins adultes infectés. *Ann. Parasitol.* 18, 180–186, 1941.
3. **Ball, G. H.**, Attempts to cultivate the mosquito phase of *Plasmodium relictum. Am. J. Trop. Med.* 27, 301–307, 1947.
4. **Ball, G. H.**, Extended persistence of *Plasmodium relictum* in culture. *Am. J. Trop. Med.* 28, 533–536, 1948.
5. **Ball, G. H.**, Prolonged contraction of mosquito digestive tract in vitro with partial development of oocysts of *Plasmodium relictum. Exp. Parasitol.* 3, 358–367, 1954.
6. **Ragab, H. A.**, Observation on the isolated gut of the mosquito. *Trans. R. Soc. Trop. Med. Hyg.* 43, 225–230, 1949.
7. **Beckel, W. E.**, Maintenance of adult mosquito tissue in a tissue culture medium. *Nature* 177, 534–535, 1956.
8. **Kitamura, S.**, The in vitro cultivation of the midgut tissue of mosquito *Culex pipiens var. pallens. Bull. Kobe Med. College* 21, 602–612, 1961.
9. **Kitamura, S.**, The in vitro cultivation of tissues from the mosquito, *Culex pipiens var. molestus.* I. Cultivation of ovary tissues in vitro. *Kobe J. Med Sci.* 10, 85–94, 1964.
10. **Peleg, J. and Trager, W.**, Cultivation of insect tissues in vitro and their application to the study of arthropod-borne viruses. *Am. J. Trop. Med. Hyg.* 12, 820–824, 1963.
11. **Peleg, J.**, Growth of mosquito tissues in vitro. *Nature* 206, 427–428, 1965.
12. **Grace, T. D. C.**, Establishment of a line of mosquito (*Aedes aegypti* L.) cells grown in vitro. *Nature* 211, 366–367, 1966.
13. **Singh, K. R. P.**, Cell cultures derived from larvae of *Aedes albopictus* (Skuse) and *Aedes aegypti* (L.). *Curr. Sci.* 36, 506–508, 1967.
14. **Mitsuhashi, J. and Maramorosch, K.**, Leafhopper tissue culture: embryonic, nymphal and imaginal tissues from aseptic insects. *Contrib. Boyce Thompson Inst.* 22, 435–460, 1964.
15. **Peleg, J.**, Behaviour of infectious RNA from four different viruses in continuously subcultured *Aedes aegypti* mosquito embryo cells. *Nature* 221, 193–194, 1969.
16. **Varma, M. G. R. and Pudney, M.**, The growth and serial passage of cell lines from *Aedes aegypti* (L.) larvae in different media. *J. Med. Entomol.* 6, 432–439, 1969.
17. **Bailey, L., Newman, J. F. E., and Poterfield, J. S.**, The multiplication of Nodamura virus in insect and mammalian cell cultures. *J. Gen. Virol.* 26, 15–20, 1975.
18. **Bhat, U. K. M.**, Establishment of new cell lines from *Aedes albopictus* larval cells in media containing calf and goat sera. *Curr. Sci.* 42, 66, 1973.
19. **Hsu, S. H., Huan, M. H., Wang, B. T., Wong, W. J., and Liu, S. N.**, Growth of some togaviruses in *Aedes albopictus* (Diptera: Culicidae) cell cultures. *J. Med. Entomol.* 14, 581–584, 1978.

20. **Mitsuhashi, J.,** A new continuous cell line from larvae of the mosquito *Aedes albopictus* (Diptera, Culicidae). *Biomed. Res.* 2, 599–606, 1964.
21. **Cahoon, B. E., Hardy, J. L., and Reeves, W. C.,** Initiation and characterization of a diploid cell line from larval tissues of *Aedes dorsalis* (Meigen). *In Vitro* 14, 255–260, 1978.
22. **Pant, U., Sudeep, A. B., Dhanda, V., Mourya, D. T., and Ganerfee, K.,** New embryonic cell line from *Aedes krombeini* (H.) (Diptera: Culicidae). *In Vitro Cell. Dev. Biol.* 28A, 567–568, 1992.
23. **Varma, M. G. R., Pudney, M., and Leake, C. J.,** Cell lines from larvae of *Aedes (Stegomyia) malayensis* Colless and *Aedes(s) pseudoscutellaris* (Theobald) and their infection with some arbovirus. *Trans. R. Soc. Trop. Med. Hyg.* 68, 374–382, 1974.
24. **Bhat, U. K. M. and Guru, P. Y.,** *Aedes novalbopictus:* establishment and characterization of larval cell lines. *Exp. Parasitol.*, 33, 105–113, 1973.
25. **Rowley, W. A., Dorsey, D. C., and Knowles, M. A.,** The replication of two California serogroup viruses in a cell line from the mosquito *Aedes triseriatus* (Diptera: Culicidae). *J. Med. Entomol.* 21, 501–505, 1984.
26. **Sweet, B. H. and McHale, J. S.,** Characterization of cell lines derived from *Cliseta inornata* and *Aedes vexans* mosquitoes. *Exp. Cell Res.* 61, 51–63, 1970.
27. **Bhat, U. K. M. and Singh, K. R. P.,** Establishment of a diploid cell line from larval tissues of *Aedes vittatus* (Bigot, 1861). *Curr. Sci.* 39, 388–390, 1970.
28. **Singh, K. R. P. and Bhat, U. K. M.,** Establishment of mosquito cell lines from larval tissues of *Aedes w-albus. Experientia* 27, 142–143, 1971.
29. **Marhoul, Z. and Pudney, M.,** A mosquito cell line (Mos. 55) from *Anopheles gambiae* larvae. *Trans. R. Soc. Trop. Med. Hyg.* 66, 183–184, 1972.
30. **Schneider, I.,** Establishment of three diploid cell lines of *Anopheles stephensi* (Diptera:Culicidae). *J. Cell Biol.* 42, 603–606, 1969.
31. **Pudney, M. and Varma, M. G. R.,** *Anopheles stephensi* var. *mysorensis:* establishment of a larval cell line (Mos. 43). *Exp. Parasitol.* 29, 7–12, 1971.
32. **Pudney, M., McCarthy, D., and Chortridge, K. F.,** Rod-shaped virus-like particles in cultured *Anopheles* cells in an *Anopheles* laboratory colony. Proc. 3rd Int. Colloq. Invert. Tissue Culture, Smolnice, 1971, 337–345, 1973.
33. **Pant, U. and Dhanda, V.,** Establishment of a cell line from *Culex bitaeniorhynchus. J. Tissue Culture Methods* 6, 61–63, 1980.
34. **Chao, J. and Ball, G. H.,** The establishment of *Culex pipiens* cell line and its adaptation to grow in three different insect tissue culture media. *In Vitro* 8, 406, 1973.
35. **Kitamura, S.,** Establishment of cell line from *Culex* mosquito. *Kobe J. Med. Sci.* 16, 41–50, 1970.
36. **Hsu, S. H., Mao, W. H., and Cross, J. H.,** Establishment of a line of cells derived from ovarian tissues of *Culex quinquefasciatus* Say. *J. Med. Entomol.* 7, 703–707, 1970.
37. **Schneider, I.,** Establishment of cell lines from *Culex tritaeniorhynchus* and *Culex salinarius* (Diptera: Culicidae). Proc. 3rd. Int. Colloq. Invert. Tissue Culture, Smolenice, 1971, 121–134, 1973.
38. **Hsu, S. H., Li, S. Y., and Cross, J. H.,** A cell line derived from ovarian tissue of *Culex tritaeniorhynchus summorosus* Dyar. *J. Med. Entomol.* 9, 86–91, 1972.
39. **Oro, G.,** Establishment of a mosquito cell line from *Haemagogus equinus* larvae. *In Vitro* 20, 153–156, 1984.
40. **Tesh, R. B.,** Establishment of 2 cell lines from the mosquito *Toxorhynchites amboinensis* (Diptera: Culicidae) and their susceptibility to infection with arboviruses. *J. Med. Entomol.* 17, 338–343, 1980.
41. **Kuno, G.,** A continuous cell line of a nonhematophagous mosquito, *Toxorhynchites amboinensis. In Vitro* 16, 915–917, 1980.
42. **Kuno, G.,** A method for isolating continuous cell lines from *Toxorhynchites amboinensis* (Diptera, Culicidae). *J. Med. Entomol.* 18, 140–144, 1981.
43. **Munderloh, U. G., Kurtti, T. J., and Maramorosch, K.,** *Anopheles stephensi* and *Toxorhynchites amboinensis:* aseptic rearing of mosquito larvae on cultured cells. *J. Parasitol.* 68, 1085–1091, 1982.
44. **Peleg, J.,** Growth of *Aedes aegypti* embryonic cells and tissues in vitro. *Experientia* 22, 555–556, 1966.
45. **Igarashi, A. and Srivastava, A. K.,** Suspension culture of *Aedes albopictus,* clone C6/36, cells and its application to the production of flavivirus. In *Invertebrate and Fish Tissue Culture,* Kuroda, Y., Kurstak, E., and Maramorosch, K., Eds., Japan Scientific Societies Press, Tokyo/Springer-Verlag, Berlin, 137–139, 1988.
46. **Greene, A. E., Charney, J., Nicols, W. W., and Coriell, L. L.,** Species identity of insect cell lines. *In Vitro* 7, 313–322, 1972.
47. **Suitor, E. C. Jr., Chang, L. L., and Liu, H. H.,** Establishment and characterization of a clone from Grace's in vitro cultured mosquito (*Aedes aegypti* L.) cells. *Exp. Cell Res.* 44, 572–578, 1966.
48. **Hsu, S. H., Liu, H. H., and Suitor, E. C. Jr.,** Further description on a subline of Grace's mosquito (*Aedes aegypti* L.) cells adapted to hemolymph-free medium. *Mosquito News* 29, 439–446, 1969.
49. **Zepp, H. D., Conover, J. H., Hirschhorn, K., and Hodes, H. L.,** Human mosquito cell hybrids induced by ultraviolet-inactivated Sendai virus. *Nature (London) New Biol.* 229, 119–121, 1971.
50. **Conover, J. H., Zepp, H. D., Hirschhorn, K., and Hodes, H. L.,** Production of human-mosquito somatic cell hybrids and their response to virus infection. *Curr. Topics Microbiol. Immunol.* 55, 85–92, 1971.

51. **Řeháček, J.**, The growth of arboviruses in mosquito cells in vitro. *Acta Virol.* 12, 241–246, 1968.
52. **Mitsuhashi, J., Grace, T. D. C., and Waterhouse, D. F.**, Effects of insecticides on cultures of insect cells. *Entomol. Exp. Appl.* 13, 327–341, 1970.
53. **Filshie, B. K. and Řeháček, J.**, Studies of the morphology of Murray valley encephalitis and Japanese encephalitis viruses growing in cultured mosquito cells. *Virology* 34, 435–443, 1968.
54. **Ball, G. H. and Chao, J.**, The cultivation of *Plasmodium relictum* in mosquito cell lines. *J. Parasitol.* 57, 391–395, 1971.
55. **Wood, D. E. and Suitor, E. C. Jr.**, In vitro development of microfilariae of *Macacanema formosana* in mosquito cell cultures. *Nature* 211, 868–870, 1966.
56. **Sweet, B. H. and Dupree, L. T.**, Growth of cells derived from *Cliseta inornata* and *Aedes vexans* in tissue culture. A preliminary note. *Mosquito News* 28, 368–373, 1968.
57. **McHale, J. S. and Sweet, B. H.**, Morphological and cultural characteristics of *Culiseta inornata* and *Aedes vexans* mosquito cell lines. *Curr. Topics Microbiol. Immunol.* 55, 42–47, 1971.
58. **Ibrahim, A. N. and Sweet, B. H.**, Antigenic relationships of mosquito cell lines as determined by immunodiffusion techniques. *Curr. Topics Microbiol. Immunol.* 55, 47–51, 1971.
59. **Cupp, E. W.**, Development of filariae in mosquito cell cultures. Workshop on Development of Filariae in Mosquitoes, U.S.-Japan Cooperative Medical Science Program, U.S. Panel on Parasitic Diseases, Los Angeles, 1972, 7–8.
60. **Cunningham, A., Buckley, S. M., Casals, J., and Webb, S. R.**, Isolation of chikungunya virus contaminating an *Aedes albopictus* cell line. *J. Gen. Virol.* 27, 97–199, 1975.
61. **Hirumi, H., Hirumi, K., Speyer, G., Yunker, C. E., Thomas, L. A., Cory, J., and Sweet, B. H.**, Viral contamination of a mosquito cell line, *Aedes albopictus* associated with syncytium formation. *In Vitro* 12, 83–97, 1976.
62. **Hirumi, H.**, Viral, microbial, and extrinsic cell contaminations of insect cell cultures. In *Invertebrate Tissue Culture, Research Applications*, Maramorosch, K., Ed., Academic Press, New York, 233–268, 1976.
63. **Gorziglia, M., Botero, L., Gil, F., and Esparza, J.**, Preliminary characterization of virus-like particles in a mosquito *(Aedes pseudoscutelaris)* cell line (Mos-61). *Intervirology* 13, 232–240, 1980.
64. **Mitsuhashi, J.**, Isolation of a continuous cell line from larval fat bodies of an Arctiid moth, *Spilarctia seriatopunctata* (Insecta, Lepidoptera, Arctiidae). *Zool. Sci.* 1, 415–419, 1984.
65. **Mitsuhashi, J. and Inoue, H.**, Obtainment of a continuous cell line from the larval fat bodies of the mulberry tiger moth, *Spilosoma imparilis* (Lepidoptera, Arctiidae). *Appl. Entomol. Zool.* 23, 488–489, 1988.
66. **Mitsuhashi, J.**, Continuous cultures of insect cell lines in media free of sera. *Appl. Entomol. Zool.* 17, 575–581, 1982.
67. **Kuno, G.**, Cultivation of mosquito cell lines in serum-free media and their effects on dengue virus replication. *In Vitro* 19, 707–713, 1983.
68. **Lee, C. H. and Schloemer, R. H.**, Identification of the antiviral factor in culture medium of mosquito cells persistently infected with Banzi virus. *Virology* 110, 445–454, 1981.
69. **Newton, S. E., Short, N. J., and Dalgarno, L.**, Bunyamwera virus replication in cultured *Aedes albopictus* cells: establishment of a persistent viral infection. *J. Virol.* 38, 1015–1024, 1981.
70. **Elliot, R. M. and Wilkie, M.**, Persistent infection of *Aedes albopictus* C6/36 cells by bunyamwera virus. *Virology* 150, 21–32, 1986.
71. **Buckley, S. M., Singh, K. R. P., and Bhat, U. K. M.**, Small and large-plaque variants of chikungunya virus in two vertebrate and seven invertebrate cell lines. *Acta Virol.* 19, 10–18, 1975.
72. **Igarashi, A.**, Characteristics of *Aedes albopictus* cells persistently infected with dengue viruses. *Nature* 280, 690–691, 1979.
73. **Kuno, G.**, Persistent infection of a nonvector mosquito cell line (TRA-171) with dengue viruses. *Intervirology* 18, 45–55, 1982.
74. **Lisak, V. M., Korolev, M. B., Chunikbin, S. P., and Izotov, V. K.**, Comparative studies of Getah virus morphogenesis in mammalian and mosquito cell cultures. *Vopr. Virusol.* No. 2, 137–142, 1979.
75. **Hsu, S. H., Wang, B. T., Huang, M. H., Wong, W. J., and Cross, J. H.**, Growth of Japanese encephalitis virus in *Culex taeniorhynchus* cell cultures. *Am. J. Trop. Med. Hyg.* 24, 881–888, 1975.
76. **Banerjee, K. and Singh, K. R. P.**, Establishment of carrier cultures of *Aedes albopictus* cell line infected with arboviruses. *Ind. J. Med. Res.* 56, 812–814, 1968.
77. **Libikova, H. and Buckley, S. M.**, Studies with Kemerovo virus in Singh's *Aedes albopictus* cell lines. *Acta Virol.* 15, 393–403, 1971.
78. **Kascsak, R. J. and Lyons, M. J.**, Attempts to demonstrate the interferon defense mechanism in cultured mosquito cells. *Arch. Gen. Virusforsch.* 45, 149–154, 1974.
79. **Carvalho, M. G. C., Frugulhetti, I. C., and Rebello, M. A.**, Marituba (Bunyaviridae) virus replication in cultured *Aedes albopictus* cells and LA9 cells. *Arch. Virol.* 90, 325–335, 1986.
80. **Ragow, R. S., Grace, T. D. C., Filshie, B. K., Bartley, W., and Dalgarno, L.**, Ross river virus replication in cultured mosquito and mammalian cells: virus growth and correlated ultrastructural changes. *J. Gen. Virol.* 21, 109–122, 1973.
81. **Davey, M. W. and Dalgarno, L.**, Semliki forest virus replication in cultured *Aedes albopictus* cells: studies on the establishment of persistence. *J. Gen. Virol.* 24, 453–463, 1974.

82. **Stalder, J., Reigel, F., Flaviana, A., and Koblet, H.,** Infection of the *Aedes albopictus* cell clone C6/36 with Semliki forest virus. *Experientia* 37, 1229, 1981.
83. **Enzmann, P. J.,** Induction of an interferon-like substance in persistently infected *Aedes albopictus* cells. *Arch. Ges. Virusforsch.* 41, 382–389, 1973.
84. **Stollar, V. and Shenk, T. E.,** Homologous viral interference in *Aedes albopictus* cultures chronically infected with Sindbis virus. *J. Virol.* 11, 592–595, 1973.
85. **Eaton, B. T.,** Sindbis virus variants from a persistently infected mosquito cell culture contain an altered E2 glycoprotein. *Virology* 122, 486–491, 1982.
86. **Stollar, V., Peleg, J., and Shenk, T. E.,** Temperature sensitivity of a Sindbis virus mutant from persistently infected *Aedes aegypti* cell culture. *Intervirology* 2, 337–344, 1974.
87. **Otova, B. and Krobova, J.,** Persistent Tahyna virus infection of stable GMK and Singh's *Aedes albopictus* mosquito cell lines. *Acta Virol.* 18, 402–406, 1974.
88. **Nicoletti, L. and Verani, P.,** Growth of the phlebovirus Toscana in a mosquito *(Aedes pseudoscutellaris)* cell line (AP-61): establishment of a persistent infection. *Arch. Virol.* 85, 35–45, 1985.
89. **Esparza, J. and Sanchez, A.,** Multiplication of a Venezuelan equine encephalitis (Mucanmbo) virus in cultured mosquito cells. *Arch Virol.* 49, 273–280, 1975.
90. **Artsob, H. and Spence, L.,** Persistent infection of mosquito cell lines with vesiculer stomatitis virus. *Acta Virol.* 18, 331–340, 1974.
91. **Maeda, S., Hashimoto, K., and Simizu, B.,** Complementation between temperature-sensitive mutants isolated from *Aedes albopictus* cells persistently infected with western equine encephalitis virus. *Virology* 92, 532–541, 1979.
92. **Singh, K. R. P. and Paul, S. D.,** Isolation of dengue viruses in *Aedes albopictus* cell cultures. *Bull. WHO* 40, 982–983, 1969.
93. **Pavri, K. M. and Ghosh, S. N.,** Complement-fixation tests for simultaneous isolation and identification of dengue viruses, using tissue cultures. *Bull. WHO* 40, 984–986, 1969.
94. **Gubler, D. J., Kuno, G., Sather, G. E., Velez, M., and Oliver, A.,** Mosquito cell cultures and specific monoclonal antibodies in surveillance for dengue viruses. *Am. J. Trop. Med. Hyg.* 33, 158–165, 1984.
95. **Hsu, S. H., Huang, W. C., and Cross, J. H.,** Isolation of Japanese encephalitis virus from Taiwan mosquitoes by mosquito cell cultures and mouse inoculation. *J. Med. Entomol.* 14, 698–701, 1978.
96. **Igarashi, A., Buei, K., Ueba, N., Yoshida, M., Ito, S., Nakamura, H., Sasao, F., and Fukai, K.,** Isolation of viruses from female *Culex tritaeniorhynchus* in *Aedes albopictus* cell cultures. *Am. J. Trop. Med. Hyg.* 30, 449–460, 1981.
97. **Sawyer, M. and Osburn, B. I.,** Processing of wild caught Culicoides and isolation of blue tongue virus using a mosquito cell line. *J. Tissue Culture Method* 11, 171–173.
98. **Digoutte, J. P., Jouan, A., Guenno, B. L., Riou, O., Philippe, B., Meagan, J., Ksiazek, T. G., and Peters, C. J.,** Isolation of the Rift valley fever virus by inoculation into *Aedes pseudoscutellaris* cells: comparison with other diagnostic methods. *Res. Virol.* 140, 31–41, 1989.
99. **Eaton, B. T.,** Heterologous interference in *Aedes albopictus* cells infected with alpha viruses. *J. Virol.* 30, 45–55, 1979.
100. **Condreay, L. D. and Brown, D. T.,** Exclusion of super infecting homologous virus by Sindbis virus-infected *Aedes albopictus* mosquito cells. *J. Virol.* 58, 81–86, 1986.
101. **Lee, C. H. and Schloemer, R. H.,** Mosquito cells infected with Banzi virus secrete and antiviral activity which is of viral origin. *Virology* 110, 402–410, 1981.
102. **Dittmar, D., Castro, A., and Haines, H.,** Demonstration of interference between dengue virus types in cultured mosquito cells using monoclonal antibody probes. *J. Gen. Virol.* 59, 273–282, 1982.
103. **Logan, K. B.,** Generation of defective interfering particles of Semliki forest virus in a clone of *Aedes albopictus* (mosquito) cells. *J. Virol.* 30, 38–44, 1979.
104. **Eckels, K. H., Dubois, D. R., Summers, P. L., and Russell, P. K.,** The use of the C6/36 clone of *Aedes albopictus* cells for the development of live-attenuated dengue virus vaccines. 6th Int. Conf. Invert. Tissue Culture, St. Augustine, Presentation, 1983.
105. **Scott, R. M., Shelton, A. L., Bancroft, W. H., and Eckels, K. H.,** Safety tests and evaluation of hypersensitivity to mosquito cells. 6th Int. Conf. Invert. Tissue Culture, St. Augustine, Presentation, 1983.
106. **Ball, G. H. and Chao, J.,** Development in vitro of isolated oocysts of *Plasmodium relictum. J. Parasitol.* 43, 409–412, 1957.
107. **Ball, G. H. and Chao, J.,** In vitro development of the mosquito phase of *Plasmodium relictum. Exp. Parasitol.* 9, 47–55, 1960.
108. **Ball, G. H. and Chao, J.,** Contribution of in vitro culture towards understanding the relationship between avian malaria and the invertebrate host. *Ann. N.Y. Acad. Sci.* 113, 322–331, 1963.
109. **Ball, G. H. and Chao, J.,** The relationship of the mosquito stomach and other organs to malaria parasites as indicated by [in vitro] culture. *Ann. Epiphyties* 14, 205–210, 1963.

110. **Ball, G. H. and Chao, J.,** The cultivation of *Plasmodium relictum* in mosquito cell lines. *J. Parasitol.* 57, 391–395, 1971.
111. **Ball, G. H. and Chao, J.,** Use of amino acids by *Plasmodium relictum* oocysts in vitro. *Exp. Parasitol.* 39, 115–118, 1976.
112. **Simons, R. W., Rosales-Ronquillo, M. C., and Silverman, P. H.,** Long term maintenance in a nutrient medium of adult organs taken from aseptically reared mosquitoes, *Anopheles stephensi.* Proc. 3rd Int. Colloq. Invert. Tissue Culture, Smolenice, 1971, 113–119, 1973.
113. **Schneider, I.,** Cultivation of *Plasmodium gallinaceum* oocysts in Grace's cell strain of *Aedes aegypti* (L.). Proc. 2nd Int. Colloq. Invert. Tissue Culture, Tremezzo, Como 1967, C. Barigozzi, Ed., Milano, 245–253, 1968.
114. **Speer, C. A., Rosales-Ronquillo, M. C., and Silverman, P. H.,** Scanning electron microscope observations of *Plasmodium berghei* ookinates in primary mosquito cell cultures. *J. Invert. Pathol.* 24, 179–183, 1974.
115. **Speer, C A., Rosales-Ronquillo, M. C., and Silverman, P. H.,** Mobility of *Plasmodium berghei* ookinates in vitro. *J. Invert. Pathol.* 25, 73–78, 1975.
116. **Rosales-Ronquillo, M. C. and Silverman, P. H.,** In vitro ookinates development of the rodent parasite, *Plasmodium berghei. J. Parasitol.* 60, 819–824, 1974.
117. **Rosales-Ronquillo, M. C., Nienaber, G., and Silverman, P. H.,** *Plasmodium berghei* ookinate formation in a non-vector cell line. *J. Parasitol.* 60, 1039–1040, 1974.
118. **Rosales-Sharp, M. C.,** In vitro culture of the sporogonic forms of a malaria parasite. *In Vitro* 15, 211–212, 1979 (abstract).
119. **Chao, J. and Ball, G. H.,** In vitro culture of the vector phase of snake hemogregarienes in mosquito cell lines. *J. Parasitol.* 58, 148–152, 1972.
120. **Chao, J. and Ball, G. H.,** The establishment of a *Culex pipiens* cell line and its adaptation to grow in three different insect tissue culture media. *In Vitro* 8, 406, 1973 (abstract).
121. **Davidson, E. W.,** Effects of *Bacillus sphericus* 1593 and 2362 spore/crystal toxin on cultured mosquito cells. *J. Invert. Pathol.* 47, 21–31, 1986.
122. **Yunker, C. E., Ormsbee, R. A., Cory, J., and Peacock, M. G.,** Infection of insect cells with *Coxiella burneti. Acta Virol.* 14, 383–392, 1970.
123. **Cory, J. and Yunker, C. E.,** Rickettsial plaques in mosquito cell monolayers. *Acta Virol.* 18, 512–513, 1974.
124. **Cory, J., Yunker, C. E., Howarth, J. A., Hokawa, Y., Hughes, L. E., Thomas, L. A., and Clifford, C. M.,** Isolation of spotted fever group and Wolbachia-like agents from field-collected materials by means of plaque formation in mammalian and mosquito cells. *Acta Virol.* 19, 443–445, 1975.
125. **Mitsuhashi, J.,** Fusion of insect cells. In *Biotechnology in Invertebrate Pathology and Cell Culture,* Maramorosch, K., Ed., Academic Press, New York, 387–400, 1987.
126. **Chiplonkar, J. M.,** Electrofusion of mosquito cells. *In Vitro Cell. Develop. Biol.* 28A, 230–232, 1992.
127. **Omar, A., Flavino, A., Reigel, F., Kohler, U., and Koblet, H.,** Syncytium formation and inhibition in Semliki Forest virus infected *Aedes albopictus* cells at low pH. In *Invertebrate Cell System Applications,* Vol. 2, Mitsuhashi, J., Ed., CRC Press, Boca Raton, FL, 147–150, 1989.
128. **Koblet, H., Omar, A., Kohler, U., and Kempf, C.,** Investigation of cell-cell fusion in Semliki Forest virus (SFV) infected C6/36 (mosquito) cells. In *Invertebrate and Fish Tissue Culture,* Kuroda, Y., Kurstak, E., and Maramorosch, K., Eds., Japan Scientific Societies Press, Tokyo/Springer-Verlag, 140–143, 1988.
129. **Randorph, V. B. and Stollar, V.,** Low pH-induced cell fusion in flavivirus-infected *Aedes albopictus* cell cultures. *J. Gen. Virol.* 71, 1845–1850, 1990.
130. **Fallon, A. M. and Durbin, J. E.,** Polybrene-mediated transfection of cultured mosquito cells: expression and regulation of transferred genes. *In Vitro* 21, 42A, 1985 (abstract).
131. **Durbin, J. E. and Fallon, A. M.,** Transient expression of chloramphenicol acetyltransferase gene in cultured mosquito cells. *Gene* 36, 173–178, 1985.
132. **Kjer, K. M. and Fallon, A. M.,** Efficient transfection of mosquito cells is influenced by the temperature at which DNA-calcium phosphate coprecipitates are prepared. *Arch. Insect Biochem. Physiol.* 16, 189–200, 1991.
133. **Mento, S. J. and Stollar, V.,** Isolation and partial characterization of drug-resistant *Aedes albopictus* cells. *Somatic Cell Gen.* 4, 179–191, 1978.
134. **Mazzacano, C. A. and Fallon, A. M.,** Thymidine kinase-deficient mutants of *Aedes albopictus* mosquito cells. *In Vitro Cell. Develop. Biol.* 28A, 455–458, 1992.
135. **Mento, S. J., Malinoski, F., and Stollar, V.,** Phenotypic characterization of ouabain-resistant *Aedes albopictus* cells. *J. Cell Physiol.* 101, 515–522, 1979.
136. **Fallon, A. M. and Stollar, V.,** Isolation and characterization of puromycin-resistant clones from cultured mosquito cells. *Somatic Cell Gen.* 8, 521–532, 1982.
137. **Sherwood, A. and Stollar, V.,** *Aedes albopictus* cells resistant to adenosine because of defect in nucleoside transport. *Somatic Cell Gen.* 8, 575–585, 1982.
138. **Nouri, N. and Fallon, A. M.,** Cycloheximide-resistant mosquito cells are temperature-sensitive. *In Vitro Cell. Dev. Biol.* 22, 14A, 1986 (abstract).
139. **Fallon, A. M.,** Role of gene amplification in insecticide resistance. *In Vitro Cell. Dev. Biol.* 23, 63A, 1987 (abstract).

Chapter 3

LEPIDOPTERAN CELL CULTURE*

James L. Vaughn

TABLE OF CONTENTS

* Names are necessary to report factually on available data; however, the USDA neither guarantees nor warrants the standard of the product, and the use of the name by USDA implies no approval of the product to the exclusion of others that may also be suitable.

James L. Vaughn, Ph.D., is the Research Leader of the Insect Biocontrol Laboratory, ARS, USDA, Beltsville, Maryland. He received his education at the University of Wisconsin, Madison obtaining his BA degree in Bacteriology in 1957 and his Ph.D. in bacteriology in 1962. Research Officer, Forestry, Insect Pathology Research Institute, Sault St. Marie, Ontario, Canada, 1961–1965; Research Microbiologist, Insect Pathology Pioneering Research Laboratory, ARS, USDA, Beltsville, MD, 1965–1979; Research Leader, Insect Pathology Laboratory ARS, USDA, Beltsville, 1979–1990; Research Leader, Insect Biocontrol Laboratory, 1990–to date. TCA; SIP; AAAS; ASM; Insect cell culture; media development; baculoviruses.

I. INTRODUCTION

Man has a keen interest in the moths and butterflies comprising the family Lepidoptera. The bright coloration and attractive markings of many adult Lepidoptera make them very visible and desirable for decorative arrangements, collections, and jewelry. Many are of economic importance; the silkworm is a source of silk thread, which fueled interest in its biology and pathology. Other Lepidoptera are serious pests of crops, forests, and ornamentals, and many studies have been done to develop methods for their control. The large size of many Lepidoptera made them useful in physiological studies. Thus, it is not surprising that the first efforts in insect cell culture were with tissues from a lepidopterous insect.

In 1915 Goldschmidt[1] observed cell division and gametogenesis *in vitro* using cecropia spermatocytes in a simple salt solution plus the hemolymph from cecropia larva. The first extended cell survival and the replication of an insect virus *in vitro* was reported in 1935 by Trager.[2] He used a medium consisting of a balanced salt solution, maltose, egg albumin digest, and silkworm hemolymph. The cells were from ovarian tissues of the silkworm and were maintained *in vitro* for several weeks. Extended culture of cells from the tissues of the silkworm were reported in 1959 by Gaw and his colleagues in China.[3] However, this report was not widely available and the results were overlooked for many years. Interested readers are referred to Aizawa[4] for a more inclusive review of these early studies.

In 1962 Grace[5] reported the indefinite culture of cells from the tissues of another lepidopteran, the large Australian emperor gum moth, *Antheraea euclypti.* These are generally considered to be the first cell lines from an invertebrate animal. This report had an immediate effect on the culture of invertebrate cells, and within 10 years there were 85 lines from 35 species of invertebrates, including 13 lines from 10 species of Lepidoptera reported.[6] There are now more than 150 cell lines reported from Lepidoptera.[7]

These cell lines are utilized not only in research on the biology and pathology of insects, but in human and veterinary medicine as well. As discussed elsewhere in this book, an active, biotechnology-based commercial field is developing using lepidopteran cells as a substrate for the manufacture of a number of biologicals. Therefore, there is widespread interest in the development of lepidopteran cell culture. This chapter will review the tissues most frequently used to establish lepidopteran cell lines, some methods used to characterize the cell lines, and then briefly explore some areas of research where additional cell lines would be desirable. For a comprehensive listing of cell lines and their sources, the reader is referred to the listings of Hink and associates[6,8-11] and the review by Oberlander and Miller.[12]

II. TISSUES OF ORIGIN

A. OVARIES

The cell lines developed by Grace were from the ovaries of diapausing pupae of *A. eucalypti.* The procedures that he used were similar to those used by others at the time; the organs were dissected from the insect, unwanted tissues removed, and the ovarian tissue either treated with a proteolytic enzyme, i.e. trypsin, or minced and placed in culture medium. However, the medium was his modification of the medium of Wyatt[13] to which he added B vitamins and choline chloride. In addition to the new medium, which quickly became a standard in the field, what had Grace done differently to achieve success where others had not during the 45 years since the work of Goldschmidt? In retrospect, two things seem to have been critical. First, he used a high ratio of tissue to medium, tissue from 4 ovaries to 0.9 ml of medium; and second, he did not rush to subculture the primary cultures. The medium on the primary cultures was changed every 7 to 14 days for 10 months until the cells began to multiply rapidly. Grace[14] was later to refer to this as "organized neglect." He observed that after 50 days many cells had begun to degenerate and by 70 days only a few cells remained healthy, at which time most workers would have discarded the primary. He noted 4 months later that the number of cells was beginning to increase rapidly and the first subculture was made 12 days after the renewed growth had begun.

Ovarian cell cultures have been established from tissues taken from all developmental stages of Lepidoptera. Stanley and Vaughn[15] found that good primary cultures depended more on the developmental stage of the ovary and not on the stage of the insect. In their studies the best primary cultures were obtained from tissues in which follicular development was evident, but the eggs had not matured. The intermediate layer contained healthy cells and the muscle was not yet contracting.

The actual source of the cells, which grew out of the primary explants, has been the subject of much speculation and some study. Trager[2] suggested that the cells that migrated from the tissue fragments in his cultures came from the connective tissue lining the ovarian tubes. Grace[16] and Jones and Cunningham[17] concluded that the cells of the intermediate layer also were the source of many of the migrating cells. Stanley and Vaughn,[15] who used minced tissue in which the membranes surrounding the ovariols were breached, concluded that under those conditions follicular cells were an important component of the cellular outgrowth in ovarian primary cultures. Whatever the tissue of origin, ovaries became a frequent choice for primary culture.

Lines derived from these primaries contained cells with a variety of morphologies, but no specific cell types have been identified. For studies on media formulation and virus replication, this mixed cell population has not presented any problems and the heterogeneity has been used to advantage by some researchers. Volkman and Summers[18] obtained a clone from Hink's *Trichoplusia ni* 368 cell line that was a more sensitive indicator in virus plaque assays than the uncloned line. Similarly, Summers and Smith[19] obtained a clone, Sf9, from the *Spodoptera frugiperda* line, IPLB-Sf-21AE,[20] that was a superior producer of expressed proteins when infected with a baculovirus expression vector.

B. HEMOCYTES

Another source of dividing cells in Lepidoptera that were easily obtained were the hemocytes. These cells were abundant in all species, required no tedious dissection to obtain, and no treatment to dissociate the tissue. In early attempts by Grace[21] to culture hemocytes from *Bombyx mori*, a syncytium was formed within 24 h of establishing the primary culture. The cells survived less than a week in this condition. Martignoni and Scallion[22] obtained polynucleated synctium and single cell monolayers with hemocyte cultures from *Peridroma saucia*. The monolayer was composed of plasmatocytes, prohemocytes, spheroidocytes, and other cells, many of which were degenerating. The monolayer was maintained for 10 to 15 days and contained predominantly plasmocytes. Mitsuhashi[23] succeeded in subculturing hemocytes from pupae of the rice borer, *Chilo suppressalis*. Of the seven types of hemocytes, only prohemocytes and plasmatocytes attached readily to the culture vessel and multiplied in the primary culture. Cell division began in both types after about 1 week in culture, and the prohemocytes became the dominant cell type as the culture was passaged. The granular hemocytes usually did not survive more than 2 days. The primary culture was maintained for 12 months by changing the medium weekly. Subculturing was first attempted after 7 months by gently scraping off parts of the monolayer. The subculture survived for 5 months and cells divided by mitosis. Chao and Ball[24] reported similar results in a line from *Samia cynthia* based entirely on the comparative morphology of the cells after several months in culture.

A more detailed study by Arnold and Sohi[25] of the cells of a hemocyte line that had been in culture for over a year compared those cells to hemocytes in fresh blood smears from the same insect, *Malacosoma disstria*. They reported that the cell line contained prohemocytes, plasmatocytes, and granular hemocytes. The cell types present in the culture were sufficiently distinctive to permit identification, but were not morphologically identical to those in fresh blood. The cells were highly polyploid and generally larger than the comparable type in fresh blood. As in the previously described cell lines, the prohemocyte-plasmatocyte complex predominated. Mitosis was seen clearly only in these cells in culture, although the authors reported "that cell division of some kind was common there also in the granular hemocytes." Thus, it appears that of the several types of cells in the hemolymph of Lepidoptera, the prohemocytes can be cultured consistently, the plasmatocytes may also grow under some conditions, but that other cell types are lost. To date, 18 hemocyte derived cell lines have been reported from Lepidoptera (Lynn, personal communication).

C. EMBRYOS

Although the two tissues discussed above have been excellent sources of cells for culture, each provides a limited number of cell types that can be cultured indefinitely. If it is correctly assumed that the cell lines derived from these tissues came only from stem cells or cells that had not yet terminally differentiated, then embryonic tissues may contain a more diverse selection of such dividing cells. Since Lepidoptera lay copious quantities of eggs, embryonic tissues in sufficient quantity to establish primary cultures are available, and a number of attempts have been made to culture these tissues. The age at the time of culture varied from 24 h old to fully formed embryos within 24 h of hatching.

The outgrowths from early stage embryos contain cells of a variety of morphologies: epithelial-like, fibroblast-like, highly refractive small spherical cells, large cells containing light or dark vacuoles, and long spindle-shaped fibrous cells.[26-30] The cells grew either as aggregates, monolayers, or free floating. The composition of the primary culture changed with time as growth patterns and cells with different morphologies appeared.

Miltenburger et al.[26] established a number of cell lines from embryos of the apple codling moth, *Cydia pomonella*. These embryos were 2 to 3 days old or older. Cells from 200 embryos were used for each primary, and in 32 out of 35 primaries cell growth reached a density that permitted subculture. A number of cell lines were established using shaking, pipetting, scraping, or enzyme treatment to remove the cells from the surface. The predominant cell morphology in the individual lines included round, round to spindle-shaped, fibroblast-like (some with fatty granules), epithelial-like (some with fatty granules), and net-like fibroblasts. The difference in predominating cell type was believed by the authors to be related to the methods used to passage the cells, e.g., scraping resulted in lines of fibroblast-like cells.

Dwyer et al.[28] obtained a very diverse outgrowth from 24-h old embryos of *Pieris rapae*. Within 2 weeks of initiating the cultures, cell aggregates appeared that often contracted, indicating the presence of muscle or nerve tissue. Multicellular fibers developed and radiated from the aggregates and formed a net-like structure over the surface of the flask. Epithelial-like cells, fibroblast-like cells, and other types developed attached to or underneath this network. Dwyer and her colleagues established over 250 sublines with different cell populations which they tested for susceptibility to *P. rapae* granulosis virus. Most lines were not susceptible, but they selected five lines with different cell morphologies and attachment characteristics that were susceptible to the granulosis virus.

Other workers, also interested in cell lines for virus studies, have used similar approaches, Sieburth and Maruniak[29] from *Anticarsia gemmatalis* and Lee and Hou[30] from the diamondback moth, *Plutella xylostella*. The primary cultures, established from 2- to 3-day-old embryos, contained the variety of cell types reported by earlier workers. The authors selected for cells that grew attached to the culture surface by removing any floating cells with the spent medium and then scraping the remaining cells from the flask with a rubber policeman.

It is not necessary to use early stage embryos as the previous workers have done. Several cell lines have been established from embryos that are fully formed and within 24 h of hatching.[31-33] The cellular outgrowth in the primary cultures was similar to that reported from younger embryos. In some primary cultures, vesicles, a single cell layer thick, formed. Some

Lepidoptera, e.g., *B. mori,* have an egg diapause, and large numbers of embryos at the same developmental stage can be obtained and stored for several months prior to culture of the cells. Inoue and Mitsuhashi[34] developed cell lines from *B. mori* embryos held at 5°C until the termination of their diapause.

Although no longer embryos, newly hatched larvae are a source of tissues that produce primary cultures very similar to those from embryos.[35] In these primary cultures, many of the newly explanted tissues died quickly due to melanization. From the few that survived, free cells and/or vesicles developed and were removed from the primary culture by periodic replacement of the growth medium. Morphology of the free cells was similar to those from embryos. Of the eight lines obtained, three contained a mixture of spindle-shaped and spherical cells, four were predominantly spindle-shaped cells, and in one cells were epithelial-like.

D. OTHER TISSUES

Although the largest number of lepidopteran cell lines have been derived from these tissues,[7] specific needs of investigators may require a cell line from known tissues. This is particularly true for research on endocrinology and development. In the past most of the *in vitro* studies were done with organ culture; for a review see Oberlander and Lynn.[36] The organs in these studies were the endocrine glands, larval epidermal systems, imaginal discs, fat body, and gonads. Other than the ovaries, very few cell lines are available from these tissues. Only two such cell lines are listed in Hink's compilations on invertebrate cell lines. One was from the testes of the larval gypsy moth.[9] It consists of spindle-shaped cells that perhaps were derived from connective tissues in the testes. It was developed by Miltenburger in 1977 and was listed by Hink as an unpublished reference. The only other line of testicular origin was developed by Lynn et al.[36] and was from the testicular sheath of *Heliocoverpa (Heliothis) virescens.* Preliminary tests on this cell line indicated that the cells did not synthesize ecdysteroids.

Imaginal discs from Lepidoptera, despite their use in organ culture,[37] have not been used to develop cell lines until recently. The first line (IAL-TND1) reported was from the wing discs of *T. ni.*[38] The same group of investigators also developed lines from the wing discs from *S. frugiperda* (IAL-SFD1) and *Plodia interpunctella* (IAL-PID2).[39] Initially cell migration occurred as sheaths or as blebs on the tissue fragments in primary cultures from *T. ni* and *S. frugiperda* discs. When these outgrowths were disrupted during the first transfer, aggregates of cells developed which formed fluid-filled vesicles after several days in culture. If these vesicles were broken into small fragments, each formed new, smaller vesicles. Although the primary cultures from *P. interpunctella* were established by apparently the same methods, the results were quite different. Cell outgrowth from these tissue fragments was as a monolayer of small spindle-shaped cells. Neither did the cells from the IAL-PID2 line synthesize new proteins in response to 20-hydroxy-ecdysone treatment as did cells of the other two lines.[39] The cells of the *T. ni* and *S. frugiperda* lines synthesized two proteins in response to hormone treatment that separated on polyacrylamide gels in a pattern remarkably similar to proteins from recently explanted whole discs. Thus, this provided additional evidence that these two lines were derived from imaginal disc cells that still maintained their biochemical capabilities and would be acceptable *in vitro* models for study.

The fat body in insects has a number of metabolic functions, e.g., synthesis of storage proteins and vitellogenins, storage of nutrient reserves, and, when infected, the replication of many insect viruses. Thus, both physiologists and pathologists are interested in cell lines derived from this tissue. Despite the wide interest in the use of fat body derived lines, there have been only a few reported, most of them from the laboratory of Dr. Jun Mitsuhashi. The first of these was from *Mamestra brassicae,* the cabbage armyworm.[40] The cells in these primary cultures were a mixture of large fat cells containing numerous small particles, epithelial-like cells, and spherical hemocyte-like cells. Subsequent primary cultures from last instar larvae of the common armyworm, *Leucania separata,* consisted of similar cells.[41] In both instances only the spherical hemocyte-like cells proliferated. Later, Inoue and Mitsuhashi[42] established five additional lines from the fat body of the cabbageworm. Again they found that only the prohemocyte-like cells multiplied, although fat cells migrated from the explants. Cells in all five lines grew in suspension.

The remaining lines derived from lepidopteran fat body were derived in the Insect Pathology (now Insect Biocontrol) Laboratory in Beltsville, MD by Goodwin from preemergent pupae of *T. ni*[43] and by Lynn from *L. dispar.*[36] Goodwin's *T. ni* fat body line grew only as a syncytium and did not passage well. The *L. dispar* cells were spherical and contained cytoplasmic vacuoles that were stained with the lipophilic stain, Oil Red O. In addition, this line contained an additional allozyme for isocitrate dehydrogenase which was found in extracts of pupal fat body, but not in any of the embryonic lines from *L. dispar.* Thus, the line appears to be of fat body origin, but the extent to which the properties of differentiated fat body have been retained is not known.

III. PRIMARY CULTURE METHODS

Despite the diversity exhibited among the members of the order Lepidoptera, the procedures for the initiation of cell cultures are remarkably similar in their broad outline. The steps involved are usually the same regardless of the choice of tissue for primary culture, the insect species, or developmental stage that will be used. The choice of what insect or what tissue to use is now more often dictated by the researcher's needs for the cell line rather than by what can be cultured. The choice of medium to be used will be covered by others in this book and will not be discussed here. Once these decisions have been made, then only three questions need be answered: how to obtain the desired tissue aseptically, how to dissociate it to free cells, and how to establish primary cultures and maintain them until sustained growth begins. Lynn[44] described in detail general methods for establishing primary cultures and the reader is referred to that paper for details. The following section will review the general procedures and discuss some of the variations that have been used in the past. As cautioned by Lynn, no general method will work in all situations and some modifications will be required to meet unusual situations.

A. SURFACE STERILIZATION

Sterilization of the exterior surfaces of Lepidoptera is most often done by immersing the eggs or the entire body in the disinfectant solution and then rinsing thoroughly in sterile distilled water or a saline solution. The most gentle treatment is to wash the animal with a saline solution containing antibiotics, e.g., penicillin and streptomycin.[26] This treatment was used by Miltenburger and his colleagues to prepare larvae for bleeding to collect hemocytes. It is probably most suitable for this type of technique where only minimum dissection is required. Other investigators used 70% ethanol for 1 min[32,45] which also has a minimum risk of damaging internal tissues. Various quaternary ammonium-based disinfectants have also been used and do not harm the internal tissues.[33,46]

The most commonly used disinfectant is NaClO, often in the form of commercial bleaches. The concentration used is 0.05%[47,27] or 0.5%[31] NaClO. It may be used alone, as in the references above, or it can be used in combination with a separate ethanol treatment.[28-30] NaClO is a strong treatment and some embryos can be killed by the treatment; so, pretesting is advisable if a new insect species is being used.

To obtain sterile tissue from neonate larvae, Sato[35] used 70% ethanol followed by 0.1% mercury to sterilize eggs. The treated eggs were then picked on the shoulder of a test tube containing medium. The newly hatched larvae migrated to the medium, and sterility could be determined within 2 to 3 days by observing the medium. Gelernter and Federici[48] obtained sterile neonate larvae for their studies by treating eggs in 5% formaldehyde. The treated eggs were then removed from the solution and allowed to dry on filter paper in a sterile hood. Within 24 h the eggs would hatch and larvae could be aseptically transferred to a salt solution for dissection.

B. TISSUE DISSOCIATION

Once sterile tissue is obtained, it is necessary to dissociate the tissue in some way to free the constituent cells or to allow them to migrate from the tissue into the culture medium. This may be as simple a procedure as removing unwanted tissues, such as trachae or fat body, from the organ and then cutting it into very small pieces. In the method provided by Lynn,[44] this latter step is carried out directly in the standing drop in which the tissues will be incubated. This is the simplest procedure and requires the least manipulation of the tissue. However, if the primary culture is to be established in flasks, the treatment must be done in small petri dishes and the tissue fragments transferred by pipette to the flask. Early stage embryos can be dissociated by gentle pipetting,[49] by crushing them through a 100-mesh screen,[28] or in a homogenizer using a Teflon plunger.[33]

More complete dissociation can be obtained by treating the tissue fragments with enzymes. The most extensive enzymatic treatment was that described by Goodwin.[50] The cleaned tissues were treated with a 1:20 dilution of the commercial tissue culture trypsin (2.5% active ingredient) and then with 0.02% hyaluronidase. Trypsin treatment tended to leave the tissue fragments sticky and difficult to manipulate, and the second treatment with hyaluronidase reduced this problem. Goodwin found that when collagenase was used for the first enzyme treatment, the cell fragments dissociated more completely in culture and an excellent monolayer was produced. Initial subculturing was required within a week as compared to several months with the trypsin-treated explants. However, he had also changed the medium used, and it is not possible to separate the effects of the two changes. Readers are cautioned that enzymatic treatments have had deleterious effects on the cells of many insects and should be used with care.

C. ESTABLISHMENT AND MAINTENANCE OF PRIMARY CULTURES

The best rules to follow in establishing primary cultures are (1) have an adequate ratio of cells and tissues to medium volume and (2) do not hurry to make the first subculture, and when making the subculture do not dilute the cell concentration too much. In most situations tissue fragments make up most of the inoculum for primary culture, and cell numbers cannot be counted accurately. Therefore, the amount of inoculum is usually given as tissue from so many insects or lightly packed tissue volume. Goodwin[50] used a tissue pellet of 0.05 to 0.1 cm^3, after centrifugation at $10 \times g$ for 10 min in 4 ml of medium to establish his primary cultures. This represented the ovaries from 15 to 18 fall armyworm pupae or 20 to 25 cabbage looper pupae. Lynn[44] recommended using 20 to 25 embryos in a standing drop culture of 0.1 ml. Gelernter and Federici[48] used 200 minced larvae in 4 ml of medium. Lynn and Oberlander[39] used 13 imaginal discs from *S. frugiperda* and 40 from *P. interpunctella* in a 0.2-ml standing drop culture for primary cultures of that tissue. Hemocytes tend to attach to surfaces readily and, thus, they are not counted when establishing primary cultures. Sohi[46] used the blood from 5 fully grown *M. disstria* larvae in 3 to 5 ml of medium. Hemocytes from fewer larvae did not survive in culture, and if more blood was used it was difficult to prevent melanization which killed the cells.

These data should provide some guidance for determining the tissue/medium ratio for those who are attempting to culture tissues from new sources. If the first attempts at primary culture fail, an increase in the ratio may be beneficial. A limited overabundance of tissue can be corrected easily by subculturing sooner.

Successful insect primary cultures all follow a general pattern of development. In the first days following initiation of the culture, many cells migrate from the tissue fragments and they may or may not attach to the growth surface. Depending upon the source of the tissue, some mitotic division may be observed in the migrated cells. This migration phase, which may last several days, is followed by a quiescent phase. During this phase little seems to happen and the number of cells may decrease. This phase can last several days or months, and the cells must be fed by the addition or replacement of the medium. In the standing drop cultures, Lynn[44] recommended that after 1 or 2 days in culture an additional 1 ml of medium be added to each 35-mm petri dish. Then at 7 to 10 day intervals, 0.5 ml of medium was added until the total volume reached 2.5 ml. When the culture needed to be fed again, 2 ml of spent medium was removed and 0.5 ml of fresh medium was added. The process was repeated until the monolayer of cells was nearly confluent, i.e., covered 90% of the dish. If there are large numbers of floating cells, these can be removed from the spent medium by gentle centrifugation, resuspended in fresh medium, and returned to the culture.

If the primary cultures are established in flasks, the cells are fed by removing half of the culture medium and replacing it with fresh medium at 7 to 10 day intervals until the monolayer is nearly confluent. Floating cells can be removed by centrifugation. If vesicles form, they can usually be broken up by gentle pipetting and returned to the original flask.

At some time in the development of a successful cell line, growth will begin. In a culture containing attached cells, small islands of very uniform cells will appear in which mitotic figures can be seen. It may be desirable to increase the feeding at this time either by increasing the frequency or the proportion of spent medium replaced. When the monolayer reaches 90% confluency or in cultures of floating cells, when the cells nearly cover the bottom of the flask, the first split (subculture) can be attempted.

Attached cells can be loosened from the surface of the culture vessel by a number of methods depending on how firmly they attach. Many lepidopteran cells attach very loosely and can be detached by gently tapping the flask, flushing with medium from a pipet, or scraping gently with a rubber policeman. Roberts[51] found that more firmly attached cells were damaged by scraping and required treatment with proteolytic enzymes to release them. In his study, 10 to 30% of the *S. frugiperda* and *S. littorals* cells were damaged when scraped from the growth surface. Only slight improvement occurred when the cells were removed with trypsin or pronase; however, with a neutral protease, dispase, 95 to 100% of the cells were undamaged. However, McIntosh et al.[47] found that trypsin in combination with ethylenediaminetetraacetic acid (EDTA) was satisfactory for releasing *H. virescens* cells.

Miltenburger and colleagues[26] established several hundred primary cultures from a variety of tissues of the apple codling moth using either physical or enzymatic methods, including trypsin and collagenase, to passage them. Based on morphology, adherence, growth rate, and isoenzyme analysis, they concluded that the lines obtained from any one tissue source had significant differences. However, lines passaged by the same techniques were somewhat similar. Thus, it may be sound practice to try more than one method of passaging cells from primary cultures from the same insect or tissue.

Almost always the first split of a primary culture is 1:2. That is two new cultures from the single primary culture. The new subcultures should contain some used medium and not be transferred with entirely fresh medium, if possible. Transfers should be made thereafter when the cells have nearly reached a confluent monolayer. A 1:2 split ratio should be maintained until the growth rate is such that splitting is required more than twice a week. At this point the split ratio can be increased.

IV. CHARACTERIZATION OF NEW CELL LINES

It is important to establish the identity of a new cell line as soon as possible. Because of the long quiescent time in the development of insect cell lines, during which they are fed several times, there is always a possibility that the primary culture could be contaminated with cells from other lines maintained in the laboratory. After a cell line has been developed, the possibility of contamination exists whenever the cultures are opened.

The first contamination of an insect cell line was reported by Greene and Charney[52] in 1971. They presented serological and isoenzyme data showing that the cells of the Grace *Aedes aegypti* line that they had were actually moth cells of the Grace *A. eucalypti* line. This was corroborated by the genetic studies of Nichols et al.,[53] reported at the same time. This finding, early in the development of insect cell culture, alerted cell culturists to the problem, and, although it has not prevented other incidents of contamination, it has resulted in the early characterization of most newly reported cell lines.

The cytogenetic studies of Nichols et al. revealed that lepidopteran cells in culture contained both micro- and macrochromosomes. This characteristic appears to be unique to the Lepidoptera and serves to differentiate these cells from those of other insects. However, the cells of all lepidopteran lines are highly polyploid, and the cytogenetics has not been useful in identifying the species of origin for lepidopteran cell lines. Disney and McCarthy[54] reported that omission of the standard colchicine treatment from the preparation of metaphase smears reduced the fragmentation of the chromosomes, suggesting that the fragmentation may be partly an artifact of the method of preparation. Today cell karyotyping is usually done as part of the characterization of a cell line, but it does not serve to identify the origin further than to order.

Krywienczyk and Sohi[55] used double diffusion to serologically characterize four cell lines from *M. disstria* and one from *Choristoneura fumiferana*. Using absorbed antiserum they were able to identify the species of origin of each cell line. Comparisons of the four *M. disstria* lines, which were from either ovarian tissue or hemocytes, revealed some antigenic differences, but the method did not identify the tissues of origin.

Aldridge and Knudson[56] conducted a more extensive study, serologically comparing five cell lines from different genera of Lepidoptera with hemagglutination, complement-fixation, immunodiffusion, and immunoelectrophoresis. The lines originated from *Spodoptera littoralis, S. frugiperda, H. zea, Laspeyresia pomonella,* and *C. fumiferana.* They could distinguish the lines from different genera, but could not distinguish between the two lines from *Spodoptera.* Thus, serology appears to be most useful only to the intergeneric level.

Tabachnick and Knudson[57] then investigated isozyme analyses as a means of identifying cell lines. This method had been shown by Greene and Charney[52] to distinguish between moth and mosquito lines. Tabachnick and Knudson included 14 lepidopteran and 2 dipteran cell lines in the study. The lepidopteran lines were from 8 taxonomic families and 11 genera. There were cell lines from two species of the genus *Spodoptera.* A total of 18 isozyme were examined; however, all the cell lines could be distinguished with only 4 enzymes, isocitrate dehydrogenase, malic enzyme, phosphoglucoisomerase, and phosphoglucomutase. Furthermore, the isozyme patterns found in the intact insect from which two dipteran and one lepidopteran cell lines were derived were the same as in the respective cell lines. They concluded that isozyme profiles would be useful in the identification of cell lines at least to the species level. Most publications on new cell lines from Lepidoptera now contain some isozyme profile in the characterization, as examples see References 27 to 30, 58.

Harvey and Sohi[59] examined the use of isozyme profiles to distinguish cell lines from the same species using a number of lines from *C. fumiferana.* These were compared with profiles from the three other Lepidoptera and a Hymenoptera. The insects and cells were examined for nine isozymes. All nine were found in the whole insect, but not all were found in every cell line. Cell lines were easily confirmed as to species of origin; however, two cell lines were found to be of a different species than labeled. Other tests confirmed that these were contaminations. No two CF cell lines were identical at all loci, suggesting that isozyme profiles, if extensive enough, may be useful in distinguishing lines from the same species.

McIntosh and Ignoffo[60] used isoelectric focusing and the single enzyme phosphoglucoisomerase to distinguish between six cell lines from five genera *(Heliothis zea* and *H. virescens* have now been placed in the genus *Helicoverpa).* They reported differentiation between lines derived from *Plutella, Spodoptera, Trichoplusia, Heliothis,* and *Helicoverpa.*

The development of commercially available kits for the development of isoenzyme profiles from insect cell lines makes this system of characterization more available to the cell culturists. Lynn and his colleagues[27] found the Corning Authentikit (Innovative Chemistry, Marshfield, MA) useful in characterizing several cell lines from *T. ni.* Cell lines from ovaries, imaginal discs, and embryos were tested. All cell lines were identical for glucose-6-phosphate, malate dehydrogenase, mannose phosphate isomerase, malic enzyme, and lactate dehydrogenase. However, the embryonic cell line differed from the others in the migration distance of aspartate aminotransferase and peptidase B. This is another indication that if the right isoenzymes are used, cell lines with different tissues of origin may be distinguished. Mitsuhashi[60] used the APIZYM kit manufactured in France by API System, S.A., Montaleau, France. This kit uses 19 enzymes, and Mitsuhashi could distinguish species of origin among 19 cell lines. He also was able to distinguish between some lines from the same species. Thus, the commercial isoenzyme kits, designed primarily for characterizing vertebrate cell lines, can be used to distinguish lepidopteran cell lines at least to species of origin.

V. FUTURE NEEDS

In an Epilogue to the Proceedings of the 5th International Conference on Invertebrate Cell and Tissue Culture in 1979, G. Echalier[62] discussed the need for additional research on invertebrate cell culture. He pointed out that there was very little variation among the many cell lines that had been developed up to that time. Most originated from similar tissues and, therefore, were composed of similar cell types, predominantly of "undifferentiated fibroblast-like Morphology." Echalier urged that new methods be developed to better study *in vitro* the physiological diversity observed in insects *in vivo.* He urged particularly that there be efforts to culture every type of differentiated tissue and that these cells should maintain their specialized functions *in vitro.* In the intervening 15 years, there has been some progress toward that objective with lepidopteran cells.

Despite the major advances in the use of cell culture for the nuclear polyhedrosis viruses discussed elsewhere in the book, there are still no easily cultured cell lines that can be used to study the granulosis virus. The work of Miltenburger et al.[26] and Dwyer et al.[28] has shown that by beginning with embryonic tissues that have a broad range of tissue types capable of *in vitro* growth, and employing a number of simple techniques, a variety of cell types with different physiological properties can be selected. Some of these lines supported granulosis virus replication for a time. The ability to produce any of the granulosis viruses economically is, perhaps, the major impediment to their use in pest management. Additional studies to develop *in vitro* systems for these systems are badly needed.

The studies of Oberlander and colleagues[37-39] on the culture of imaginal discs demonstrated that it is possible to culture cells from specific tissues that do retain the capability *in vitro* to carry on at least some of their *in vivo* functions. All these cell lines were developed from wing imaginal discs and represent less than 1% of the lepidopteran cell lines available. Despite the obvious usefulness of *in vitro* systems for studies on genetics, physiology, differentiation, and development of Lepidoptera, there has been only modest advancement in this area since 1979. It is in these areas that lepidopteran cell culture has the greatest promise for the future development.

REFERENCES

1. **Goldschmidt, R.,** Some experiments on spermatogenesis in vitro. *Proc. Natl. Acad. Sci. USA,* 2, 220, 1915.
2. **Trager, W.,** Cultivation of the virus of grasserie in silkworm tissue cultures. *J. Exp. Med.,* 61, 501, 1935.
3. **Gaw, Z. Y., Liu, M. T., and Zie, T. U.,** Tissue culture methods for the cultivation of virus grasserie. *Acta. Virol. Engl. Ed.,* 3, 55, 1959.
4. **Aizawa, K.,** Cell culture of Lepidoptera. In *Invertebrate Tissue Culture,* Vol. I, Vago, C., Ed., Academic Press, New York, 1971, 211.
5. **Grace, T. D. C.,** Establishment of four strains of cells from insects in vitro. *Nature,* 195, 788, 1962.
6. **Hink, W. F.,** A catalog of invertebrate cell lines. In *Invertebrate Tissue Culture,* Vol. II, Vago, C., Ed., Academic Press, New York, 1972, 363.
7. **Lynn, D. E.,** Establishing invertebrate cells in culture: continued need for new cell lines. In Proc. 8th Int. Conf. on Invertebr. and Fish Tissue Culture. Fraser, M. J., Jr., Ed., Tissue Culture Assoc., Columbia, MD, 1991, 1.
8. **Hink, W. F.,** A compilation of invertebrate cell lines and culture media. In *Invertebrate Tissue Culture: Research Applications,* Maramorosch, K., Ed., Academic Press, New York, 1976, 319.
9. **Hink, W. F.,** The 1979 compilation of invertebrate cell lines and culture media. In *Invertebrate Systems In Vitro,* Kurstak, E., Maramorosch, K., and Dubendorfer, A., Eds., Elsevier/North Holland, Amsterdam, 1980, 553.
10. **Hink, W. F. and Bezanson, D. R.,** Invertebrate cell culture media and cell lines. In *Techniques in The Life Sciences, Cell Biology,* Vol. 1, *Techniques in Setting Up and Maintenance of Tissue and Cell Cultures,* Kurstak, E., Ed., Elsevier, County Clair, Ireland, 1985, Chapter C111.
11. **Hink, W. F. and Hall, R. L.,** Recently established invertebrate cell lines. In *Invertebrate Cell System Applications,* Vol. 2, Mitsuhashi, J., Ed., CRC Press, Boca Raton, FL, 1989, 269.
12. **Oberlander, H. and Miller, S.,** Lepidopteran cell lines: tools for research in physiology, development and genetics. In *Advances in Cell Culture,* Vol. 5, Maramorosch, K., Ed., Academic Press, New York, 1987, 187.
13. **Wyatt, S. S.,** Culture in vitro of tissue from the silkworm Bombyx mori L. *J. Gen. Physiol.,* 39, 841, 1956.
14. **Grace, T. D. C.,** Development of insect cell cultures. In *Invertebrate Cell Culture Applications,* Maramorosch, K. and Mitsuhashi, J., Eds., Academic press, New York, 1982, 1.
15. **Stanley, M. S. M. and Vaughn, J. L.,** Origin of migrating cells in cultures of moth ovarian tissue based on developmental stages producing optimum outgrowth. *Ann. Entomol. Soc. Am.,* 61, 1067, 1968.
16. **Grace, T. D. C.,** Effects of various substances on growth of silkworm tissues in vitro. *Aust. J. Biol. Sci.,* 11, 407, 1958.
17. **Jones, B. M. and Cunningham, I.,** Growth by cell division in insect tissue culture, *Exp. Cell Res.,* 23, 386, 1961.
18. **Volkman, L. E. and Summers, M. D.,** Nuclear polyhedrosis virus detection: relative capabilities of clones developed from Trichoplusia ni ovarian cell line TN-368 to serve as indicator cells in a plaque assay. *J. Virol.,* 16, 1630, 1975.
19. **Summers, M. D. and Smith, G. E.,** *A Manual of Methods for Baculovirus Vectors and Insect Cell Culture Procedures,* Texas Agricultural Experiment Station, College Station, TX, Bull. no. 1555, 1987.
20. **Vaughn, J. L., Goodwin, R. G., Tompkins, G. J., and McCawley, P.,** The establishment of two insect cell lines from the insect Spodoptera frugiperda (Lepidoptera: Noctuidae). *In Vitro,* 13, 213, 1977.
21. **Grace, T. D. C.,** Culture of insect tissues, *Nature,* 174, 187, 1954.
22. **Martignoni, M. E. and Scallion, R. H.,** Preparation and uses of insect hemocyte monolayers "in vitro". *Biol. Bull.,* 121, 507, 1961.
23. **Mitsuhashi, J.,** Tissue culture of the rice stem borer, Chilo suppressalis Walker (Lepidoptera: Pyralidae). II. Morphology and in vitro cultivation of hemocytes. *Appl. Entomol. Zool.,* 1, 5, 1966.
24. **Chao, J. and Ball, G. H.,** A cell line isolated from hemocytes of Samia cynthia pupae. *Curr. Topics Microbiol. Immunol.,* 55, 28, 1971.
25. **Arnold, J. W. and Sohi, S. S.,** Hemocytes of Malacosoma disstria Hubner (Lepidoptera: Lasiocampidae): morphology of the cells in fresh blood and after cultivation in vitro. *Can. J. Zool.,* 52, 481, 1974.
26. **Miltenburger, H. G., Naser, W. L., Harvey, J. P., Huber, J., and Huger, A. M.,** The cellular substrate: a very important requirement for baculovirus in vitro replication. *Z. Naturforsch.,* 39c, 993, 1984.
27. **Rochford, R., Dougherty, E. M., and Lynn, D. E.,** Establishment of a cell line from embryos of the cabbage looper, Trichoplusia ni (Hubner). *In Vitro,* 20, 823, 1984.
28. **Dwyer, K. G., Webb, S. E., Shelton, A. M., and Granados, R. R.,** Establishment of cell lines from Pieris rapae embryos: characterization and susceptibility to baculoviruses. *J. Invertebr. Pathol.,* 52, 268, 1988.
29. **Sieburth, P. J. and Maruniak, J. E.,** Growth characteristics of a continuous cell line from the velvetbean caterpillar, Anticarsia gemmatalis Hubner (Lepidoptera: Noctuidae). *In Vitro Cell. Dev. Biol.,* 24, 195, 1988.
30. **Lee, S.-H. and Hou, R. F.,** Establishment of a cell line derived from embryos of the diamondback moth, Plutella xylostella (L.). *J. Invertebr. Pathol.,* 59, 174, 1992.

31. **Hink, W. F. and Ellis, B. J.,** Establishment and characterization of two new cell lines (CP-1268 and CP-169) from the codling moth, Carpocapsa pomonella (with a review of culture of cells and tissues from Lepidoptera). *Curr. Topics Microbiol. Immunol.,* 55, 19, 1971.
32. **Pant, U., Mascarenhas, A. F., and Jagannathan, V.,** In vitro cultivation of a cell line from embryonic tissue of potato tuber moth Gnorimoschema operculella (Zellev). *Indian J. Exp. Biol.,* 15, 244, 1977.
33. **Tsang, K. R., Ward, G. B., Mardan, A. H., Harien, P. K., Brooks, M. A., and Jacobson, L.,** Establishment and characterization of a cell line from embryos of the Indian meal moth, Plodia interpunctella. *J. Invertebr. Pathol.,* 46, 180, 1985.
34. **Inoue, H. and Mitsuhashi, J.,** A new Bombyx continuous cell line from embryos and its characterization. In *Invertebrate Cell System Applications,* Vol. 2, Mitsuhashi, J., Ed., CRC Press, Boca Raton, FL, 1989, 199.
35. **Sato, T.,** Establishment of eight cell lines from neonate larvae of tortricids (Lepidoptera) and their several characteristics including susceptibility to insect viruses. In *Invertebrate Cell System Applications,* Vol. 2, Mitsuhashi, J., Ed., CRC Press, Boca Raton, FL, 1989, 187.
36. **Lynn, D. E., Dougherty, E. M., McClintock, J. T., and Loeb, M.,** Development of cell lines from various tissues of Lepidoptera. In *Invertebrate and Fish Culture,* Kuroda, Y., Kurstak, E., and Maramorosch, K., Springer-Verlag, Berlin, 1988, 239.
37. **Oberlander, H. and Lynn, D. E.,** Morphogenesis in insect cell culture. In *Advances in Cell Culture,* Vol. 2, Maramorosch, K., Ed., Academic Press, New York, 1982, 237.
38. **Lynn, D. E., Miller, S. G., and Oberlander, H.,** Establishment of a cell line from lepidopteran wing discs: induction of newly synthesized protein by hydroxyecdysone. *Proc. Natl. Acad. Sci. U.S.A.,* 79, 2589, 1982.
39. **Lynn, D. E. and Oberlander, H.,** The establishment of cell lines from imaginal wing discs of Spodoptera frugiperda and Plodia interpunctella. *J. Insect Physiol.,* 29, 591, 1983.
40. **Mitsuhashi, J.,** Establishment and some characteristics of a continuous cell line derived from fat bodies of the cabbage armyworm (Lepidoptera: Noctuidae). *Dev. Growth Differ.,* 23, 63, 1981.
41. **Mitsuhashi, J.,** A continuous cell line derived from fat bodies of the common armyworm, Leucania separata (Lepidoptera: Noctuidae), *Appl. Entomol. Zool.,* 18, 533, 1983.
42. **Inoue, H. and Mitsuhashi, J.,** Further establishment of continuous cell lines from larval fat bodies of the cabbage armyworm, Mamestra brassicae (Lepidoptera: Noctuidae), *Appl. Entomol. Zool.,* 20, 496, 1985.
43. **Goodwin, R. H., Vaughn, J. L., Adams, J. R., and Louloudes, S. J.,** The influence of insect cell lines and tissue-culture media on baculovirus polyhedra production. *Misc. Publ. Entomol. Soc. Am.,* 9, 66, 1973.
44. **Lynn, D. E.,** Methods for the development of cell lines from insects. *J. Tissue Cult. Methods,* 12, 23, 1989.
45. **Mitsuhashi, J.,** Establishment of cell lines from the pupal ovaries of swallowtail, Papilio xuthus L. (Lepidoptera: Papilionidae). *Appl. Entomol. Zool.,* 8, 64, 1973.
46. **Sohi, S. S.,** In vitro cultivation of hemocytes of Malacosoma disstria Hubner (Lepidoptera: Lasiocampidae). *Can. J. Zool.,* 49, 1355, 1971.
47. **McIntosh, A. H., Andrews, P. A., and Ignoffo, C. M.,** Establishment of two continuous cell lines of Heliothis virescens (F.) (Lepidoptera: Noctuidae). *In Vitro,* 17, 649, 1981.
48. **Gelernter, W. D. and Federici, B. A.,** Continuous cell line from Spodoptera exigua (Lepidoptera: Noctuidae) that supports replication of nuclear polyhedrosis viruses from Spodoptera exigua and Autographa californica. *J. Invertebr. Pathol.,* 48, 199, 1986.
49. **Eide, P. E., Caldwell, J. M., and Marks, E. P.,** Establishment of two cell lines from embryonic tissue of the tobacco hornworm, Manduca sexta. *In Vitro,* 11, 395, 1975.
50. **Goodwin, R. H.,** Insect cell culture: improved media and methods for initiating attached cell lines from the Lepidoptera. *In Vitro,* 11, 369, 1975.
51. **Roberts, P. L.,** The use of proteolytic enzymes for harvesting insect cell lines which support nuclear polyhedrosis virus replication. *FEMS Microbiol. Lett.,* 29, 189, 1985.
52. **Greene, A. E. and Charney, J.,** Characterization and identification of insect cell cultures. *Curr. Topics Microbiol. Immunol.,* 55, 51, 1971.
53. **Nichols, W. W., Bradt, C., and Bowne, W.,** Cytogenetic studies on cells in culture from the Class Insecta. *Curr. Topics Microbiol. Immunol.,* 55, 64, 1971.
54. **Disney, J. E. and McCarthy, W. J.,** A modified technique for the improved characterization of lepidopteran chromosomes from cells in culture. *In Vitro,* 21, 563, 1985.
55. **Krywienczyk, J. and Sohi, S. S.,** Serologic characterization and identification of four lepidopteran cell lines. *Can. J. Zool.,* 54, 1559, 1976.
56. **Aldridge, C. A. and Knudson, D. L.,** Characterization of invertebrate cell lines. I. Serologic studies of selected lepidopteran lines. *In Vitro,* 16, 384, 1980.
57. **Tabachnick, W. J. and Knudson, D. L.,** Characterization of invertebrate cell lines. II. Isozyme analyses employing starch gel electrophoresis. *In Vitro,* 16, 392, 1980.
58. **Corsaro, B. G. and Fraser, M. J.,** Characterization of clonal populations of the Heliothis zea cell line IPLB-Hz-1075. *In Vitro,* 23, 855, 1987.

59. **Harvey, G. T. and Sohi, S. S.,** Isozyme characterization of 8 hymenopteran and 20 lepidopteran cell lines. In *Invertebrate Cell System Applications,* Vol. 1, Mitsuhashi, J., Ed., CRC Press, Boca Raton, FL, 1989, 71.
60. **McIntosh, A. H. and Ignoffo, C. M.,** Characterization of lepidopteran cell lines by isoelectric focusing and phosphoglucoisomerase. In *Invertebrate Cell System Applications,* Vol. 1, Mitsuhashi, J., Ed., CRC Press, Boca Raton, FL, 1989, 65.
61. **Mitsuhashi, J.,** Comparison of zymograms of some insect cell lines by means of APIZYM system. *Appl. Entomol. Zool.,* 25, 535, 1990.
62. **Echalier, G.,** Epilogue; Necessity of radically new insect cell culture methods. In *Invertebrate Systems In Vitro,* Kurstak, E., Maramorosch, K., and Dubendorfer, A., Eds., Elsevier/North Holland, Amsterdam, 1980, 589.

Chapter 4

GRASSHOPPER CELL CULTURE

Ulrike G. Munderloh, Timothy J. Kurtti, Yan Liu, and Chunsheng Chen

TABLE OF CONTENTS

0-8493-7642-4/94/$0.00+$.50

Timothy J. Kurtti, Ph.D., is Associate Professor of Entomology at the University of Minnesota, St. Paul, MN. He received his B.A. in Microbiology in 1965, and his Ph.D. in Entomology in 1974, both from the University of Minnesota. In 1977 he joined the International Laboratory for Research on Animal Diseases in Nairobi, Kenya, and then returned to the US (Waksman Institute of Microbiology, Rutgers University, NJ) in 1979. Since 1985, he has directed a program on Insect Microbiology at the Department of Entomology at the University of Minnesota. Sigma Xi, NY Academy of Sciences, American Men and Women of Science; Entomological Soc. of America; Soc. Protozoologists, ASM, American Soc. Parasitologists, Soc. for Invertebrate Pathology, Tissue Culture Assoc.; Chair, Invertebrate Div. of the TCA, 1992–1994. Invertebrate cell culture, biology of insect-pathogenic microsporidia, vector-borne animal and human pathogens

Ulrike G. Munderloh, D. Med. Vet., is a Research Associate at the Department of Entomology of the University of Minnesota. She received a degree in veterinary medicine at the University in Munich, Germany, in 1975, and the doctorate in veterinary medicine in 1977 from the same university. She held postdoctoral fellowships at the International Laboratory for Research on Animal Diseases in Nairobi, Kenya (1978–1979), and the Waksman Institute for Microbiology (1979–1984). 1985–1986: Assistant Research Prof., Rutgers University; 1986–present: Research Associate, University of Minnesota. American Men and Women of Science; AAAS; Soc. for Vector Ecology. Insect and tick cell culture, vector-borne pathogens *in vitro;* molecular biology of *Borrelia* and microsporidia.

I. INTRODUCTION

Grasshoppers and their kin, migratory locusts (*Locusta migratoria* and *Schistocerca gregaria*), have attracted the attention of people in a way few other insects can. While they do not spread fear as do African killer bees or share the splendor of the Morpho butterfly, their peculiar behavior of aggregating into huge swarms in times of food shortage, descending on lush farm lands and leaving devastation in their wake, has earned them disrepute since biblical times. Because locusts are in close contact only periodically, horizontal pathogen transmission, which occurs most effectively under crowded conditions, is reduced. In addition, the semi-arid nature of their natural habitat has hampered the use of moisture-loving fungi and nematodes as biological control agents. The search has thus focused on *Nosema locustae*,[1] a natural pathogen of grasshoppers and locusts that can endure the dryness of the desert.[2] This microsporidian, which in its vegetative state is an obligatory intracellular parasite, readily infects locusts and grasshoppers and multiplies to high numbers within their fat bodies;[3] but it is not an effective killer.[4] Another group of pathogens, the entomopox viruses, is just beginning to be explored. While they will not instantaneously kill insects,[5] they, too, hold the promise of keeping locusts grounded (thus preventing them from migrating), because the fat body, which is the target organ of the viruses,[6] is the primary source of the lipids that provide the fuel for sustained flight.[7]

Whichever pathogen may eventually be favored for controlling locusts and grasshoppers, they all need improvement before becoming the useful tool that we desire. Insect cell culture has traditionally, and successfully, been employed to elucidate the basic biology of insect pathogens at the cellular and molecular level. Especially techniques such as plaque purification from cell layers have led to the identification of genetically distinct clones of insect viruses.[8] Such important, basic tools have not, until now, been available to orthopterists. Although Tsang et al.[9] have reported the isolation of cell lines from the lesser migratory grasshopper, *Melanoplus sanguinipes*, they are apparently no longer available, and their susceptibility to grasshopper pathogens was not ascertained. Primary cultures of *L. migratoria* and *S. gregaria*, which were initiated by Raina and Khurad,[10] seemingly failed to develop into established lines, with Khurad et al.[9] citing the unavailability of host insect cell lines as the reason that prompted them to grow *N. locustae* in noctuid cells.

To fill this need, we set out to isolate new cell lines from *M. sanguinipes*, to develop them into useful tools for the type of studies outlined above, and to characterize them by karyology and isozyme electrophoresis in sufficient detail so as to distinguish them from cell lines of other orthopterans, most notably cockroaches. We chose *M. sanguinipes* because it is of considerable importance in the midwestern regions of North America, finding undisturbed breeding grounds on federally managed range land, and invading adjacent, well irrigated, green farms. It is also susceptible to the same pathogens as the migratory locusts that devastate marginally profitable farms in Africa and Asia.[4] Therefore, cell lines derived from this insect lend themselves to studies that should be widely applicable to other acridid species as well.

The following is an account of the methods we employ in our laboratory for the cultivation and characterization of grasshopper cells. Where appropriate, we discuss the significance of the approaches or materials used by us, as they may deviate from those traditionally associated with insect cell culture. This, we hope, will encourage other researchers interested in, e.g., the *in vitro* analysis and manipulation of obligate intracellular parasites of grasshoppers, such as viruses and microsporidia, to use the cell lines available now for their work.

II. CELL CULTURE

A. CULTURE MEDIUM — GENERAL CONSIDERATIONS

The basal medium used in our laboratory for the cultivation of all cell lines is L-15B,[12] but with 80 m*M* glucose, instead of 12.43 m*M*. It is modified to suit the requirements of any particular cell line, e.g., it is diluted to 67% with H_2O for mosquito cells, and additionally buffered using MOPS (3-*N*-Morpholino-propanesulfonic acid) or PIPES (Piperazine -*N,N'*-bis 2-ethanesulfonic acid) for moth and grasshopper cells. Established cell lines grow well with just 1 to 5% of fetal bovine serum (FBS), and (except for mosquito cells) 0.1 to 0.25% of a bovine lipoprotein concentrate (BLC) (ICN, Irvine, CA), although a few (e.g., young tick cell lines) also require tryptose phosphate broth (TPB). This medium is based on Leibovitz's[13] L-15 medium originally formulated for the culture of mammalian cells at an osmotic pressure of ~320; the additives that make L-15B bring this value to ~425 mOsm/l. The hemolymph pH of insects is buffered by high concentrations of amino acids and organic acids, a condition mimicked by L-15B. Cell lines from mosquitoes, moths, ticks, cockroaches, grasshoppers, and mice have been either isolated in this medium or successfully been adapted to it. In addition, *Crithidia fasciculata,* a trypanosomatid parasitic in mosquitoes, and *Phytomonas serpens,* a pathogen of latex-bearing plants and transmitted by a bug,[14] can be successfully multiplied in L-15B (Munderloh, unpublished). This unsurpassed versatility and the fact that a major portion of L-15B, L-15, is commercially available made L-15B the medium of choice for our grasshopper embryo primary cell cultures.

The primary cultures were initiated in a medium with high concentrations of 3 antibiotics (L-15B pH ~6.8, supplemented with 20% FBS, 10% TPB, 250 µg/ml gentamycin, 500 IU/ml penicillin, and 500 µg/ml streptomycin) to suppress antibiotic-resistant bacteria which often are associated with laboratory colonies of insects. However, antifungal compounds were not used as these are quite toxic for cells.[15] One month after initiation, the grasshopper cells in the cultures were metabolizing the medium vigorously enough to demand the addition of a buffer. Subsequently, PIPES (Research Organics, Cleveland, OH), which has a pKa of 6.6, was added at a concentration of 25 m*M,* to guard against the harmful effects of excessive acidification of the medium. Penicillin and streptomycin were omitted after 2 months, leaving only gentamycin. This antibiotic was subsequently deleted as well, and all cell lines were maintained in antibiotic-free medium.

During the following year, as the lines developed very slowly, we introduced several changes in the culture conditions in an attempt to improve cell growth. These included varying the depth of the culture medium (i.e., the volume per flask), reduction of the FBS concentration from 20 to 5%, addition of lipoprotein (bovine lipoprotein concentrate, BLC; ICN) and bovine serum albumin (BSA, fraction V). All supplements should be cell culture grade, and tested on grasshopper cells before purchase and general use. An important ingredient in L-15B is glutamic acid which should be kept desiccated, and used no longer than ~6 months, as its growth supporting quality deteriorates.

Last, and certainly not least, the quality of the water can determine success or failure. The suitability of laboratory water for grasshopper cell culture is a combination of the quality of the general water supply and the efficiency of the purification method or apparatus employed to furnish tissue culture grade water. As such, there probably is no "one and only" way to prepare water suitable for invertebrate cell culture. As a rule, the purest obtainable grade water which has been filtered to remove pyrogens should be used. Water purification systems must be constantly monitored and regularly serviced to consistently yield high quality water. If the suitability of a water source is doubtful, purchasing cell culture grade water from a biological supply company may be a good alternative.

B. INITIATION OF PRIMARY CULTURES

The source of the grasshopper cell lines reported herein was a colony of the lesser migratory grasshopper, *M. sanguinipes,* maintained at the Rangeland Insect laboratory of the USDA in Bozeman, MT. Two batches of eggpods from the 74th laboratory generation were generously supplied by Dr. D. A. Streett. They were shipped in vermiculite, and, upon receipt, incubated at 31°C for 7 days. Embryos from the first shipment were used before dorsal closure, and when only the posterior margin of the eyes was pigmented. In these embryos, a cuticle had already formed. The second shipment was used while the embryos were still younger: before formation of an embryonic cuticle, and before any signs of eye pigmentation appeared. In each case, individual eggs were teased from the egg pods and carefully examined microscopically to ensure they were viable, and not obviously contaminated with bacteria or fungi. Any eggs accidentally damaged were also discarded. The 100 healthy eggs of each batch were individually cleaned with a soft paint brush, removing the outer chorion, and stored on moist filter paper overnight at 4°C. The eggs were then surface sterilized by gentle agitation for 5 min each in 0.1% benzalkonium chloride and 70% ethanol with 1 drop of Tween 80 per 100 ml. They were rinsed in 3 changes of sterile water and collected into 2.5 mg/ml gentamycin, 5000 IU/ml penicillin, and 5 mg/ml streptomycin in water. Embryos were removed from their eggs into a drop of L-15B medium,[12] pH ~6.8, supplemented with 20% heat-inactivated FBS, 10% TPB, 500 µg/ml gentamycin, 1000 IU/ml penicillin, and 1 mg/ml streptomycin. They were subsequently pooled into 10 ml L-15B with FBS and TPB as above and half the stated concentration of antibiotics, and centrifuged at 150 g for 10 min. The pellet was resuspended using a Pasteur pipette to disrupt the embryos, and the first batch of older embryos seeded into three 25-cm^2 flasks (Nunc, Roskilde, Denmark) in 5 ml of medium (see above) per flask. Thus, each flask received a tissue amount equivalent to approximately 33 embryos. Because the embryos from the second shipment were smaller, 50 of them were pooled into one flask. The high concentrations of antibiotics used to initiate cultures reduced the incidence of contamination with antibiotic-resistant bacteria which had caused the loss of all cultures in a previous attempt to isolate grasshopper cell lines (Munderloh and Kurtti, unpublished).

C. ISOLATION OF CELL LINES

The primary cultures initiated from the older embryos were coded MSE1, 2, and 3, and those from the second, younger batch MSE4 and 6. MSE2 and MSE6 failed to develop, and were eventually discarded after nearly 2 years. Hence, these lines will not be discussed further. After 3 to 7 days post initiation, 3 ml of each culture was replaced by L-15B supplemented as above, but containing only 50 µg gentamycin/ml, 100 IU penicillin/ml, and 100 µg streptomycin/ml. Care was taken not to remove any unattached tissue fragments. In this manner, the cultures were fed once a week until the growth surface was completely covered with cells.

From the second day a significant proportion of cells attached, covering 5 to 10% of the growth area (confluency), while abdominal segments and malpighian tubules remained afloat and continued to contract actively. By 1 week, hollow, multicellular balls (termed vesicles) started to emerge from suspended tissues. These would subsequently comprise 10 to 70% of the cell types present in a culture, but this varied significantly with the line. Vesicles were most prominent in lines MSE1 and MSE4, where they persisted until passage about 10, when they gradually disappeared. After 4 weeks, one fourth of the culture fluid containing unattached tissues from MSE1 and 2 were transferred to a new culture assigned the code MSE7. One week later, all of the remaining unattached cells and fragments from MSE2 and 3 were pooled into a new flask and designated MSE8.

D. APPEARANCE AND MAINTENANCE OF ESTABLISHED LINES

The phase contrast appearance of four MSE cell lines is shown in Figure 1. MSE1 comprises predominantly polygonal, epithelial cells (panel A), line MSE3 cells (panel B) are long and bipolar, while MSE4 cells resemble hemocytes (panel C). MSE8 cultures (panel D) resemble line MSE1, and MSE7 (not shown) is morphologically identical to MSE3. In all lines, except MSE4, small groups of pulsating muscle cells are regularly present, and in MSE1, 3, and 7, tubular structures reminiscent of tracheoles can be observed. The differentiation of "tracheoles" *in vitro* may be characteristic of acridid cells, as these were also seen by Tsang et al.[9] and Goodwin[16] in *M. sanguinipes* and by Raina and Khurad[10] in locust cell

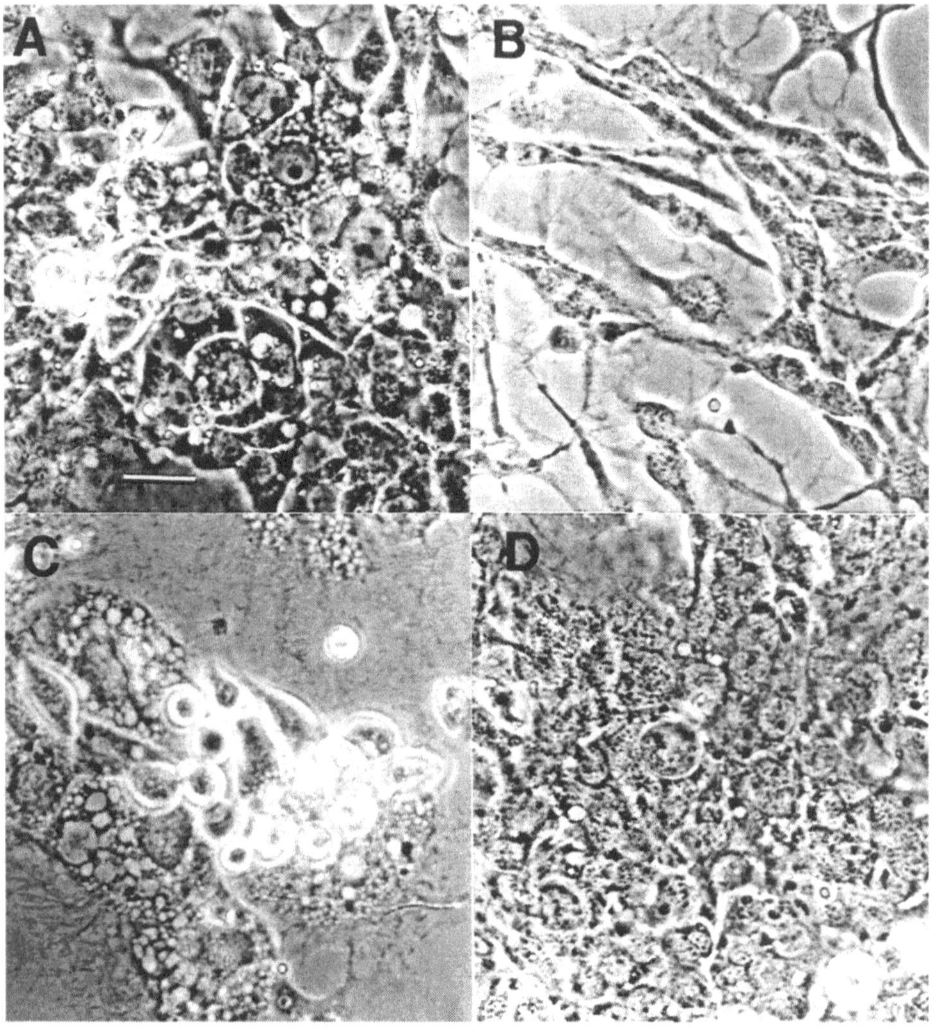

FIGURE 1. Phase contrast appearance of grasshopper *(Melanoplus sanguinipes)* cell lines. All cultures were photographed using the same magnification. Panel A, cell line MSE1; panel B, cell line MSE3; panel C, cell line MSE4; panel D, cell line MSE8. Bar in panel A represents 50 μm.

cultures. While muscle tissue commonly arises in young cultures from many arthropods (our unpublished observation), "tracheoles" have been reported only in acridid cell cultures.

A high level of serum is commonly used for the initiation of invertebrate primary cultures, and usually serves as the only source of lipids. In invertebrate hemolymph, as in vertebrate blood, lipids are transported bound to carrier proteins. The rapid rate of lipid oxidation by locust muscle during sustained flight requires, however, concentrations of lipoprotein in the hemolymph that exceed those in mammalian blood.[7] Thus, even 20% FBS may not meet the needs of grasshopper cells in culture. Furthermore, Goodwin[16] found that additional lipid supplementation improved the growth of differentiated *M. sanguinipes* embryonic cells *in vitro.* The medium for MSE cultures was therefore modified to comprise L-15B with 5% FBS, 10% TPB, 0.25% BLC, 0.5% BSA (prepared in L-15B), and was buffered to pH 6.8 with 25 m*M* PIPES. For routine maintenance, all cell lines are kept in 25-cm^2 flasks at a temperature of 31°C. Between subcultures, the growth medium is replaced once a week, and, due to the high metabolic rate of MSE cells, the use of "conditioned" medium is not beneficial. In our experience, MSE cells favor a relatively shallow culture medium overlay of 0.2 ml/cm^2 (5 ml per 25-cm^2 flask), regardless of the size of culture vessel used. Even though it is tempting to provide an actively metabolizing culture with a larger volume of medium, doing so will cause rapid development of anoxic conditions in the vicinity of the cells,[17] forcing them to adopt anaerobic metabolic pathways which favor the production of lactic acid. To improve attachment,[16] MSE1, 3, 7, and 8 are cultured in Falcon Primaria® (Becton Dickenson Co., Oxnard, CA) plastic flasks. This culture substrate markedly enhances adhesion and spreading, especially of MSE1 and 8, but does not apparently affect line MSE4 which is maintained in Nunc flasks. Cells of all lines except MSE4 adhere very tightly to the growth surface and have to be dislodged using a cell scraper. When doing this, it is important to keep the cell layer submerged, so that cells that are being detached can float up freely into the medium without being "run-over" by the scraper. We have tried to disperse cells with trypsin, but have found that it severely damages the cells, and retards attachment following subculture. Line MSE4 does not adhere to the same degree, and it is sufficient to flush the cells off their substrate using a stream of medium from a 14-gauge 4-inch laboratory cannula (with its tip bent to facilitate access to all corners) fitted to a "Luer-Lok" type syringe. A Pasteur pipette may also be used, but is less effective. The currently used subculture schedules are listed in Table 1. However, these regimens are not static, because, as the cells adapt to the culture conditions provided, and as we learn how to best handle them, subculture intervals will shorten and split ratios will become smaller.

All MSE cell lines have been successfully frozen in liquid nitrogen (LN) using 10% dimethyl sulfoxide (DMSO) in growth medium as above, except that 20% FBS must be used. The freezing medium should be completely prepared before addition of the cells to be frozen, to ensure that the endothermic reaction induced by DMSO has subsided. Freezing medium and equipment should be kept at room temperature before insertion into the neck of the nitrogen tank. We routinely use a Union Carbide freezing tray and 35 VHC tank adjusted to achieve a cooling rate of –1 to –2°C/min., but other models that achieve this cooling rate can be used as well. The cells are transferred into the liquid or vapor phase of LN after at least 90 min. To regenerate a culture, the tube containing the cell suspension is thawed rapidly in a 37°C water bath, and its contents pipetted gently into a 5- to 10-ml volume of growth medium in a culture flask. After an overnight incubation at 31°C, the culture overlay is replaced with complete growth medium, and the cells thereafter fed weekly as outlined above.

III. CHARACTERIZATION

A. KARYOTYPING

Chromosome spreads of lines MSE3, 4, and 8 were obtained by modifying the techniques outlined by Verma and Babu[18] and Sessions.[19] Cell cultures were treated with 0.2 µg/ml of colcemid (GIBCO, Grand Island, NY) for 24 h at 34°C. Suspended cells were lysed in 75 m*M* KCl for 30 min at 34°C, and then fixed 3 times for 30 min in a freshly prepared mixture of 3 parts methanol and 1 part glacial acetic acid. The fixed lysate was dropped onto slides, air dried, and stained with 10% Giemsa's stain for 30 min. The number of chromosomes per cell was determined from counts of at least 100 metaphase sets for each cell line.[20]

Lines MSE4 and MSE8 display the chromosome number and acrocentric karyotype characteristic of *M. sanguinipes*.[21,22] They can thereby be readily distinguished from cockroach cell lines, whose chromosomes, although identical in number, are clearly metacentric.[23] Male sets have one sex chromosome (XO) and 22 autosomes (2n = 23), whereas female sets possess 2 sex chromosomes (XX) (2n = 24) (Figure 2). In MSE3, an apparent fusion between two autosomes created a metacentric chromosome that was identifiable by its banding pattern (C. Chen, unpublished). Thus, the karyotype of this line is reduced to 2n = 21 for male cells and 2n = 22 for female cells. While lines MSE3 and MSE4 present an overwhelmingly female karyotype, a large proportion of MSE8 cells are male (Table 2).

TABLE 1
Lightmicroscopic Appearance and Maintenance Schedule of Existing Grasshopper Cell Lines

Cell line	Predominant appearance	Current passage #	Subculture interval	Split ratio (cell dilution)	Subculture method
MSE1	Epithelial, myocytes, fibroblastic	13	2 months	1:2	Scraping
MSE3	Fibroblastic, myocytes	9	6–8 weeks	1:2 to 1:3	Scraping
MSE4	Hemocytic	15	3–4 weeks	1:5 to 1:10	Flushing
MSE7	Fibroblastic, myocytes	5	months	1:2 to 1:3	Scraping
MSE8	Epithelial, myocytes fibroblastic	12	4–6 weeks	1:2 to 1:3	Scraping

B. ENZYME ELECTROPHORESIS

We chose isoelectric focusing of isozymes in native polyacrylamide gels because of the high resolution of enzyme bands that can be achieved with this technique. Gels (5% polyacrylamide with 3% crosslinker) were cast and run on a BioRad (Hercules, CA) model 111 mini isoelectric focusing (IEF) cell as per the manufacturer's specifications using Sigma's (St. Louis, MO) Ampholine® ampholytes (pH 3.5 to 9.5). The following 6 isozymes were assayed: lactate dehydrogenase (LDH, EC 1.1.1.27.), malate dehydrogenase (MDH, EC 1.1.1.37.), phosphogluconate dehydrogenase (PGD, EC 1.1.1.4.), malic enzyme (ME, EC 1.1.1.40.), acid phosphatases (ACP, EC 3.1.3.2.), and esterases (EST, EC 3.1.1.1. or EC 3.1.1.2.). Cultures were prepared by washing the monolayers twice with Ca^{2+}-, Mg^{2+}-free Dulbecco's phosphate buffered saline (DPBS). The cells were then suspended in DPBS and centrifuged at 275 g at 4°C for 10 min. Cell pellets were resuspended in an equal volume of a 1% glycine solution, and stored at –20°C until needed. Thawed extracts were clarified by centrifugation at 15,600 g in the cold, and 5 to 10 μl volumes applied to squares of filter paper placed ~1 cm from the cathode. Enzyme bands were focused at 4°C in a three-stage program for 15 min at 100 V, another 15 min at 200 V, followed by 450 V for 1 h. Specific isozymes were visualized using the staining protocols of Pasteur et al.[24] The distance of migration, from the anode, of major and minor bands was measured directly on the dried gels. A mixture of eight standard proteins with known pI values ranging from 4.65 to 9.6 (BioRad) was separated under identical conditions and stained with BioRad's silver stain. A linear regression plot was derived from measurements obtained from the standard proteins, and used to calculate the pI values of the isozymes in cell or tissue extracts. MSE samples were also compared with extracts from *M. sanguinipes* tissues (femoral muscle and integument), as well as with cell lines BGE1 and BGE5β from embryos of *B. germanica.*[23]

The isoelectric focusing analysis of several enzymes has confirmed the *M. sanguinipes* identity of the new MSE cell lines described here. The pattern of major and minor bands agreed with those found when grasshopper tissues were examined, but differed from those of cockroach cell lines, as well as lines from other arthropods (mosquitoes, moths, and ticks) maintained in this laboratory (data not shown). Figure 3 is a diagrammatic summary of multiple analyses of all six enzymes, while Figures 4 to 6 are photographs of gels showing typical analyses of ACP, MDH, and EST using lines MSE4 and MSE8 as well as adult *M. sanguinipes* tissue extracts in comparison with the cockroach cell lines BGE1 and BGE5β.

The usefulness of IEF to characterize insect cell lines has been established by McIntosh and Ignoffo.[25,26] Using this technique and a single enzyme, phosphoglucoisomerase (PGI), the authors were able to distinguish between lines isolated from two closely related *Helicoverpa* species, *H. zea* and *H. virescens.* Similarly, cloned *H. zea* cell lines were identifiable by IEF.[25,27] The isozyme patterns of the *Helicoverpa* cell lines remained stable after extended serial passages,[26] and, when compared with extracts derived from larval hemolymph and pupal ovaries, confirmed their species identity. In contrast, an analysis of five isozymes (isocitrate

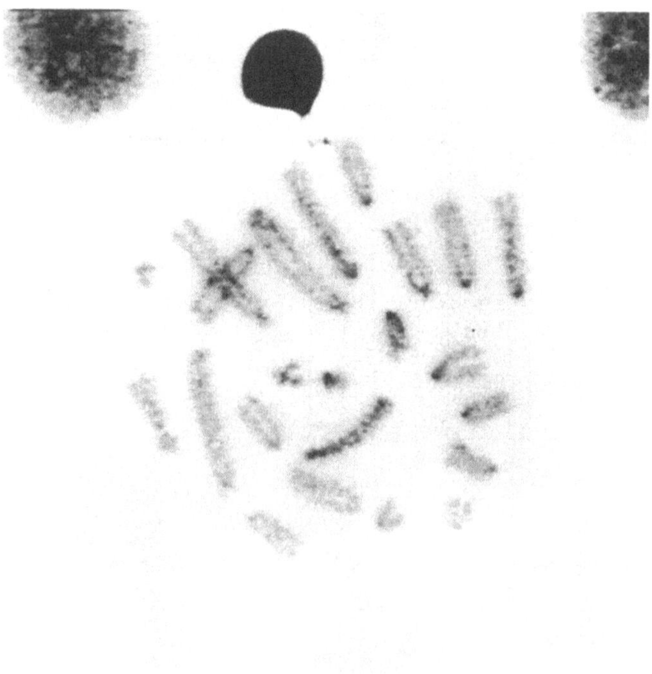

FIGURE 2. Set of metaphase chromosomes of grasshopper cell line MSE4. Male (XO) complement of 23 chromosomes. Note acrocentric appearance of chromosomes, which is typical for *M. sanguinipes*.

TABLE 2
Karyotypes of Three Established Grasshopper Cell Lines

No. chromosomes per set	Cell lines: MSE3 Frequency	MSE3 %	MSE4 Frequency	MSE4 %	MSE8 Frequency	MSE8 %
20	1	0.93				
21	13	12.15	4	3.85	12	11.88
22	87	81.31	1	0.96	21	20.79
23	5	4.68	19	18.27	23	22.77
24	1	0.93	77	74.04	36	35.65
25			3	2.88	6	5.94
26					2	1.98
46					1	0.99
Total	107	100	104	100	101	100

dehydrogenase, PGI, phosphoglucomutase, malic enzyme, and tetrazolium oxidase) by acetate electrophoresis did not allow separation of these two lepidopteran species.[25] In our hands, analysis of LDH alone distinguishes between three MSE cell lines, and separates them from cockroach cells (Table 3).

IV. CONCLUSIONS

We have isolated five new cell lines from embryonated eggs of the migratory grasshopper, *M. sanguinipes*. Their growth medium, L-15B,[12] is based on a commercially available

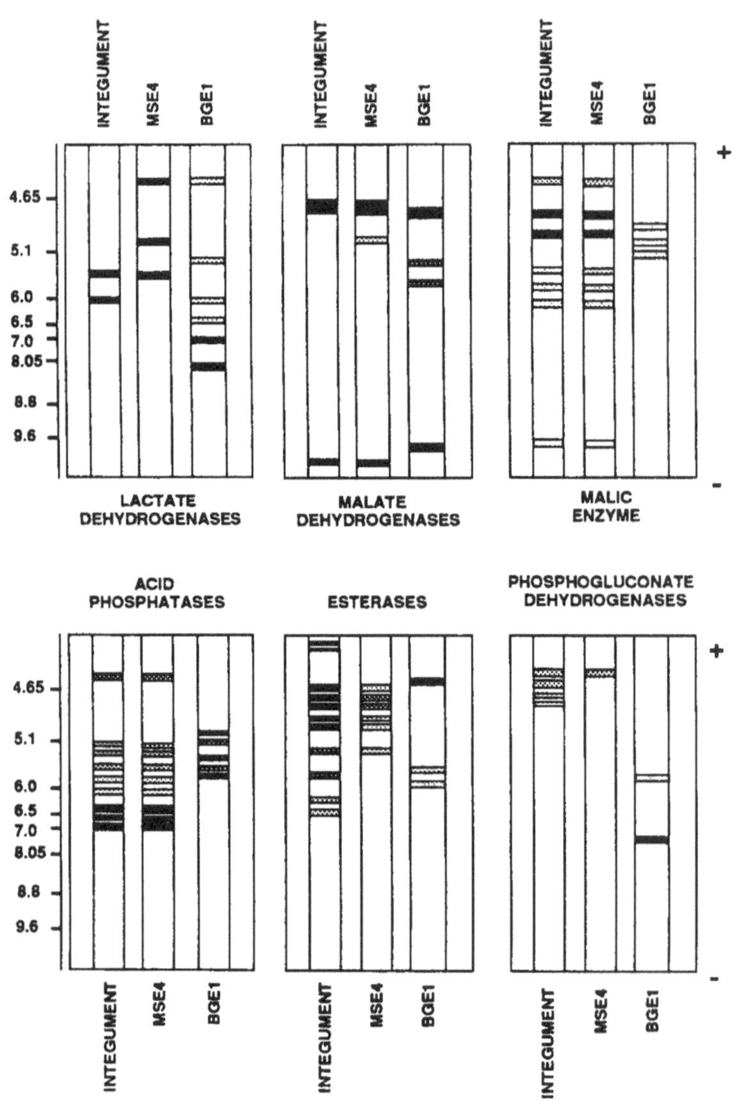

FIGURE 3. Summary of isoelectric focusing of six enzymes (see text) using extracts of *M. sanguinipes* cell line MSE4 in comparison with extracts from *M. sanguinipes* integument and cell line BGE1 from the cockroach, *Blattella germanica*.

mammalian cell culture medium, L-15,[13] and requires little effort to prepare. This is in contrast to the media described previously for the cultivation of orthopteran cells,[28,29,16] which are very complicated, and time-consuming to prepare. The relatively high concentrations of lipids apparently favored by acridid cells[16] are conveniently supplied in the form of water-soluble lipoprotein preparations, namely BSA and BLC, which are quite stable when stored desiccated as a powder at 4°C (BSA) or at –20°C (BLC). Our medium formulation has maintained the presence of contractile muscle and "tracheole"-forming cells in MSE1, 3, 7, and 8 to this date, through several passages. In addition, it allowed the isolation of cell lines from a small inoculum of six to ten embryos per ml of culture. Instead of aiming to seed primary cultures with a single-cell suspension[9] or disrupting them in protein-free saline,[16] we used serum-supplemented growth medium and minimal manipulation in an effort to preserve maximal viability of embryonic fragments. In our experience with tick cells, it is not necessary

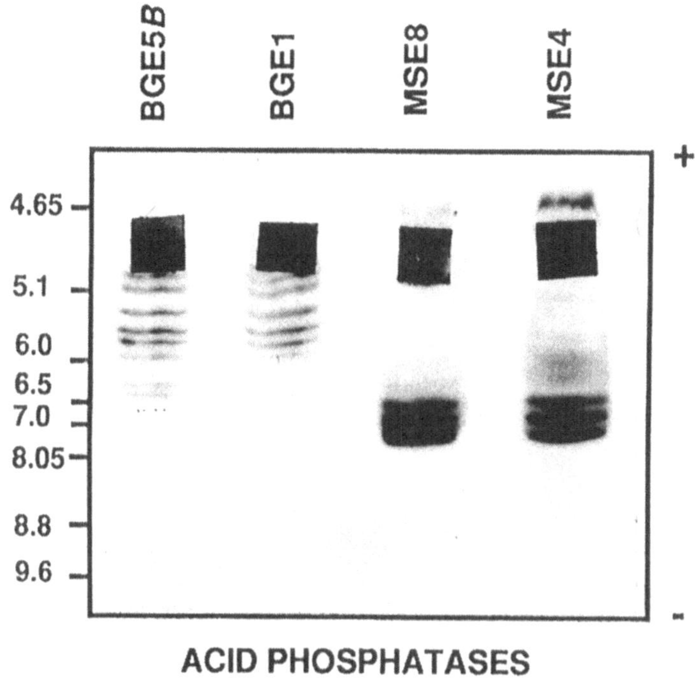

FIGURE 4. Isoelectric focusing analysis of acid phosphatases (ACP) from extracts of cockroach *(B. germanica)* cell lines BGE5β and BGE1 and *M. sanguinipes* cell lines MSE8 and MSE4.

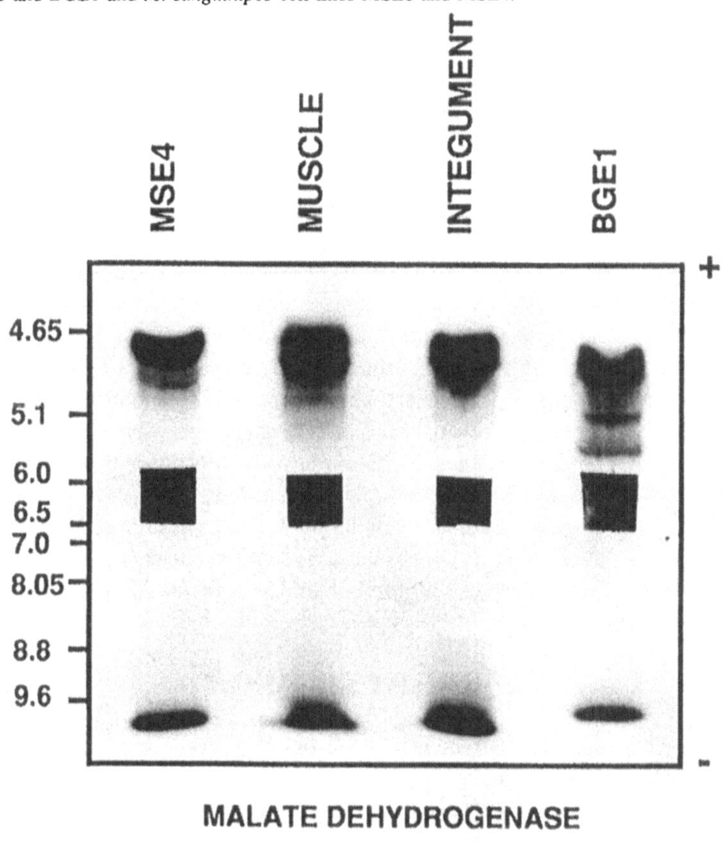

FIGURE 5. Isoelectric focusing analysis of malate dehydrogenase (MDH) from extracts of grasshopper *(M. sanguinipes)* cell line MSE4, *M. sanguinipes* adult muscle, *M. sanguinipes* adult integument, and cockroach *(B. germanica)* cell line BGE1.

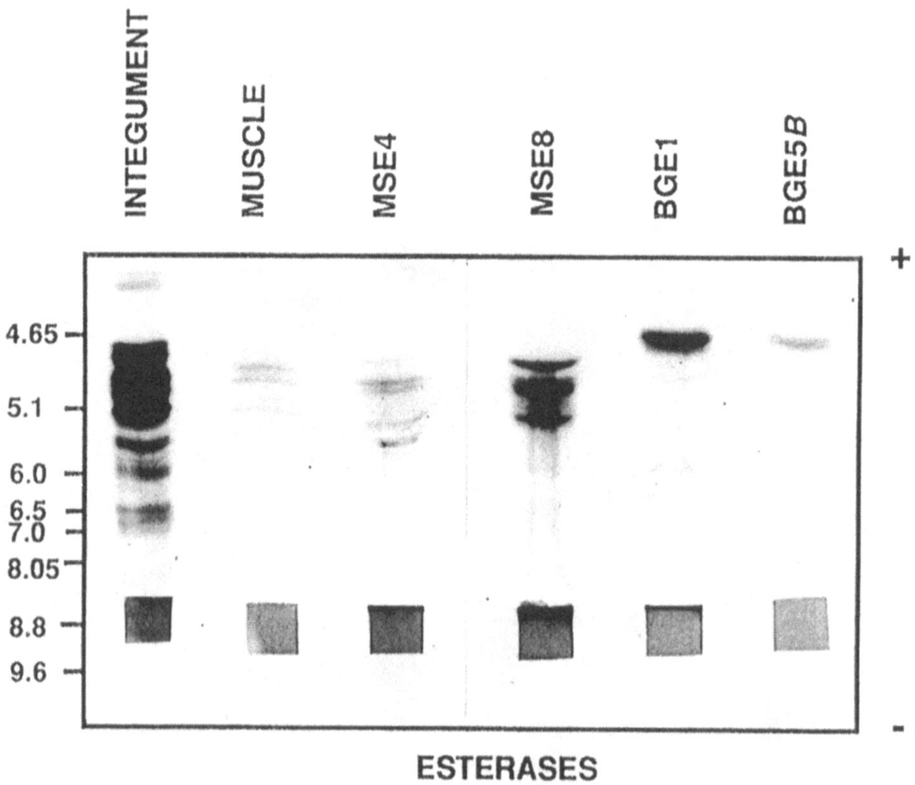

FIGURE 6. Isoelectric focusing analysis of esterases (EST) from extracts of grasshopper *(M. sanguinipes)* adult integument, *M. sanguinipes* adult muscle, *M. sanguinipes* cell lines MSE4 and MSE8, and cockroach *(B. germanica)* cell lines BGE1 and BGE5β.

to completely remove egg yolk (unpublished); in fact, egg extract can be used to replace FBS and stimulate growth in tick cell culture.[30]

The culture materials and techniques employed to grow our MSE lines are those used in most laboratories for the cultivation of vertebrate or invertebrate cells. Contrary to many mammalian cells, insect cells in culture, including the MSE lines, do not require a CO_2 atmosphere, which allows cultivation of cells in tightly closed flasks and reduces the risk of contamination. Thus, expensive, and sometimes unreliable, CO_2 incubators are not needed to control pH, so that MSE and other invertebrate cell lines can be grown in simple convection-type incubators. Two of the MSE cell lines, MSE1 and 8, are fully susceptible to *M. sanguinipes* entomopox virus,[31] and MSE4 supports the replication of a grasshopper-pathogenic microsporidian, *Nosema acridophagus* (Munderloh and Kurtti, unpublished). We hope that the availability of these cell lines, as well as the conventional culture techniques used for their propagation, will encourage their use in research applications that require the availability of established grasshopper cell lines.

ACKNOWLEDGMENTS

Supported in part by the USDA, APHIS, PPQ Grasshopper IPM Project, grant No. 12-34-81-0140-6R, in part by funding approved by the Minnesota Legislature as recommended by the Legislative Commission on Minnesota Resources from the Minnesota Environment and Natural Resources Trust Fund. This is paper No. 20457 of the Minnesota Agricultural Experiment Station.

TABLE 3
Comparison of Major Isoenzymes from *Melanoplus sanguinipes* Tissues and MSE Cell Lines with the *Blattella germanica* Cell Line BGE1

Approx.			MSE cell lines			Cockroach line
pI	Integument	Muscle	4	7	8	BGE1
3.8			LDH	LDH		
3.9						EST
4.1	EST			EST		
4.4	MDH	MDH	MDH	MDH	MDH	
4.5	PGDH	PGDH	PGDH	PGDH	PGDH	
4.5	ME	ME	ME			MDH
4.6	EST	EST		EST	EST	
4.9	ME	ME	ME			
5.0		LDH	LDH	LDH		
5.0	EST	EST		EST	EST	
5.0	PGDH					
5.4						AP
5.5						MDH
5.7	LDH	LDH	LDH		LDH	
5.8						AP
6.0						PGDH
6.0						MDH
6.3	LDH	LDH			LDH	
6.7	AP		AP	AP	AP	
7.0			AP	AP	AP	
7.2						LDH
7.3	AP		AP	AP	AP	
7.4						PGDH
7.8						LDH
8.1						ME
9.8						MDH
10.0	MDH	MDH	MDH	MDH	MDH	

Note: LDH: lactate dehydrogenase; EST: esterases; MDH: malate dehydrogenases; PGDH: phosphogluconate dehydrogenases; ME: malic enzyme; and AP: acid phosphatases.

REFERENCES

1. **Henry, J. E. and Oma, E. A.,** Protozoa: Pest control by *Nosema locustae,* a pathogen of grasshoppers and crickets, in *Microbial Control of Pests and Plant Disease 1970-1980,* Burges, H. D., Ed., Academic Press, New York, 1981, pp. 573-586.
2. **Undeen, A. H. and Epsky, N. D.,** In vitro and in vivo germination of *Nosema locustae* (Microspora: Nosematidae) spores, *J. Invertebr. Pathol.,* 56, 371-379, 1990.
3. **Raina, S. K., Rai, M. M., and Khurad, A. M.,** Grasshopper and locust control using microsporidian insecticides, in *Biotechnology in Invertebrate Pathology and Cell Culture,* Maramorosch, K., Ed., Academic Press, New York, 1987, chap. 21.
4. **Henry, J. E.,** Experimental application of *Nosema locustae* for control of grasshoppers, *J. Invertebr. Pathol.,* 18, 389-394, 1971.
5. **Woods, S. A., Streett, D. A., and Henry, J. E.,** Temporal patterns of mortality from an entomopoxvirus and strategies for control of the migratory grasshopper (*Melanoplus sanguinipes* F.), *J. Invertebr. Pathol.* 60, 33-39, 1992.
6. **Henry, J. E., Nelson, B. P., and Jutila, J. W.,** Pathology and development of the grasshopper inclusion body virus in *Melanoplus sanguinipes, J. Virol.,* 3, 605-610, 1969.
7. **Goldsworthy, G. J.,** Hormonal control of flight metabolism in locusts, in *Biology of Grasshoppers,* Chapman, R. F. and Joern, A., Eds., John Wiley & Sons, New York, 1990, chap. 8.

8. **Fraser, M. J. and Hink, W. F.**, Comparative sensitivity of several plaque assay techniques employing TN-368 and IPLB-SF 21 AE insect cell lines for plaque variants of *Galleria mellonella* nuclear polyhedrosis virus, *J. Invertebr. Pathol.*, 40, 89-97, 1982.
9. **Tsang, K. R., Freeman, F. A., Kurtti, T. J., Brooks, M. A., and Henry, J. E.**, New cell lines from embryos of *Melanoplus sanguinipes* (F.) (Orthoptera: Acrididae), *Acrida*, 10, 105-111, 1981.
10. **Raina, S. K. and Khurad, A. M.**, Primary cultures of ovarian tissue from locusts, *Locusta migratoria* and *Schistocerca gregaria* (Orthoptera: Acrididae), in *Invertebrate and Fish Tissue Culture*, Kuroda, Y., Kurstak, E., and Maramorosch, K., Eds., Springer Verlag, Berlin, 1988, pp. 3-6.
11. **Khurad, A. M., Raina, S. K., and Pandharipande, T. N.**, In vitro propagation of *Nosema locustae* using fat body cell line derived from *Mythimna convecta* (Lepidoptera: Noctuidae), *J. Protozool.*, 38, 91S-93S, 1991.
12. **Munderloh, U. G. and Kurtti, T. J.**, Formulation of medium for tick cell culture, *Exp. Appl. Acarol.*, 7, 219-229, 1989.
13. **Leibovitz, A.**, The growth and maintenance of tissue-cell cultures in free gas exchange with the atmosphere, *Am. J. Hyg.*, 78, 173-180, 1963.
14. **Freymuller, E., Milder, R., Jankevicius, J. V., Jankevicius, S. I., and Camargo, E. P.**, Ultrastructural studies on the trypanosomatid *Phytomonas serpens* in the salivary glands of a phytophagous hemipteran, *J. Protozool.*, 37, 225-229, 1990.
15. **Schaffner, C. P.**, Animal and plant tissue culture decontamination, in *Practical Tissue Culture Applications*, Maramorosch, K. and Hirumi, H., Eds., Academic Press, New York, 1979, chap. 14.
16. **Goodwin, R. H.**, The effect of lipids on the subculture of differentiated cells from primary cultures of grasshopper embryonic tissues, *In Vitro Cell. Dev. Biol.*, 24, 388-400, 1988.
17. **McLimans, W. F.**, The gaseous environment of the mammalian cell in culture, in *Growth, Nutrition, and Metabolism of Cells in Culture*, Rothblat, G. H. and Cristofalo, V. J., Eds., Academic Press, New York, 1972, Vol. 1, chap. 5.
18. **Verma, R. S. and Babu, A.**, *Human Chromosomes: Manual of Basic Techniques*, Pergamon Press, New York, 1989.
19. **Sessions, S. K.**, Chromosomes: molecular cytogenetics, in *Molecular Systematics*, Hillis, D. M. and Moritz, C., Eds., Sinauer Associates, Sunderland, MA, 1990, pp. 156-203.
20. **Dawson, M.**, Characterizing cells in culture, in *Cell Culture Labfax*, Butler, M. and Dawson, M., Eds., BIOS Scientific Publishers, Oxford, 1992, pp. 43-58.
21. **Smith, D. S.**, Oviposition, development of eggs, and chromosome complement of eggs of virgin females of the grasshopper, *Melanoplus sanguinipes*, *Can. Entomol.* 101, 23-27, 1969.
22. **Chapco, W. and Bidochka, M. J.**, Genetic variation in prairie populations of *Melanoplus sanguinipes*, the migratory grasshopper, *Heredity*, 56, 397-408, 1985.
23. **Kurtti, T. J. and Brooks, M. A.**, Isolation of cell lines from embryos of the cockroach, *Blattella germanica*, *In Vitro*, 13, 11-17, 1977.
24. **Pasteur, N., Pasteur, G., Bonhomme, F., Catalan, J., and Britton-Davidian, J.**, *Practical Isozyme Genetics*, John Wiley & Sons, New York, 1988.
25. **McIntosh, A. H. and Ignoffo, C. M.**, Characterization of five cell lines established from species of *Heliothis*, *Appl. Entomol. Zool.*, 18, 262-269, 1983.
26. **McIntosh, A. H. and Ignoffo, C. M.**, Characterization of lepidopteran cell lines by isoelectric focusing and phosphoglucoisomerase, in *Invertebrate Cell System Applications*, Vol 1, Mitsuhashi, J., Ed., CRC Press, Boca Raton, FL, 1989, chap. 6.
27. **Rice, W. C., McIntosh, A. H., and Ignoffo, C. M.**, Yield and activity of the *Heliothis zea* single nuclear polyhedrosis virus propagated in cloned and uncloned lines of *Heliothis* cells, *In Vitro Cell. Dev. Biol.*, 25, 201-204, 1989.
28. **Landureau, J.-C. and Jollès, P.**, Études des exigences d'une lignée de cellules d'insectes (souche EPa). I. Acides aminées, *Exp. Cell. Res.*, 54, 391-398, 1969.
29. **Kurtti, T. J.**, Phenotypic variations of cell lines from cockroach embryos, in *Invertebrate Tissue Culture: Research Applications*, Maramorosch, K., Ed., Academic Press, New York, 1976, pp. 39-56.
30. **Samish, M., Louloudes, S. J., Vaughn, J. L., Kurtti, T. J., and Munderloh, U. G.**, Promotion of tick cell growth by proline and fractions from tick eggs, *Int. J. Parasitol.*, 15, 21-26, 1985.
31. **Munderloh, U. G., Kurtti, T. J., Liu, Y., Chen, C., Krueger, D. E., and Prioux, V.**, Grasshopper cells in vitro: Tools for entomological research, paper presented at Orthopterists' Society Informal Conference, ESA 1992 Annual Meeting, Baltimore, MD, December 6-10.

Chapter 5

TICK CULTURE: APPLICATIONS IN BIOLOGY

Josef Řeháček

TABLE OF CONTENTS

0-8493-7642-4/94/$0.00+$.50

Josef Řeháček, Ph.D., D.Sc., is Senior Researcher of the Department of Rickettsiae at the Institute of Virology, Slovak Academy of Sciences, Bratislava, Slovak Republic. He received his Ph.D. and D.Sc. degrees in virology at the Institute of Virology in Bratislava. He is a member of the International Advisory Board on Invertebrate Tissue Culture, honorary fellow of the Australian National University, Canberra, advisor of WHO in Q fever and head of the ecology section, European Working Group on Rickettsia, Coxiella and Ehrlichia. American Society of Rickettsiologists, Parasitological Society, Slovak Zoological and Slovak Entomological Society.

I. INTRODUCTION

For many years my major interest has been the study of tick-pathogen interactions. In 1955 I decided to cultivate tick cultures *in vitro.* The major reason was that ticks are second only to mosquitoes as the most important vectors of human and animal-pathogenic viruses, rickettsiae, spirochetes, and protozoan parasites. Tick organs, tissues, and cells *in vitro* could bring new dimensions to the study of tick-borne pathogens and of their interactions with tick vectors. *In vitro* studies could elucidate parasite life cycles in arthropod hosts and help design strategies for eventual control. Furthermore, tick cultures *in vitro* could be used to recover tick-borne agents from nature, otherwise difficult or impossible to isolate and propagate in vertebrates *in vivo* and *in vitro,* since ticks, the natural hosts, would be the most susceptible substrates. Close relationships between pathogens and their tick hosts suggest the absence of cytopathic effects in the hosts and retention of full virulence for mammalian hosts. The latter characteristic is very important for the preparation of appropriate antigens and highly effective vaccines.

In general, propagation of pathogenic agents in arthropod organ, tissue, and cell cultures *in vitro* could substantially contribute to the investigations of growth cycles, mechanisms of pathogenicity at the cellular level, biological properties, studies of antigen variations, morphology, metabolic aspects, biochemical properties, various developmental stages, strain variations, and screening of therapeutic agents. Most researchers prefer easily prepared substrates for cultivation of microorganisms to obtain sufficient numbers of highly sensitive cells in a cost-efficient way.

In our country when somebody starts something new we say that he was present at the source or cradle; I was in that position nearly 40 years ago. Impetus to begin developing the new subject of tick tissue culture came at the beginning of 1955 following a lecture by my colleague in the Institute of Virology, Bratislava, who had just introduced a tissue culture method of vertebrate cells, as a tool for virological studies at our Institute. Immediately after the lecture I visited some libraries to find out whether somebody had cultivated invertebrates *in vitro,* as I intended to examine tick cultures *in vitro* for the cultivation of the EEE virus, the relation of which to ticks was the subject of my interest and also of my Ph.D. thesis.[1] The first reference I found was Goldschmidt's paper on insect culture *in vitro,* published in *Zentralblatt f. Bakteriologie* in 1916 which provided the first data on this attractive problem. I began to cultivate explants from various organs of adult *Ixodes ricinus, Dermacentor reticulatus,* and *D. marginatus* ticks using hanging drop preparations and various mixtures of media used at that time for the cultivation of vertebrate cells *in vitro.* I was pleased by the partial success of a few days survival of excised pieces of organs accompanied by limited migration of some fibroblast-type and epithelial-type cells. Encouraged by this result, I explanted small pieces of excised tissues of developing adults from metamorphosing engorged nymphs of *D. marginatus* ticks in which the migration of cells and the limited cell multiplication became quite evident.[1-4] At that time a film demonstrating the cell growth from *D. marginatus*-explanted tissues had also been prepared in my laboratory. At this same time a famous U.S. virologist who visited our institute remarked on the high scientific standard of our workers and highlighted two topics which particularly impressed him, i.e., interferon and tick tissue cultures. This encouraged me very much to continue in this difficult subject.[139] I had been working on this problem until 1970.

At the beginning of my work on tick cultures *in vitro,* I believed I was the first who introduced this subject to science; however, subsequently the priority was given to Professor F. Weyer who attempted, even if unsuccessfully, to cultivate fragments from louse guts infected with *Rickettsia prowazekii* and *R. typhi* together with small pieces of various organs of *Rhipicephalus bursa* ticks in an effort to transmit rickettsiae from louse to tick organs. He used a culture medium consisting of rabbit plasma, spleen, testicles, and extract of lice and the hanging drop method.[5] Thus, the endeavor to establish tick tissue cultures *in vitro* had ricketsiological implications, which to me as a rickettsiologist gave great pleasure.

For the sake of convenience, let me divide tick cultures *in vitro* into four categories, i.e., hemocyte cultures, organ cultures, primary tissue and cell cultures, and continuous cell lines. However, before one reads this paper I would like to emphasize that this topic has already been brilliantly reviewed by several workers.[6-19] In these reviews, all necessary practices and procedures concerning tissue cultures *in vitro* are presented in more detail.

II. HEMOCYTE CULTURES

Hemolymph, in particular hemocytes, of arthropods have several important functions. In addition to some physiological functions associated with metabolic processes, they are the most important target for the distribution and development of pathogens, particularly of viruses and bacteria transmitted through hematophagous arthropods to vertebrates. Some types/forms of hemocytes having phagocytic properties may play a defense role, which, in the case of ticks, would be interesting to investigate *in vitro*. Therefore, it is not surprising that many researchers interested in pathogen-tick relationships included tick cultures *in vitro* in their research programs. And again as in other types of *in vitro* tick tissue cultures their principal involvement was the cultivation of pathogenic microorganisms, in particular arboviruses.

A special technique was devised for collecting and culturing large numbers of tick hemocytes using microculture vessels, such as 96-well tissue culture plates or Rose chambers or capillary tubes and small screw-cap glass vials.[20] The cultivation of tick hemocytes *in vitro* is relatively easy to perform.

However, tick hemocytes *in vitro*, in contrast to organ, primary cell cultures, and cell lines, have been rarely employed in biological investigations. The reason for this could be the limited number of cells available and their relatively short survival time. In some cases it was yet to be found that they could survive for long periods,[21-23] as, for example, from the argasid tick *Ornithodoros moubata* up to 8 months[24] and from *O. coriaceus* up to 9 months,[23] but this was survival without any significant multiplication.[21,25] Sometimes, during the relatively short time of hemocyte survival, an alteration in cell population was evident.[23,24] Slight changes in the cell ultrastructure could also occur.[26,27] Cultivating hemocytes of third nymphal stage of *Alveonasus lahorensis* at room temperature, it was observed that the various cellular processes connected with the forming of different structures, i.e., mainly membranes.

Surviving hemocytes *in vitro* of several argasid ticks, i.e., *A. lahorensis, O. papillipes,* and *Argas persicus,* were used for the study of their phagocytic function to various bacteria species.[28] The viability and suspected high sensitivity of tick hemocytes *in vitro* were examined in the cultivation of tick-borne encephalitis virus in hemocytes of *Hyalomma dromedarii* and of Colorado Tick Fever virus in hemocytes of *Dermacentor andersoni* ticks *in vitro*.[25,29] The production of viruses compared to that in primary tissue cultures was in both cases lower. This could be explained by differences in the number of cells producing viruses in both examined substrates. Further cultivation of Colorado Tick Fever virus in hemocytes of *D. andersoni in vitro* showed that the growth curves of this virus were similar to those in primary tissue cultures of tick viscera.[21]

The single experiment employing tick hemocytes for cultivating babesiae compared the longevity of viability of uninfected hemocytes from nymphs and females of *Boophilus annulatus* with those from ticks infected with *Babesia bigemina*. It was shown that uninfected cells survived for 8 to 11 days and infected ones for 3 to 4 days only.[30]

Although the preparation and maintenance of hemocyte cultures from various tick species is relatively easy and without any major problems, their limited use today does not reflect their potential application in various branches of biology. I am convinced that with the renaissance in research on tick-borne zoonoses, i.e., in particular borrelioses, rickettsioses, anaplasmosis, etc., researchers will consider hemocyte cultures which could contribute considerably to the elucidation of host-pathogen relationships including defense mechanisms of ticks.

III. ORGAN CULTURES

Very little attention has been devoted until now to the preparation and use of organ cultures from ticks for studying various problems in biology. In my opinion tick organ cultures, which require a little more laborious preparation by an experienced worker, are suitable substrates for many special investigations, i.e., observations of some chemicophysiological processes, e.g., osmoregulation, salivation, elimination of feces, digestion, oogenesis, spermatogenesis, functional changes of organelles, etc., and in particular for studying some pathogenic processes following the infection with some pathogens, i.e., for studies of their developmental cycles, virulence, etc.

One valuable application, especially of salivary glands, is as a source of antigens that provoke an immune response against subsequent tick infestation and for developing vaccines against ticks using immunogens derived from the salivary glands *in vitro.* Organ cultures might also have applications in the study of hemoparasites especially of *Theileria* and *Babesia* which reach a part of their life cycles, i.e., infective stages, in the salivary glands; they could thus contribute considerably to a basic understanding of parasite-host relationships. Almost nothing is known about the physiological and biochemical barriers that determine the mechanisms of host specificity and adaptation of a parasite to a certain host species.

With regard to the latter phenomenon, I suppose organ culture could provide some explanation of tick-pathogen specificity in the same habitats with the same spectrum of host animals. This could be a goal contributing to the solution to one of the basic biological problems of vector-borne disease ecology.

The advantage of organ cultures compared to cell cultures is that they preserve better the original properties of host organisms and this would be expressed in the sensitivity of such cultures to pathogenic microorganisms.

Two types of organ cultures have been developed, i.e., complete explanted organs or small organ pieces, and the so-called "backless tick explants" which were introduced to science.[31] The latter cultures have ticks cultivated *in vitro* with their dorsal integument removed; however, this technique does not completely isolate the organs from each other. The backless tick explants technique is simpler and generally supplies better results than cultures of the excised individual organs, and it provides maintenance conditions closely related to those in ticks themselves.

In 1977 tissues were supported with stainless steel wire mesh,[32] while more recently agarose gel was being used to support the organs.[33] In 1982 microchambers were introduced for the cultivation of tick organs.[34]

The viability of organ cultures exhibiting sometimes only poor cell migration and no evidence of growth has been proved either by histological criteria[22,30,35] or by contractions and peristaltic-like waves for several days after the cultures were initiated.[35] It was also assessed by measuring fluid secretory competence.[32,33] It was reported that when female salivary glands of *Amblyomma hebraeum* were cultured at 26°C, they secreted fluid at 65% on day 2, 59% on day 4, and 42% on day 6.[33] Salivary glands from fed males cultured for 4 days lost only 5% of their fluid secretion, whereas glands from males 4 days post-removal lost 63% of their fluid secretion.

As mentioned in the introductory part of this chapter, explants of tick organs *in vitro* represent the first attempt of tick cultures *in vitro.* Professor Weyer used them unsuccessfully for the cultivation of typhus group rickettsiae.[5] Further use of organ cultures was directed also to the cultivation of tick-borne pathogens, i.e., of "filterable" particles of *Coxiella burnetii.* At that time Dr. Kordova investigated the possible existence of so-called "ultrafilterable forms" of *C. burnetii* and asked me to prepare surviving fragments of *Ixodes ricinus* and *D. marginatus* ticks *in vitro* to investigate the existence of those forms. Although this effort reached the aimed target and "filterable forms" were transformed to normal rickettsiae after 12 to 16 days survival in tick explants,[36,37] these results need further confirmation by the use of modern advanced methods.

Following the above experiments on surviving fragments of tissues of various tick organs of *D. marginatus, D. reticulatus,* and *I. ricinus,* ticks were used for the cultivation of EEE virus.[1] Survival of these fragments in Eagle's medium for a few days in *in vitro* conditions was proved by staining and in particular by rhythmical movement of ovarial pieces, pieces of digestive tract, and Malpighian tubes. EEE virus not being tick-borne demonstrated clear propagation only in explants from connective tissues. The multiplication was evident especially with the administration of small inocula, the virus increment being of the order of 4 log units of TCID per ml of virus. Other tissues of these ticks and of *D. marginatus* caused only gradual virus inactivation.[1,38]

The cultivation method of explants from the whole developing imagoes of ticks was elaborated in 1981, and tick-borne encephalitis and Powassan viruses were cultivated in explants prepared from *H. anatolicum anatolicum* and *H. dromedarii* ticks.[39] Both viruses multiplied in this substrate over 200 days after the infection. Dual infection with both viruses did not result in the inhibition of 1 of those viruses within 1 to 2 months but did in the following months.[40,41]

The greatest benefit of tick organ culture has been obtained with hemosporidians, *Theileria* and *Babesia,* which have a complex life cycle developing in various target tissues of arthropod hosts involving gametogony and sexual fusion in the gut, formation of motile kinetes, and sporogeny in the salivary glands. Hadani, in 1981, presented a review on development of these parasites in tick organs and tissues *in vitro.*[42] Longevity of maintenance of uninfected and infected nymph and adult organs of several tick species at 28°C with *B. bigemina* showed longer values for uninfected organs, e.g., intestinal tract 17 to 25, ovaria 28 to 58, Malpighian tubes 13 to 17, salivary glands 26 to 50 days, respectively.[30,43] Of particular interest was that *B. bigemina* did not develop in cultures of salivary glands, although in cultures of intestinum, ovary, Malpighian tubes, and hemocytes of nymphs and females. *B. annulatus* schizonts and merogoits at 28°C could be identified. Parasitized organs survived less than 1 to 2 weeks, releasing kinetes into the medium.[30] In 1978 the maturation was reported of kinetes of *Nuttallia danii (Babesia merionis),* a rodent babesia. Kinetes *in vitro* turned into infective particles in salivary glands of partially fed *H. anatolicum excavatum* nymphs.[44] Larvae fed on infected gerbils hatched to nymphs, and their cultured salivary glands demonstrated parasite development at a rate similar to that observed under normal conditions. McCall studied the effect of salivation stimulants on *T. parva* — infected salivary glands of unfed adult *R. sanguineus* ticks.[45,46]

The backless tick explant method[31,47] did not find wider application in practice; however, McCall observed development of *T. parva* in the culture of salivary glands of *R. appendiculatus.* Maturation *in vitro* of *T. parva* in backless tick explants was compared with that in cultured excised salivary glands. After 3 to 7 days at 30°C, salivary glands from backless tick explants and excised salivary glands showed similar numbers of infected acini per infected tick. After 12 days at 28°C, backless tick explants showed 20 to 30 times as many infected acini per tick as excised salivary glands. Later on McCall noticed that the whole nymph explants of *H. anatolicum anatolicum* infected with *T. annulata* released large numbers of *T. annulata* kinetes into the medium. Some of these kinetes subsequently infected the salivary glands of the explants. Backless tick explant cultures from infected adult ticks supported the maturation of theileriae from sporoblasts to sporozoites.[48]

Kurtti et al.,[49] demonstrated the infectivity of *T. parva* in salivary glands taken from 2-day fed *R. appendiculatus* and cultured for 3 days at 32°C; glands from unfed infected ticks cultured for 5 days failed to yield infective sporozoits.

Tick organ culture technique has been used mainly for maintenance *in vitro* of excised salivary glands and for various physiological studies.[50,51] The reason for this is that ticks of the family Ixodidae employ their salivary glands as an osmo- and volume-regulatory system, i.e., during a blood meal, when excess fluid is secreted directly back into the host's circulation system. It was in particular Professor Kaufman and his school who devised a method for the study of salivary secretion under *in vitro* conditions and tested various salivary stimulants.[52,53] Salivary glands *in vitro* in TC 199 medium secreted at a rate of 73% on day 3, 25% on day 7, and 8% on day 14.[32]

It was found that dopamine and spiperone, which are supposed to be natural transmitter substances in the tick salivary glands, stimulated fluid secretion in isolated salivary glands from *Amblyomma hebraeum in vitro.*[54] In another study the same authors tested the effectiveness of butaclamol and sulpiride in blocking secretion stimulated by a number of antagonists using the same substrate, i.e., salivary glands of *A. hebraeum in vitro.*[55] However, salivary glands of *A. americanum* degenerated *in vitro* when cultivated with a variety of ecdysteroid analogs but not when incubated with a number of "vertebrate" steroids, which, on the contrary, improved fluid secretory competence by 22 to 52% after 4 days in culture.[56,57]

Fluid secretory competence *in vitro* can be used as an index of salivary gland degeneration in *A. hebraeum.* The glands dissected from fed males which had been left off the host for 4 days lost 62% of their fluid secretory competence compared to glands dissected shortly after the males were removed.[58] The significant reduction of fluid secretory competence of *A. hebraeum* salivary glands *in vitro* was found under influence of some ecdysteroids and their analogs.[59]

Regarding organ cultures, other than salivary glands only the maturation of spermatozoa of *O. moubata* and *D. variabilis* ticks *in vitro* had been observed.[60]

Not only zoologists and physiologists, but also microbiologists by their ignorance or reluctance to include tick organ cultures in their studies, have missed the chance of contributing comprehensively to the elucidation of many basic biological problems.

IV. PRIMARY TISSUE CULTURES

Primary tick tissue cultures could be considered as a more or less unsuccessful attempt to develop a cell line which for various reasons stopped short of achieving the goal. This type of culture includes the cultivation of migrating cells from tissues of donors when a significant multiplication of cells occurs for but a limited period (Figure 1).

The methods for the preparation of primary tick tissue cultures have been brilliantly reviewed.[14] Primary cultures are derived either from embryos removed from eggs or from developing adults within metamorphosing engorged nymphs. Embryonic cultures from eggs, e.g., from *H. asiaticum*[61] from embryonic tissues of *A. americanum*[62] and from B. *microplus,*[63] In these cultures the cells were passaged several times before they died off. Growth of spindle-shaped fibroblast-like cells was observed from embryonic cells of *H. rufipes rufipes* within 7 to 10 days. During the following 2 to 3 weeks cells grew extensively, forming a network which resembled muscle cells.[64] The technique of the preparation of embryonic cells from ticks has been summarized.[65]

Eggs which were surface sterilized or eggs deposited by surface-sterilized females were incubated aseptically and cracked open by gentle pressure with a glass rod or syringe plunger. Embryonal fragments were washed, broken mostly by pipetting, centrifuged and seeded into cultivating flasks. The removal of egg shells at that time had probably only a cosmetic importance. One of the main problems concerning the preparation of cells from developing eggs is that of frequent bacterial and especially fungal contamination. It seems that the technique of obtaining sterile eggs from surface sterilized preovipositing females is simpler and more reliable than surface sterilization of oviposited eggs.[63]

Unfortunately, preparation of most primary tick tissue cultures is more time-consuming than that from developing eggs, for example, cultures from developing tissues of adults within nymphs undergoing metamorphosis.[2-4,35,63,66-75]

Surface sterilization of engorged nymphs is achieved successfully by rinsing in 70% ethanol or 10% benzalkonium chloride followed by washing in sterile distilled water.[7] The nymphs are dissected while immersed in sterile saline.[4] The best source of a high number of growing cells is the nymphs in which the developing adults, i.e., the frontal parts of the

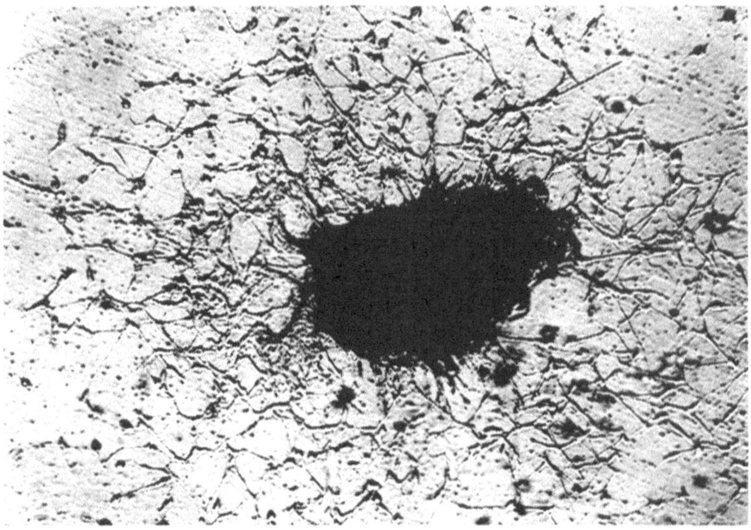

FIGURE 1. Six-day-old primary culture from *Dermacentor reticulatus.*

imaginal body, are clearly visible. The preimaginal tissues in early development as well as in later stages, after a cuticle starts to develop, i.e., hardening of cuticle, signal lesser ability of the explants to yield growth.[67,76,77] The pooled tissues are dissociated either by enzymatic treatment/trypsinization, which results in considerable amount of cells but may have a detrimental effect on the growth of cells/or better, by repeated pipetting, or by simply teasing the tissues apart by use of scissors or forceps. After several washings in saline or in the culture media, the cells are centrifuged for about 10 min at about 1500 rpm and resuspended in the medium for subsequent seeding.[4] The material consists in this case of cells and hemocytes of the whole preimaginal tick with the exception of cells and tissues of Malpighian tubes and the digestive tract, which are discarded during the dissection. The success of cultivation thus depends on the selection of tick species, optimal stage of development for the dissection of nymphs, and, of course, on the growth medium. Leibovitz L-15 medium supplemented with tryptophase phosphate broth and fetal calf serum seems to be the best. Under these conditions, several cell passages could be performed without any problems.[78]

The third convenient source is cells from ovarial tissues of adult females. When small pieces of ovaries were removed from partially engorged females of *D. reticulatus* and *D. marginatus* migration of fibroblast-type cells occurred during first days of cultivation, followed by the production of small discrete colonies of epithelial — type cells by day 7 to 10. The cells further slowly increased in number to cover the surface of culture flasks in 2 to 4 weeks. These cultures were subcultured 5 to 9 times (Figure 2), but subsequently the cells ceased to multiply, became very large and granular, and died off.[67] Similar cultures were also developed from *Haemaphysalis inermis* ticks (Řeháček, unpublished results). In these cultures, multiplication of cells often occurs in the form of discrete foci, with cells piled on top of each other; such cultures resemble a hilly country and these foci are clearly visible to the naked eye.

One of the most important factors for successful cultivation is the use of an appropriate medium. Different culture media have been employed. At first hemolymph composition of ticks was used as a guide in an effort to create physiological culture conditions for tick cells.[69,70,79,80] The media were supplemented with extracts from eggs of ticks, spiders, sera of different beetle

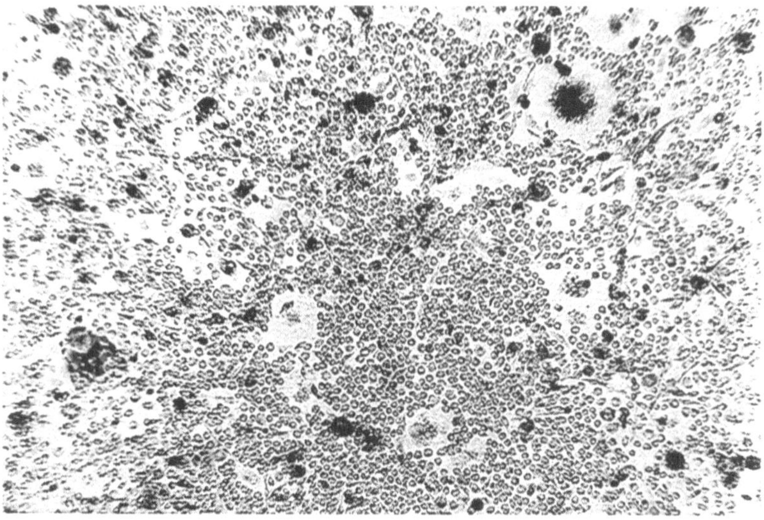

FIGURE 2. *D. reticulatus* cells from explanted ovary; 7th passage.

larvae and adults, hemolymph from various arthropod species, i.e., cockroaches, beetles, etc., but the resulting media did not promote the continued cultivation of the cells nor result in the establishment of a cell line. The analysis of hemolymph taken from ticks partially engorged on vertebrate animals focused on proteins, carbohydrates, and ionic composition. Based upon analysis of the amino acids, sugars, and cations in the hemolymph of *B. microplus, Argas lagenoplastis,* and *H. dromedarii,* a medium was designed in which the cells of *R. sanguineus* grew and multiplied for 9 months.[79] We investigated the utilization of amino acids and sugars by the cells of *R. sanguineus* ticks grown for a 10-day period. Aspartic acid, glutamic acid, threonine, proline, methionine, leucine, and phenylalanine decreased in concentration. Valine, cystine, and isoleucine remained at about the same concentration during this period, while serine, alanine, tyrosine, lysine histidine, glycine, and arginine increased in amount. Both sugars in the medium, glucose and inositol, were extensively utilized.[80] Uptake of glucose as a possible indicator of growth and metabolism of the tick cells *in vitro* was observed also.[81] When judging these attempts on improvement of the chemical composition of cultivating media, it is now obvious that a suggestion was considered at that time, i.e., that the optimal cultivating media should reflect the natural conditions of host environs, i.e., the composition of hemolymph which creates the natural environment for arthropod organs. Second, in an attempt to obtain as much hemolymph as possible, partly engorged female ticks were used. It is possible that host serum components including globulins and albumins penetrated the gut wall of ticks and contaminated the hemolymph.[82-85] At that time nobody knew or considered this possibility. Future successful growth of tick cells will probably depend on the above factors and on the amount of cells to initiate cultures. The major factor in my opinion is the tick species involved.

For experimental studies, the preparation of a sufficient number of primary cultures is too laborious, e.g., preparation of 20 vials from developing imagoes within engorged nymphs with cells will take 1 day of intensive work. This was probably the main obstacle for the broader application of primary tick cultures. Despite this obstacle, at the beginning of the 1960s the status of tick tissue culture appeared to be reasonable for study of a broad spectrum of various biological problems connected with the cultivation of pathogens, in particular arboviruses.[86]

Primary tissue cultures from developing tick adults were successfully used for the first time for the propagation of tick-borne encephalitis virus. The cell cultures prepared from developing imagoes within engorged *D. marginatus* and *H. asiaticum* nymphs propagated the virus up to 5 log units.[25] The importance of that substrate was emphasized by high susceptibility and the ability to detect small amounts of virus, i.e., 0.3 to 0.03 IFD_{50} in higher percentage, i.e., 76 and 18 within tick cells, compared to 34 and 3 in chick embryo cell cultures.[87] This finding was applied for the isolation of this virus from field material, i.e., blood samples and brains of small mammals and from ticks. The isolation experiments were carried out on primary cultures of *H. dromedarii* ticks and simultaneously also with chick embryo cell cultures. Of 187 samples examined, 5 strains of the virus, namely 1 from the blood of *Talpa europaea,* 1 from the blood of *Apodemus flavicollis,* and 3 from *I. ricinus* ticks, were isolated. All these strains were also isolated in chick embryo cell cultures. The results demonstrated that the method of isolating the tick-borne encephalitis virus in tick primary cells was equal to that in chick embryo cell cultures, and therefore we recommended that it should be used for virus isolations from nature in addition to suckling mice and chick embryo cell cultures.[141]

This pioneering work on the application of tick tissue culture for virus propagation was followed by experiments to establish the spectrum of this substrate to various viruses, using *H. dromedarii* primary cultures. It was shown that most of arboviruses, irrespective of whether they were mosquito or tick-borne, propagated well in this culture exhibiting mostly a 10^4- to 10^5-fold increase in virus levels until the 8th day of cultivation. However, nonarboviruses, i.e., encephalomyelitis, polio, vaccinia, vesicular stomatitis, NDV, and pseudorabies, with the exception of LCM virus which multiplied in tick cells, did not multiply in this substrate.[88] These results indicated that tick tissue cultures may represent a suitable medium not only for the propagation of some arboviruses but also for a biological characterization of viruses.

Other studies on primary tissue cultures from *H. dromedarii* ticks confirmed the high sensitivity of this system for *in vitro* cultivation, i.e., with Langat and Quaranfil viruses[89] and with the Tribec virus.[90] The latter virus multiplied, reaching titers up to $10^{4.5}$ mouse intracerebral LD_{50} per 0.01 ml during 4 to 14 days post-infection. The fluorescent antibody technique revealed the viral antigen in the form of confluent and brightly fluorescing granules in the cytoplasm of infected cells starting from the second day.

The growth of Colorado Tick Fever virus in primary tissue cultures of its vector, *D. andersoni,* again confirmed tick cultures to be a very sensitive system for the detection of an arbovirus. After a 6 to 10 day latent period following inoculation, the virus multiplied intensively reaching a 4.5 log unit increase of extracellular virus at 4 to 5 weeks. Afterwards, virus was detected up to 166 days in cultivated tissues. It was shown that less than 0.1 of a suckling mouse intracerebral LD_{50} can be taken up and propagated to higher levels by tick tissues *in vitro.* Further observation on certain characteristics of the Florio-2 strain showed that this virus remained unchanged after 98 and 124 days *in vitro,* i.e., difference from the other strains of Colorado Tick Fever virus in its ability to kill adult mice after intraperitoneal infection remained unchanged. In addition, the average survival time — 5 days — of intracerebrally inoculated suckling mice was not altered,[29] and the titer of mouse brain stock prepared from virus that had been in tick tissue cultures for 124 days remained approximately equal. This is very important because this cultivation system could be used for keeping the pathogenic agents when a change in their basic properties, i.e., virulence, is not desirable. However, such studies need confirmations as it was shown in repeated experiments of tick-borne encephalitis virus in tick tissue cultures that the virus — western subtype — when serially passaged at weekly intervals during a 1 year period persisted for 253 days in passaged cultures with titers between 10^3 to $10^{6.5}$ log units per ml, indicating that practically no change in virulence had occurred. It can be concluded that it is possible to maintain tick-borne encephalitis virus in tick tissue cultures as a virus stock for experiments.[16]

High sensitivity of *H. dromedarii* tick cells *in vitro* to tick-borne encephalitis virus and Lipovnik virus (Kemerovo group) was also used to study whether tick-borne encephalitis virus, acting as an excellent interferon inducer in vertebrate cells, could do the same in invertebrate cells *in vitro.* However, when tick cultures were inoculated separately with both viruses, in the first virus 3 to 15 days after the other, no interferon-mediated or other interference was observed. On the contrary, if both viruses were inoculated into the culture simultaneously, marked mutual inhibition appeared as documented by growth curves and by a decreased percentage of infected cells.[91,92]

In all experiments with the arboviruses mentioned above, the appearance of tick tissue cultures was not affected by viral infections as shown by the absence of cytopathic effects and inclusions, which probably limits their wider use in virology.

Primary tick cell cultures were also shown to be a reasonable substrate for the cultivation of rickettsiae. This was first demonstrated with *H. asiaticum* cells inoculated with *Coxiella burnetii.* After 6 days cultivation at 28°C, this rickettsia reached a titer of 10^6 ID_{50} per ml in chick embryos. Comparison of *C. burnetii* propagation in tick cultures with propagation in other substrates showed that the tick cells were the most susceptible.[93] In addition to *C. burnetii,* other species of rickettsiae also have been propagated in primary tick tissue cultures prepared from *H. asiaticum* ticks. Simultaneous titration of these rickettsiae in tick cell cultures and in chick embryos revealed higher sensitivity of tick cultures in the case of *R. conorii, R. akari,* and *R. prowazekii* (Figure 3) formed long chains and threads, and *R. typhi* segmented chains in the cells (Figure 4) both going around cell vacuoles. Both rickettsiae were occasionally found also in cell nuclei. The high susceptibility to rickettsiae of tick cells *in vitro* permitted investigations of various aspects, e.g., the isolation of rickettsiae from different sources in nature. The various patterns of propagation of rickettsiae also offer the basis for a preliminary diagnosis and differentiation between *C. burnetii* and other rickettsiae and between spotted fever group rickettsiae and typhus group rickettsiae, i.e., single rods being characteristic for the first, and chains and threads for the second.[94] Additionally, multiplication of *R. slovaca* in tick primary cultures from *H. dromedarii* (Figures 5 and 6) demonstrated that multiplication of this rickettsia was much greater in tick cells than in vertebrate cells *in vitro* and *in vivo.* However, experiments carried out on the comparison of the detection of small amounts of these rickettsiae demonstrated almost the same results in yolk sacs and in tick cells *in vitro.*[95]

It has been reported that primary cell cultures from *A. hebraeum* ticks became infectious for 9 days after being inoculated with blood infected with *Cowdria ruminantium.*[96]

Chlamydia psittaci, the agent of enzootic abortion of sheep, strain PL-377, was cultured for 30 to 45 days in embryonal cells of *Hyalomma genus* ticks *in vitro.*[97]

Only two papers dealt with the cultivation of protozoan parasites in primary tick tissue cultures. Cultures from *B. microplus* were used for the cultivation of erythrocytic forms of *B. ovis.* Babesiae invaded the cells and multiplied up to 48 h after the inoculation. Their structure resembled that of blood forms rather than that of gut and salivary gland forms in ticks.[98] Primary cultures from *Anocentor nitens* embryonal cells infected with *B. caballi* probably caused their death.[99]

Recently, resistance of ixodid tick infestation induced by administration of whole extracts of ticks, tick salivary gland extracts, and suspensions prepared from tick viscera has been investigated. Tick tissue cultures were shown to be efficient sources of antigen for induction of host resistance to ixodid ticks infestation using primary tissue culture and cells from developing larvae *A. americanum* administered to guinea pigs. A dose of 10^6 cells inoculated subcutaneously on day 0, 7, and 21 induced resistance in host animals on day 35 as evidenced by challenging by *A. americanum* and *D. andersoni* ticks. This was expressed by reduced engorgement weight of females, reduced oviposition by the females, and by death of ticks at the attachment site.[100]

Although primary tick tissue cultures at the time of their establishment have provided a useful tool for solving some biologic problems, their laborious preparation and the establishment of continuous tick cell lines from various tick species directed the attention of researchers to other systems.

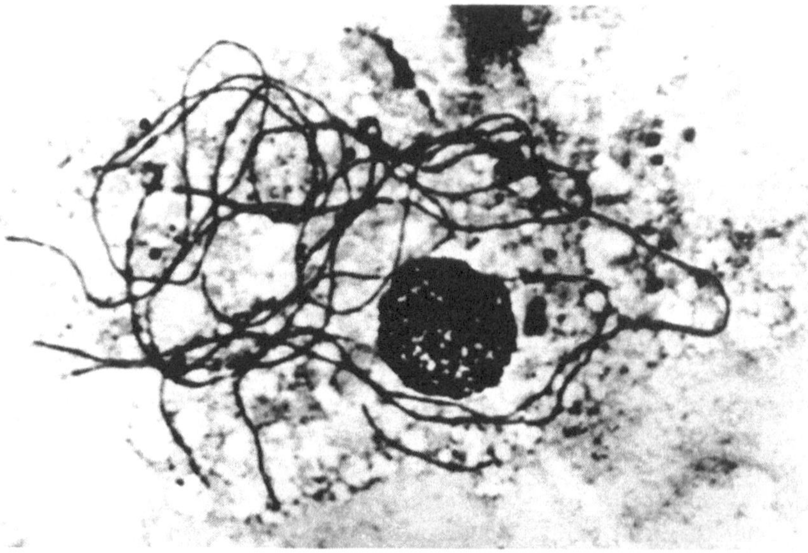

FIGURE 3. Eight-day-old culture of *R. provazekii* in *H. dromedarii*.

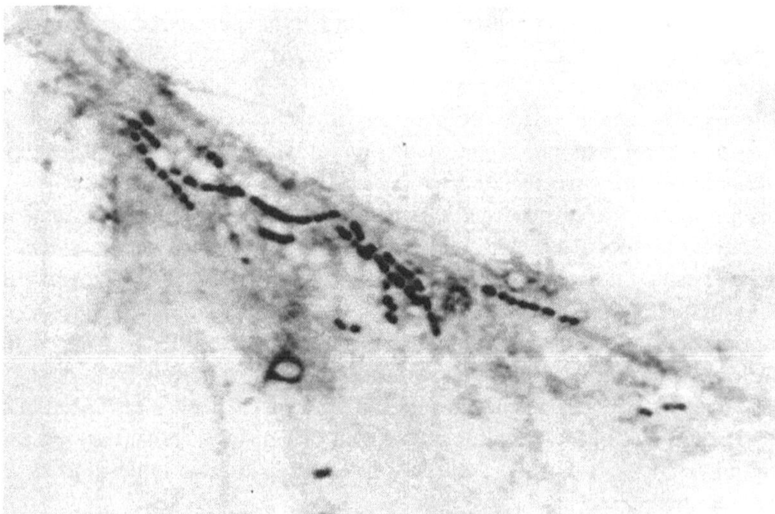

FIGURE 4. Eight-day-old culture of *H. dromedarii* with *R. typhi*.

V. CONTINUOUS CELL LINES

Most of the attention of researchers employing tick cells *in vitro* has been directed towards the development of continuous cell lines, since they are easy to handle, can be grown in relatively standardized conditions, and have an almost uniform cell type which is not the case with organ and primary cultures. The first three continuous cell lines were established from developing adult tick *R. appendiculatus*. This opened the door for the development of many other cell lines. The impetus for the development of the above cell lines was the preparation of a reasonable substrate for the cultivation of arboviruses, and this was immediately accomplished by the successful cultivation of Langat, Quaranfil, West Nile, and louping-ill viruses.[101]

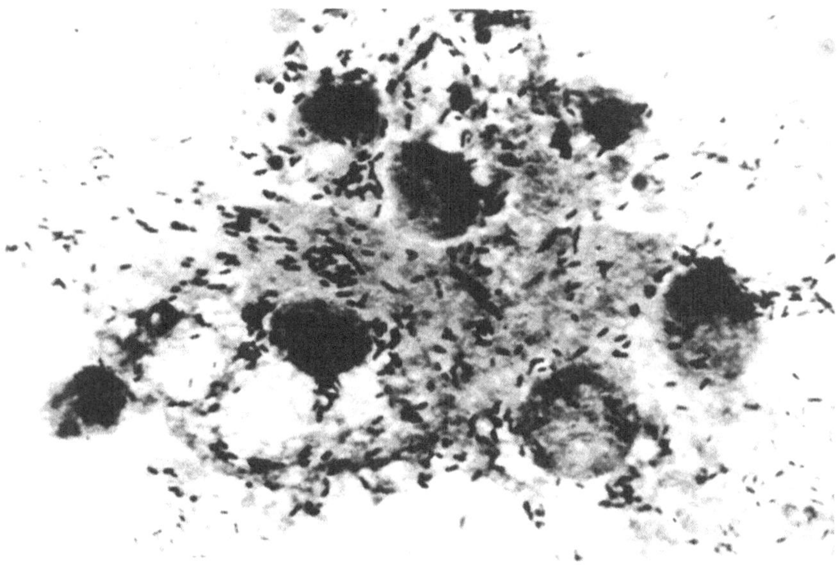

FIGURE 5. Eight-day-old culture of *R. slovaca* in *H. dromedarii.*

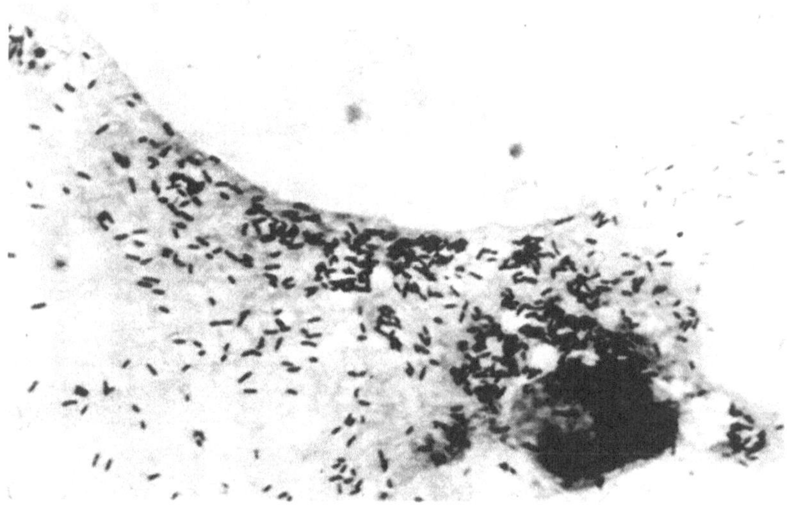

FIGURE 6. Eight-day-old culture of *R. slovaca.*

Many factors appear to be crucial for the development of continuous tick cell lines, and it is impossible to mention them all in detail in this review; they are reported in several papers describing not only the development patterns but also the cell morphology, karyology, growth requirements,[9,10,14,77,78,102,103-105,138] growth rates, etc.

Let me briefly emphasize some of the most important factors responsible for the development and successful maintenance of continuous tick cell lines. Usually the establishment of a tick cell line is determined by the availability of tick species which may not be the vector of the pathogenic microorganism intended to be studied. However, it seems that some genera of ticks are not suitable for the establishment of cell lines, e.g., *Ixodes.* There also may be differences within some genera, e.g., genus *Dermacentor,* where cell lines of some species, i.e., *D. parumapertus* and *D. variabilis*, have been established, although none from *D. andersoni*, despite enormous efforts devoted toward the creation of a cell line from this species, an important vector of several human pathogens. Next to tick species, the density of cells with mitotic activity available — namely, from embryonic and preimaginal tissues of ticks — and the physiological status of the ticks are important factors in successful seeding and propagation of cells, e.g., it is necessary to estimate the correct time of the embryonal development in eggs and of preimaginal tissues in engorged nymphs. In the latter case, the most suitable time is when the anterior part of the developing imago is clearly visible in the frontal part of the nymphal body (see Section IV). In selecting tick species for the cell cultivation, laboratory colonies are preferable to field-collected specimens to eliminate undesirable microbial contamination. The development of adults in metamorphosing nymphs could be slowed down by maintenance at +4°C for several weeks, until convenient for investigations.[72]

Surface sterilization of ticks is necessary prior to their preparation for cell cultures, i.e., either surface sterilization of engorged nymphs, eggs, or engorged females before oviposition. The latter method is probably the most important. This has been done by first removing gross contaminations like host's hair, blood, feces, etc. by washing in running tap water, thereafter by immersion in 70% ethyl-alcohol or other chemosterilants such as benzalkonium chloride. Subsequently, the ticks are rinsed in sterile distilled water with, or without, antibiotics before further processing. It is, of course, impossible to avoid internal contamination by microorganisms living especially in the digestive tract and hemolymph of ticks, but, fortunately, these contaminations are infrequent. All the processes of the preparation of tick material for culturing, i.e., dissection of tissues, seeding of cells, etc., must be done under aseptic conditions (air laminar flow cabinet) because the contamination, in particular with bacteria, mycoplasmas, yeasts, and fungi, may destroy all attempts to set up and maintain the cultures.

A suitable medium is probably the most decisive factor, after tick species involved, in the establishment and maintenance of a cell line. Basic media, such as Eagle's, Hanks, and Leibovitz L-15 with some adjustments, have been used for this purpose. Despite the described spectrum of the utilization of aminoacids and sugars by *R. sanguineus* primary cell cultures,[80] the nutritional requirement of tick cells are still undefined.[8] Addition of various substances into the above-mentioned media, such as tick extract, demonstrated a beneficial effect on tick cells *in vitro*,[106] and resulted in an increase of the splitting rate from 1.5- to 2.4-fold during a period of 30 days; addition of L-proline to the basal medium increased the splitting rate 1.2- to 1.9-fold or nearly to the same degree as crude tick egg extract.[107] An attempt was made to standardize the tick cell culture medium to assure continuous cell proliferation.[102] The L-15 medium was supplemented by the addition of aspartic acid, glutamic acid, proline, alpha-ketoglutaric acid, glucose, trace minerals, and additional vitamins. L-15 medium supplemented with 10% tryptophase broth and 5 to 20% of fetal bovine serum is at present the medium of choice for widest application. The effectiveness of bovine cholesterol (10 µ/ml) can reduce the high level, i.e., 10 to 20% of fetal bovine serum, to a 5% or lower concentration.[102,108]

The final pH of the medium should be adjusted from 6.5 to 7.0, 6.8 being the most commonly used. It should be emphasized that the heat inactivation of fetal calf serum to eliminate undesirable toxins should last about 1/2 to 1-1/2 h at 56°C. Each new lot of medium additive must be screened on limited cell cultures, to establish its toxic effect before broader use.

Many other factors concerning tick cell cultivation may play an important role also, not only in the development, but also in the maintenance, e.g., the type of culture vessels, as some plastic material might be toxic: (1) pretreatment of culture vessels by the cultivating media, (2) enrichment and use of conditioned media, (3) keeping primary cultures for several months before attempting further passages (this I found also to be a very important factor), (4) reseeding of cells into the same flasks which apparently accelerates the rate of cell attachment and proliferation, and (5) the releasing of cells from their congested foci from surface of culture vessels by scraping with a rubber policeman, pipetting, etc.[104]

Tick cells *in vitro* multiply slowly, which is to some extent influenced by temperature. Cells have been successfully cultivated mostly at 28°C, but also at 30°C and in some cases at 37°C. The osmotic pressure of media used for tick cell cultures lies around 320 mOsm/l.[8] However, exact conditions for the cultivation of tick cells *in vitro* have not yet been established, i.e., their tolerance to various factors, such as pH of the media, ion ratio, osmotic pressure, gaseous compositions of culture fluids on the tick cells. Variations in these probably influence cell growth of different tick species and type of cultures produced. Cells could be stored in liquid nitrogen in a medium containing 10% dimethyl sulfoxide or 10% glycerol as a cryoprotectant.[77,104]

Growth rates of tick cells is usually low (split 1.2 to 2.3 weekly) at 28°C.[14] Most of the lines are diploid, but some aneuploid lines exist with either the male or the female complement predominating.[8] The origins of cells or tissues from which they arose are unknown. The major cell types, i.e., epithelial-like cells and fibroblast-like cells, are the basic types of established cell lines. In addition to them, a possible third type — a muscle cell which rarely occurs in the cell cultures — was described in cultured embryonic cells of *R. appendiculatus* by use of electron microscopy.[109] With the help of electron microscopy, serology, cytochemistry, and biochemistry, we might gain more information on the cultivated cells.[110] The different phases of development, namely, growth, maturation, and aging of cells, have been described.[8] It is hoped that untrastructural studies will reveal the protein synthetic activity relevant to the antigenic properties of the cells. It was suggested that in *R. appendiculatus* cells *in vitro*, cultures express a protein which is immunologically related to vitellin, normally synthesized in a sex and stage-specific manner. This protein is synthesized by the cells at a very low level and probably is not secreted into the medium.[111] It was found that the dependence on the concentration and length of exposure to 20-HE (invertebrate hormone 20-hydroxyecdysone) and JH III (juvenile hormone III) has an influence on the morphology, growth, and adhesiveness of cultured *R. appendiculatus* and *A. nitens* cells.[112] To distinguish the cell lines from different arthropod species, including tick cell lines, cellular isozymes have been employed, using a cellulose-acetate electrophoretic technique.[113] The isozyme profiles of tick cell lines 257 from *R. appendiculatus, B. microplus,* and *A. nitens* have been determined.[6,10]

More than 20 continuous cell lines from 9 species of ixodid ticks have been developed,[10,14,114,115] but no lines exist from argasid ticks. A list of cell lines developed from ixodid ticks is in Table 1.

VI. USE OF CONTINUOUS TICK CELL LINES FOR THE CULTIVATION OF PATHOGENIC MICROORGANISMS

A. VIRUSES

The greatest impetus for the development of tick cell lines came through their potential use in the research on arthropod-borne pathogens, particularly arboviruses, bacteria, and parasitic protozoa to study host-parasite relationships and to produce tick cells in reasonable quantity for different experimental studies. Until recently the greatest progress in the application of tick cell lines has been associated with the cultivation of arboviruses. After the basic findings proved that the first established tick cell lines from *R. appendiculatus* propagated arboviruses,[101] Langat, Quaranfil, West Nile, and louping-ill (with others using other cell lines and viruses) verified these important and interesting results. The application of embryonal cell lines for the cultivation of pathogens was directed to the cells of *B. microplus,* which supported the replication of Dugbe virus. A chronic infection was established in these cells without visible inclusions or a cytopathic effect.[116] Almost all tick cell lines developed to date have been used for the cultivation of arboviruses. The results of these experiments are reviewed in Table 2 based upon several papers.[10-13,89,103,104,114,117,118] About 80 different viruses have been tested for growth in tick cell cultures; over 70 represented different arboviruses both of tick

TABLE 1
List of Continuous Cell Lines Derived from Ticks

Tick species	Cell line code	Source	Ref.
Anocentor nitens	ANE 58	Embryo	112
Boophilus microplus	BM-256 (TC-156)	Embryo	63
	BM-270	Embryo	89
	IX	Embryo	138
	VIII-633	Embryo	105
Dermacentor parumapertus	RML-14	Embryo	104
	RML-16,17	Embryo	103
D. variabilis	RML-15,18,19,20	Embryo	103
Haemaphysalis obesa	ATC-304,310	Preadult	78
H. spinigera	ATC-281/282, 309	Preadult	78
Hyalomma anatolicum a.	HAE/CTVM 7,8,9,10,11	Embryo	115
Rhipicephalus appendiculatus	RA-243 (TC-243)	Preadult	101
	RAE-25	Embryo	106
		Embryo	136
R. sanguineus	ATC-307,308	Preadult	78
	RSE-8, RSE-13	Embryo	106
	RML-21,22	Embryo	9

Compiled from papers by Varma, Pudney, Yunker, and Kurtti.

TABLE 2
Cultivation of Viruses in Continuous Tick Cell Lines

Viruses	Vectorship	No. replicating/ no. tested
Togaviridae, Alphavirus	Mosquito-borne viruses	9/10
Flaviviridae, Flavivirus	Tick-borne viruses	8/8
	Mosquito-borne	3/5
Bunyaviridae	Tick-borne	9/13
	Mosquito-borne	1/7
	Phlebotomine transmitted	0/1
Reoviridae, Orbivirus	Tick-borne	6/6
	Mosquito-borne	1/1
Rhabdoviridae	Tick-borne	2/2
	Phlebotomine transmitted	1/1
Unclassified	Tick-borne	10/11
Nonarboviruses, Arenavirus		
Picornaviridae (human polio 1)		0/1
Poxviridae (vaccinia, ectromelia, fowl pox, Orf. v.)		0/4
Herpesviridae (Aujeszky virus)		0/1
Adenoviridae (bovine adenovirus type 3)		0/1

and mosquito-borne origin. In general, viruses of ixodid tick origin grew readily and to higher titers than those from argasid ticks, with the exception of the Hughes virus which grew well in *D. parumapertus* cells.[137] The majority of tick-borne viruses were able to grow in one or more cell lines. No cytopathic effect was produced with any of these viruses. Though normally tick-borne viruses with few exceptions do not multiply in mosquito cell lines, certain mosquito-borne viruses may multiply in some tick cell cultures.[11,117] Of the mosquito-borne viruses tested in tick cell cultures, 60% multiplied, whereas mosquito cells supported the propagation of only 31% of tick-borne viruses.[114]

The most interesting finding was the isolation of the Cascade virus from *D. occidentalis* ticks.[103] This virus was also isolated on *Xenopus laevis* (XL-2) cells as well as in *D. variabilis* (RML-15) cell cultures, but not in mosquito cells, laboratory animals (baby mice and hamsters), chicken eggs, or vertebrate cell cultures. Thus, poikilothermic cells may enable the isolation of this temperature-sensitive arbovirus. Tick cell cultures became of particular value for the isolation and propagation of unclassified and new tick-borne viruses. Ten out of eleven unclassified tick-borne viruses multiplied in a cell line from *R. appendiculatus, B. microplus,* and *D. parumapertus,* but none grew in mosquito cells.[114]

Several arboviruses were persistently maintained in cell cultures, i.e., in *R. appendiculatus* "243" cells with louping-ill virus; the cells were subcultured through 90 weekly subcultures, producing 4 to 6 dex PFU per (ml) week without producing small plaque variants.[12] In these cultures, heterologous Semliki Forest virus grew to normal levels, which suggests the absence of an interferon-like antiviral defense system in these cells. Langat virus was subcultured 12 times over 98 days in *B. microplus* cells before the culture was lost due to contamination. No apparent alteration was seen in the plaque morphology of cells infected with this virus.[11] Nairobi Sheep Disease virus yielded high virus titers ranging between 10^5 to 10^6 during 35 passages in *R. appendiculatus* cells; sheep inoculated with the 26th passage material failed to develop viremia as did the 2nd and 3rd passage material; however, the titers of virus were the same as the 2nd, 4th, and 26th passages, which indicated that apparently no loss of virulence for mice did occur.[119] Indirect immunoperoxidase technique was used for the detection of Nairobi Sheep Disease virus in a tick cell line from *R. appendiculatus* (TTC-243).[120,121] Good results had been obtained between 3 and 6 days post-infection, but the test seemed less sensitive in tick cells than in BHK cells or sheep kidney cells.

B. RICKETTSIAE

The possibility of using tick cell lines for the cultivation of rickettsiae appeared to be poor although *R. rickettsii, R. akari, R. typhi,* and a new species of rickettsia grew well in cell lines from *R. sanguineus* and *D. variabilis.*[9] These cell lines may show promising potential in the study of rickettsial organisms, i.e., in the primary isolation of assay of rickettsial and rickettsia-like organisms associated with animals. The RML-15 cell line derived from *D. variabilis* supported growth of *Anaplasma marginale,* used as a tool for studying various biologic properties of this pathogen and eventually for the development of an *Anaplasma vaccine.*[122] *A. marginale* from homogenized midguts dissected from experimentally infected *D. variabilis* ticks was maintained through at least 22 passages over 11 months in a tick cell line derived from *D. variabilis* embryos. Growth of *A. marginale* in the cell line began in the area of the nuclear membrane at approximately 4 days after inoculation; thereafter the organism was observed in inclusions scattered throughout the cytoplasm of host cells. Maximal growth occurred at 7 to 14 days. Growth of *A. marginale* in tick cells did not appear to cause deleterious effects on host cells. The anaplasmas retained infectivity for host cells and antigenicity as evidenced by reaction with specific monoclonal antibodies in indirect fluorescent antibody procedures.[123] Inoculation of splenectomized calves with cell cultures infected with *A. marginale* did not result in detectable disease. This indicated that the organism growing in tick cell culture was in a stage that was not infective for cattle. The experiment also demonstrated that the organism in culture retained all the major antigens previously characterized. This finding substantiates the immunizing potential of infected cell cultures.[124]

C. PROTOZOAN PARASITES

In contrast to organ cultures tick cell lines do not provide a convenient substrate for the cultivation of protozoan parasites. Probably during prolonged continuous passages tick cells lose their original organ-like functions.

When a *B. microplus* cell line was challenged with *B. bovis*, the parasites were quickly phagocytized, but instead of multiplying they were destroyed along with the erythrocytes.[125] Propagation of *B. bovis* in a *B. microplus* cell line resulted in the appearance of sexual stages which appeared both intra- and extracellularly.[126]

It seems that tick cell cultures may differ in their ability to support babesial vector stages independent of the species of origin of the line. *B. caballi* kinetes from hemolymph and organs of *A. nitens* cocultivated with cell lines from *R. sanguineus, R. appendiculatus,* and *A. nitens* cells that were penetrated by, or infected with, *B. caballi,* rounded off, degenerated, and lysed.[99] More kinetes were associated with *R. appendiculatus* cells where the medium contained *A. nitens* egg extract. Ninety percent of the *B. caballi* kinetes were always found within or in close contact with cells from *A. nitens*. The parasites were club shaped, amoeboid, or round. Intracellular babesiae were frequently lying adjacent to the host cell nucleus. Although *B. caballi* kinetes actively invade tick cell cultures, there was no evidence of subsequent replication. The same results were noticed by Ronald and Cruz (1981) when a *B. microplus* cell line infected with *B. bovis* kinetes was cultivated.[127]

D. SPIROPLASMAS

Spiroplasma is a helical, motile, wall-less prokaryote (class Mollicutes). A new spiroplasma from *Ixodes pacificus* ticks collected in Oregon (U.S.) was isolated in a cell line from *D. variabilis* ticks.[128] Thus, the use of arthropod cell lines appears to have value in the primary isolation of arthropod-derived mollicutes. Six spiroplasma strains were recovered in the RML-19 line and 5 strains were isolated in the RML-15 cell line, from *D. variabilis* ticks. Tick cell cultures RML-15, 16, and 23 propagated *Spiroplasma mirum* with a cytopathic effect. One of the spiroplasma isolates was maintained through four consecutive weekly passages of the *D. variabilis* cell line and for three passages of *R. sanguineus* cells.[20]

E. BORRELIAE

Borrelia burgdorferi, strain 297, was cultivated in tick cell lines from *R. appendiculatus, R. sanguineus, A. nitens, D. variabilis,* and *B. microplus* in tick cell culture medium L-15B, modified by the addition of gelatin and *N*-acetylglucosamine, showed that spirochetes passaged axenically in this medium had longer population doubling times (27.1 + 4.5 h) than those grown in BSK medium (11.7 + 2.2 h) which was unsuitable for tick cells. Borreliae had the greatest affinity for *D. variabilis* cells. Scanning electron microscopy revealed numerous spirochetes attached to foci at the surface of round bodies. High spirochetal concentrations (over 10^7 per ml) elicited cytopathic effects by loss of surface membrane, extensions, rounding up, detachment, and lysis. Tick cell lines could prove valuable in the isolation of *B. burgdorferi.*[129,130]

F. LEISHMANIAE

The life cycle of Leishmania species include sandfly (genera Phlebotomus and Lutzomyia) and mammalian hosts. In the insects, parasites exist as flagellated promastigotes and in the vertebrate as nonmotile intracytoplasmic amastigotes. *Leishmania major* and *L. donovani* were propagated in embryonic tick cells from *R. appendiculatus* and *Rhipicephalus evertsi* grown at 37°C. Tick cells demonstrated intracytoplasmic amastigotes; by 24 h only 5% of tick cells harbored amastigotes, by 48 h, 50%, and it remained at this level up to 120 h. No difference in the capacity to infect tick cells was detected in the tested two species of *Leishmania.*[136]

G. BESNOITIAE

Besnoitia besnoitii (Apicomplexa, Sarcocystidae, Toxoplasmatinae) a tissue cyst-forming coccidium causing besnoitiosis, a disease of cattle, was cultivated in two cells lines, i.e., RML-15 and RA-243. After 15 continuous passages during 7 months there was no alteration of the parasite.[131] Further, this organism was cultivated in other cell lines derived from three tick (*Boophilus, Dermacentor* and *Rhipicephalus*) genera. There was no difference in parasite multiplication in the various cell lines during 4 days at 28°C or at 36°C. The yield of parasites 4 days post-inoculation was lower in tick cells than in Vero cells. In tick cells, *B. besnoitii* had the tendency to change from banana shape to a rounded form.[132]

H. MICROFILARIAE

Microfilariae of *Oncocerca volvulus* and *O. gutturosa* were very active in the RA-243 cell system, but by 2 to 3 days, the larvae were seen to slow down and became shorter and thicker. By 4 to 5 days, development *in vitro* reached a stage of development equivalent to the 48-h stage in *Simulium.* Many microfilariae reached this young sausage stage and continued to late first stage where the development of a gut could be clearly seen. A few appeared to attempt molting, but no complete molts have been seen.[133,134] Microfilariae of the bovine parasite *O. lienalis* were maintained *in vitro* in the tick cell line RA-243 where they underwent development to the late sausage stage. The culture conditions which favor parasite survival and development provide further insight into the physiological requirement of filariae.[135]

VII. PROTECTION OF ANIMALS AGAINST TICKS BY VACCINES FROM TICK CELLS *IN VITRO*

In addition to examination of primary tissue cultures for immunization of animals against tick infestation,[100] experiments with the RA-243 cell line did not give consistent results, although there was a slight reduction of feeding activity in immunized animals.[6] The author suggests that continuation of such experiments is more than desirable.

VIII. CONCLUSIONS

Four types of tick tissue culture, i.e., hemocytes, organ cultures, primary tissue cultures and continuous cell lines have been established. These in the author's opinion may supply a broad spectrum of reasonable substrates to solve various problems in biology. Their application has not yet reached the same level as insect and vertebrate tissue culture. How can we characterize these cultures?

In the first type, i.e., hemocyte cultures, the main problem is obtaining a sufficient quantity of cells. It is necessary to employ numerous ticks because hemocytes do not multiply regularly *in vitro.* The collection of ticks, their preparation for hemocyte collection, etc. is very laborious. However, cultivation of tick hemocytes may contribute to many basic biological problems, in particular the dissemination of pathogens in the arthropod body and the defense system of ticks against undesirable microorganisms entering their bodies.

The preparation of the second type of cultures, i.e., organ cultures, presents the same problem. The dissection of ticks, although not particularly difficult, is time consuming and requires experienced laboratory workers. In addition, elimination of undesirable microbial contamination is not always easy. The advantage of these cultures is evident. They can be used for the cultivation of various microorganisms such as babesiae and theileriae in some organs, in particular salivary glands, where the organisms undergo part of their developmental cycle. Excised organs, namely salivary glands, are probably more convenient for this purpose. They maintain their original charter, while cultured cells undergo great number of passages and become modified. This is at present an unconfirmed hypothesis. Organ cultures can substantially contribute to the explanation of interrelationships between ticks and protozoan parasites, i.e., uncovering developmental cycles of pathogens in tick vectors, survival mechanisms within tick hosts, and the effect of the parasites on tick organs. Tick organ culture can elucidate various properties of microorganisms, such as virulence, antigenicity, etc. Cocultivation of protozoan parasites with vector organs *in vitro* might provide information on the affinity of some tissues to parasites, the tropism to tick vector organs and tissues, as well as discover surface receptors for adhesion of pathogens to organs.

The second benefit of tick organ culture would be in the study of the physiology of ticks, to elucidate metabolic phenomena in organ cultures and endocrinology, i.e., nutritional and biochemical studies of tick organs, the production of enzymes, hormones, control of osmotic pressure by salivation, etc. The backless tick explants would particularly help to gain information on the interaction of tissues and organs towards their microorganisms and elucidate the metabolic pathways of hosts and parasites. They could also facilitate studies on the action of various compounds such as drugs or hormones on arthropod hosts and their parasites, as well as of other environmental factors such as temperature, pH, osmotic pressure, etc.

The third type of cultures, i.e., primary cell cultures, have probably the least applications in biology. The reason for that is that compared to other types of cultures their preparation is difficult, in addition to the handling of ticks and mastering the dissection technique, which is time consuming. Although primary cells are the reasonable substrate for the cultivation of arboviruses and rickettsiae, they have been superseded by cell lines because of their relatively difficult preparation. However, at the time of their development, they played an important role[141] because they demonstrated that tick tissue and cell systems *in vitro* have their place in biology, and thus primary cell cultures became a necessary link towards the development of tick cell lines.

The last type of tick cultures *in vitro,* i.e., continuous cell lines, numbering about 20 from 9 tick species seems to be the most productive for studying different biological problems. They have a broad application in the cultivation of various pathogenic microorganisms of vertebrates, as well as of invertebrates. Their high degree of sensitivity to viruses could be used for the isolation of these agents from field material and for preliminary identification of unknown viruses, to indicate their arboviral status. Propagation of some arboviruses in tick cell lines could become a source of diagnostic reagents, such as antigens, etc. The long-term cultivation of tick-borne pathogens in tick cell cultures could provide attenuated strains with vaccine potential. In some cases the reverse result would be preferred, e.g., *B. burgdorferi* cultivated in liquid BSK medium loses its virulence, but in tick cells *in vitro* it might perhaps remain virulent.

In addition to arboviruses, the continuous tick cell lines could be applied in research with other arthropod-borne pathogens, i.e., rickettsiae, rickettsia-like organisms and parasitic protozoa. Cell lines may find application in many branches of biology, i.e., physiology, biochemistry, molecular biology, pathology, immunology, diagnostics, disease prophylaxis, in studies of their metabolic requirements, host-parasite-symbiont-relationships, host specificity, chemotherapy, production of antigens, vaccines, etc. They could also contribute to basic biological problems such as, e.g., phylogenesy of pathogens to ticks.

A. ADVANTAGES AND DISADVANTAGES OF TICK CELLS *IN VITRO*

The advantage of relatively easy handling and, in particular, their high sensitivity to various pathogens make tick cells a simple and rapid system for isolation of agents from field material and for their cultivation. Cultivation of tick-borne pathogens in tick cell cultures may provide either attenuated or virulent strains with vaccine potential. Varma suggested that since vertebrates are mainly infected with a pathogen through tick bites, vaccines prepared from pathogens cultivated in tick cells might provide better protection against tick-borne agents.[6] One major potential application is an immunological approach to tick control, by the induction of host immune resistance directed towards tick infestation of animals, which also protects animals against pathogens transmitted by these arthropods. Using this control method, both environment and animals are saved from undesirable toxic chemicals used in chemical control.

The disadvantages consist in the nonavailability of continuous cell lines from important vectors of arthropod-borne pathogens, e.g., species of *Ixodes* genus remain to date refractory to attempts to establish a cell line. Furthermore, a cell line is a mixture of cells of different origin, and no clones have yet been produced from tick cells *in vitro*. Cloned tick cells could define the sensitivity to certain viruses. The lack of cytopathic effects and plaques is a limiting factor to the broader use of tick cell cultures. Currently, the multiplication of a virus can be visualized only by the fluorescent antibody method or by the employment of other media such as body mice or other vertebrate species or vertebrate cell cultures. Using tick cells *in vitro*, the natural contamination with some agents, in particular with viruses, should be considered. The observation of a viral contamination would be monitored by electron microscopic examination, as obvious visible effects on the contaminated cells may not be apparent. The possibility of viral contamination of tick cells must be kept in mind at all times particularly when tick cells are used for virological studies. This the author would like to document by finding reovirus-like particles (genus Orbivirus) in some organs of *D. reticulatus* ticks,[140] and in *R. appendiculatus* cells *in vitro*,[121] respectively.

B. HOW TO MODIFY TICK CELL CULTURES INTO MORE EFFECTIVE RESEARCH TOOLS

There is a need to select uniform cell types with increased susceptibility to pathogens; cloning of cells types may contribute to the creation of a cytopathic effect. Hybrid cell lines from fused mosquito or other hematophagous arthropods and tick cells may change the growth spectrum and sensitivity to a wider range of pathogens than tick or mosquito cells alone. Determination of isozyme profiles of tick cells in culture would permit the detection of cross-contamination between tick and other cell lines. To date the isoenzyme profile has been determined only for the RA-257 *(R. appendiculatus)* cell line.[6]

Despite the establishment of reasonable tick culture *in vitro*, many scientists, especially microbiologists, routinely prefer the use of laboratory animals or vertebrate cell cultures. This preference, I suppose, is due to a certain conservatism. Patience and time are required to overcome this reluctance. I believe the broader application of tick cell cultures in biology is possible and desirable.

ACKNOWLEDGMENT

I wish to express my sincere thanks to Professor M.G.R. Varma for the critical reading of this manuscript and for his comments.

REFERENCES

1. **Řeháček, J.:** Biology of the EEE virus in ticks *Ixodes ricinus* L., *Dermacentor marginatus* Sulzer. Ph.D. thesis, Bratislava, 1958.
2. **Řeháček, J.:** Preliminary report on tick tissues cultures. *Acta Virol.* 2, 253–254, 1958.
3. **Řeháček, J.:** An improved method of tick tissue cultures. *Acta Virol.* 6, 188, 1962.
4. **Řeháček, J.:** Preparation of tissue cultures from the tick *Hyalomma dromedarii* Koch. *J. Med. Entomol.* 2, 161–164, 1965.
5. **Weyer, F.:** Explantationsversuche bei Läusen in Verbindung mit der Kultur von Rickettsien. *Zentralhl. Bakteriol. Parasitenkd. Infektionskr. Hyg. Abt. Orig.* 159, 13–22, 1952.
6. **Varma, M. G. R.:** Progress in the study of human and animal pathogens in primary and established tick cell lines. *Invertebrate Cell Systems Applications*, Vol. 2, Mitsuhashi, Ed., CRC Press, Boca Raton, FL, 2, 119–128, 1989.
7. **Kurtti, T. J., Buscher, G.:** Trends in tick cell culture. In *Practical Tissue Culture Applications*, Maramorosch, K. and Hirumi, H., Eds., Academic Press, New York, 351–371, 1979.

8. **Kurtti, T. J., Munderloh, U. G.:** Tick cell cultures: Characteristics, growth requirements, and applications to parasitology. In *Invertebrate Cell Culture Applications*, Maramorosch, K. and Mitsuhashi, J., Eds., Academic Press, New York, 195–232, 1982.
9. **Yunker, C. E., Cory, J., Meibos, H.:** Tick tissue and cell cultures. Application to research in medical and veterinary acarology and vector-borne disease. *Acarologia* 6, 1082–1088, 1984.
10. **Kurtti, T. J., Munderloh, U. G., Ahlstrand, G. G.:** Tick tissue and cell culture in vector research. In *Advances in Disease Vector Research*, Vol. 5, Harris, K. F., Ed., Springer-Verlag, New York, 87–109, 1988.
11. **Leake, C. J.:** Comparative growth of arboviruses in cell lines derived from Aedes and Anopheles mosquitoes and from the tick Boophilus microplus. In *Arboviruses in Arthropod Cells In Vitro*, Vol. 1, Yunker, C. E., Ed., CRC Press, Boca Raton, FL, 25–42, 1987.
12. **Leake, C. G., Pudney, M., Varma, M. G. R.:** Studies on arboviruses in established tick cell lines. In *Invertebrate Systems In Vitro*, Kurstak, E., Maramorosch, K., Dubendorfer, A., Eds., Elsevier, Amsterdam, 327–335, 1980.
13. **Pudney, M.:** Tick cell lines for the isolation and assay of arboviruses. In *Arboviruses in Arthropod Cells in Vitro*, Vol. 1, Yunker, C. E., Ed., CRC Press, Boca Raton, FL, 87–101, 1987.
14. **Yunker, C. E.:** Preparation and maintenance of arthropod cell cultures: Acari, with emphasis on ticks. In *Arboviruses in Arthropod Cells In Vitro*, Vol. 1, Yunker, C. E., Ed., CRC Press, Boca Raton, FL, 35–51, 1987.
15. **Řeháček, J.:** Use of invertebrate cell culture for study of animal viruses and rickettsiae, *Invertebrate Tissue Culture*, Vol. 2, Vago, C., Ed., Academic Press, New York, 279–320, 1972.
16. **Řeháček, J.:** Maintaining of tick-borne encephalitis (TBE) virus, western subtype in tick cells in vitro. Proc. 3rd Int. Colloq. Invertebrate Tissue Culture, Řeháček, J., Blaskovic, D., Hink, W. F., Eds., Veda, Bratislava, 439–443, 1973.
17. **Řeháček, J.:** Organ, tissue and cell cultures of haematophagous arthropods and their application in virology and rickettsiology. *Entomol. Probl.*, 11, 98–155, 1973.
18. **Řeháček, J.:** Tick tissue cultures and arboviruses. In *Invertebrate Tissue Culture, Applications in Medicine, Biology and Agriculture*. Kurstak, E. and Maramorosch, K., Eds., Academic Press, New York, 21–33, 1976.
19. **Řeháček, J.:** Arthropod cell cultures in studies of tick-borne Toga viruses and Orbiviruses in Central Europe. In *Arboviruses in Arthropod Cells In Vitro*, Vol. 1, Yunker, C. E., Ed., CRC Press, Boca Raton, FL, 115–132, 1987.
20. **Yunker, C. E., Tully, J. G., Cory, J.:** Arthropod cell lines in the isolation and propagation of tick-borne spiroplasmas. *Current Microb.* 15, 45–50, 1987.
21. **Cory, J., Yunker, C. E.:** Primary cultures of tick hemocytes as systems for arbovirus growth. *Ann. Entomol. Soc. Am.* 64, 1249–1254, 1971.
22. **Hoffmann, G., Schein, E., Jagow, M.:** Untersuchungen an extirpierten und in der Kulture gehaltenen Zeckengeweben. *Z. Tropenmed. Parasitol.* 21, 46–61, 1970.
23. **Moulton, J. E.:** Preparation of primary cultures of tick cells. *Am. J. Vet. Res.* 39, 1558–1564, 1978.
24. **Fujisaki, K., Kitaoka, S., Morii, T.:** Hemocyte types and their primary cultures in the argasid tick, Ornithodoros moubata Murray (Ixodoidea). *Appl. Entomol. Zool.* 10, 30–39, 1975.
25. **Řeháček, J.:** Propagation of tick-borne encephalities (TE) virus in tick tissue cultures. *Ann. Epiphyt.* 14, 199–204, 1963.
26. **Sidorov, V. E.:** On the basic substances formation in hemocytes of Alveonasus lahorensis Neumann (Argasidae, Ixodoidea) in vitro study. *Zh. Obshch. Biol.* 38, 934–939, 1977a. In Russian.
27. **Sidorov, V. E.:** The formation of membranes of the basic substance by haemocytes of Alveonasus lahorensis (Argasidae). *Parazitologia* 6, 480–483, 1977b. In Russian.
28. **Sidorov, V. E.:** Body cavity of Argas ticks as a habitat of spirochetes and Brucella. *Zh. Mikrob. Moscow* 6, 91–97, 1960. In Russian.
29. **Yunker, C. E., Cory, J.:** Growth of Colorado Tick Fever (CTF) virus in primary tissue cultures of its vector, Dermacentor andersoni Stiles (Acarina: Ixodidae) with notes on tick tissue culture. *Exp. Parasitol.* 20, 267–277, 1967.
30. **Hoffmann, G.:** Haltung babesieninfizierter Zeckengewebe in künstlichen Nährmedien. *Z. Angew. Entomol.* 71, 26–34, 1972.
31. **Bell, J. L.:** Organ culture of Rhipicephalus appendiculatus with maturation of Theileria parva in tick salivary glands in vitro. *Acta Tropica*, 37, 319–325, 1980.
32. **Kaufman, W. R., Barnett, S. F.:** Dermacentor andersoni: culture of whole salivary gland. *Exp. Parasitol.* 42, 106–114, 1977.
33. **Kaufman, W. R.:** Organ culture of ixodid-tick salivary glands. *Exp. Appl. Acarol.* 9, 79–86, 1990.
34. **Kruchechnikov, V. N., Shcherbakov, S. V.:** Cultivation of organs of Ornithodoros papillipes (Argasidae). *Parazitologia* 16, 253–255, 1982.
35. **Martin, H. M., Vidler, B. O.:** In vitro growth of tick tissues (Rhipicephalus appendiculatus Neumann, 1901). *Exp. Parasitol.* 12, 192–203, 1962.

36. **Kordova, N., Řeháček, J.:** Experimental infection of ticks in vivo and their organs in vitro with filterable particles of Coxiella burnetii. *Acta Virol.* 3, 201–209, 1959.
37. **Kordova, N., Řeháček, J.:** Development of Coxiella burnetii from their ultrafilterable particles in ticks in vivo, their organs in vitro and in tick tissue cultures. Proc. XII Int. Congr. Entomol. London, 1964, p. 777, 1965.
38. **Řeháček, J., Pešek, J.:** Propagation of Eastern Equine Encephalomyelitis (EEE) virus in surviving tick tissues. *Acta Virol.* 4, 241–245, 1960.
39. **Chunikhin, S. P., Kochetova, G. A., Stefutkina, L. F., Korolev, M. B.:** Characteristics of the reproduction of tick-borne encephalitis and Powassan viruses in the explants from Dermacentor silvarum hatched from nymphal cuticule. *Med. Parazitol. Parazit. Bolezni* 4, 61–64, 1981. In Russian.
40. **Chunikhin, S. P., Khozinskaya, G. A., Stefutkina, L. F., Korolev, M. B.:** Mono and mixed infection of explants of tissues of the ticks of the genus Hyalomma with viruses of tick-borne encephalitis and Powassan viruses. *Parazitologia* 18, 116–122, 1984. In Russian.
41. **Chunikhin, S. P., Leonova, G. N.:** Ecology and geographical distribution of arboviruses. *AMS-USSR-Medicina*, Moscow, 127 pages, 1985. In Russian.
42. **Hadani, A.:** In vitro development of Babesia and Theileria parasites in organs and tissues of ticks. In *Babesiosis*, Ristic, M. and Kreier, J. P., Eds., Academic Press, New York, 225–266, 1981.
43. **Hoffmann, G., Köhler, G.:** Haltung von Zeckengeweben in künstlichen Nährmedien. *Ztsch. Parasitenkd.* 31, 8, 1968.
44. **Hadani, A., Kaufman, W., Barnett, S. F.:** Development of Nuttellia danii (Protozoa, Babesidae) within tick salivary glands cultured in vitro. *J. Parasitol.* 64, 501–503, 1978.
45. **McCall, H.:** Effect of salivary stimulants on isolated preparations of Theileria parva infected and non-infected salivary glands of Rhipicephalus appendiculatus. In *Proceedings of the International Conference on Tick-Borne Disease and Their Vectors*, Wilds, J. H. K., Ed., Tonbridge, Lewis Reprints, Edinburgh, 343–350, 1976.
46. **McCall, H.:** Studies of *Theileria parva* in the Tissues of *Rhipicephalus appendiculatus* in vitro. M. Phil. thesis, University of Edinburgh, 1977.
47. **Bell, J. L.:** Development of Theileria in Tick Tissue Cultures. M. Phil. thesis, Edinburgh University, 1983.
48. **Bell, J. L.:** Tick tissue culture techniques in the study of arthropodborne protozoa: the development of Theileria annulata in organ cultures of Hyalomma anatolicum anatolicum. *Acarology VI*, Griffiths, D. A. and Bowman, C. E., Eds., Horwood, Chichester, 1089–1095, 1984.
49. **Kurtti, T. J., Munderloh, U. G., Irvin, A. D., Buscher, G.:** Theileria parva: early events in the development of bovine lymphoblastoid cell lines persistently infected with macroschizonts. *Exp. Parasitol.* 52, 280–290, 1981.
50. **Needham, G. R., Sauer, J. R.:** Control of fluid secretion by isolated glands of the lone star tick. *J. Insect Physiol.* 21, 1893–1898, 1975.
51. **Sauer, J. R., Frick, J. H., Hair, J. A.:** Control of ^{36}Cl uptake by isolated salivary glands of the lone star tick. *J. Insect Physiol.* 20, 1771–1778, 1974.
52. **Kaufman, W. R., Philips, J. E.:** Ion and water balance in the ixodid tick Dermacentor andersoni I. Routes of ion and water excretion. *J. Exp. Biol.* 58, 523–536, 1973.
53. **Kaufman, W. R.:** The influence of various factors on fluid secretion by in vitro salivary glands of ixodid ticks. *J. Exp. Biol.* 64, 727–742, 1976.
54. **Wong, D. L.-P., Kaufman, W. R.:** Potentiation by spirone and other butyrophenones of fluid secretion by isolated salivary glands of ixodid ticks, *Eur. J. Pharmacol.* 73, 163–173, 1981.
55. **Kaufman, W. R., Wong, D. L.-P.:** Evidence for multiple receptors mediating fluid secretion in salivary glands of ticks. *Eur. J. Pharmacol.* 87, 43–52, 1983.
56. **Kaufman, W. R.:** The effect of steroids and azadirachtin on the salivary gland and ovary in ixodid ticks. *J. Insect Physiol.* 34, 721–723, 1988.
57. **Lindsay, P. J., Kaufman, W. R.:** Action of some steroids on salivary gland degeneration in the ixodid tick, A. americanum L. *J. Insect Physiol.* 34, 351–359, 1988.
58. **Kaufman, W. R.:** Effect of 20-hydroxyecdysone on the salivary glands of the male tick, *Amblyomma hebraeum*. *Exp. Appl. Acarol.* 9, 87–95, 1990.
59. **Kaufman, W. R.:** Further investigations on the action of ecdysteroids on the salivary glands of the female tick *Amblyomma americanum*. *Exp. Appl. Acarol.* 10, 259–265, 1991.
60. **Shepherd, J., Levine, S., Hall, J. D.:** Maturation of tick spermatozoa in vitro. *Int. J. Invertebr. Reprod.* 4, 311–321, 1982.
61. **Medvedeva, G. I., Beskina, S. R., Grokhovskaya, I. M.:** Culture of ixodid embryonal cells, *Med. Parasitol. Moscow*, 41, 39–40, 1972. In Russian.
62. **Eide, P. E., Caldwell, J. M.:** A method for obtaining primary cultures of dispersed embryonic tissues from the lone star tick, *Amblyomma americanum*, *Ann. Entomol. Soc. Am.* 66, 891–893, 1973.
63. **Pudney, M., Varma, M. G. R., Leake, C. J.:** Culture of embryonic cells from the tick *Boophilus microplus* (Ixodidae). *J. Med. Entomol.* 10, 493–496, 1973.

64. **Rahman, H.:** The culture of embryonic cells from the tick *Hyalomma rufipes rufipes* (Koch), 1844. *Ann. Trop. Med. Parasitol.* 75, 573–574, 1981.
65. **Yunker, C. E., Meibos, H.:** Culture of embryonic tick cells (Acari:Ixodidae). *Tissue Cult. Assoc. Manual,* 5, 1015–1017, 1979.
66. **Řeháček, J.:** Ticks as experimental animals in investigations with arboviruses. Aktualnye problemy virusnych infekcij. Materialy XII Nauchnoj sessii Inst. Poliomyelita i virusnych encefalitov, 459, 1965 (in Russian).
67. **Řeháček, J.:** Present status of tick tissue culture. *Curr. Topics Microbiol.* 55, 32–41, 1971.
68. **Řeháček, J.:** Tissue culture of blood sucking arthropods and their use for the cultivation of viruses and rickettsiae. *Ann. Parasitol.* 46, 197–231, 1971.
69. **Řeháček, J., Hana, L.:** Cultivation of tissue explants from ticks. In Proc. Biol. of Viruses of the Tick-Borne Encephalitis Complex, Libíková, H., Ed., Smolenice near Bratislava, October 11–14, 1960, 185–187, 1961.
70. **Řeháček, J., Hana, L.:** Notes on tick tissue cultures. *Acta Virol.* 5, 57–61, 1961.
71. **Varma, M. G. R., Wallers, W.:** An improved method for obtaining, in vitro, uniform cell monolayer sheets from tissues of the tick, *Hyalomma dromedarii* (Ixodidae). *Nature (London),* 208, 602–603, 1965.
72. **Yunker, C. E., Cory, J.:** Effectiveness of refrigerated nymphs in tick tissue culture experiments. *J. Parasitol.* 51, 686, 1965.
73. **Yunker, C. E.:** Observations on tick tissues in vitro. *Proc. 1st Int. Congr. Parasitol., Roma, 1964,* 2, 1036–1037, 1966.
74. **Zukowski, K., Bitkowska, E.:** Investigations on the suitability of some chosen media for the culture of tissues from *Hyalomma dromedarii* Koch and *Dermacentor reticulatus* (Fabr.) Acarina, Ixodides. *Wiad. Parazytol.,* 23, 49–51, 1977. In Polish.
75. **Kamio, T., Mitsuhashi, J., Fujisaki, K.:** Primary culture of developing adult tissue in a nymphal tick, *Haemaphysalis longicornis* (Acarina: Ixodidae). *Appl. Entomol. Zool.* 21, 394–398, 1986.
76. **Varma, M. G. R., Pudney, M.:** Culture of cells from the tick *Rhipicephalus appendiculatus.* Ixodoidea. Ixodidae. Proc. 3rd Int. Colloquium Invert. Tissue Culture, Řeháček, J., Blaskovic, D., Hink, W. F., Eds., Smolenice near Bratislava, June 22–25. 1971, 135–140, 1973.
77. **Pudney, M., Varma, M. G. R., Leake, C. J.:** Establishment of cell lines from ixodid ticks. *Tissue Cult. Assoc. Manual* 5, 1003–1007, 1979.
78. **Guru, P. Y., Dhanda, V., Gupta, N. P.:** Cell cultures derived from the developing adults of three species of ticks, by a simplified technique. *Indian J. Med. Res.* 64, 1041–1045, 1976.
79. **Řeháček, J., Brzostowski, H. W.:** A tick tissue culture medium based on analysis of tick haemolymph. *J. Insect Physiol.* 15, 1431–1436, 1969.
80. **Řeháček, J., Brzostowski, H. W.:** The utilization of amino acids and sugars by tick cells cultivated in vitro. *J. Insect Physiol.* 15, 1683–1686, 1969.
81. **Varma, M. G. R., Pudney, M.:** Tick cell culture: A review of recent work. Proc. 2nd Int. Congr. Acarology, 1967, Sutton Bonington, 1967, Budapest, Akadémai Kiado, p. 637, 1969.
82. **Ackerman, S., Clare, F. B., McGill, T. W., Sonenshine, D. E.:** Passage of host serum components, including antibody, across the digestive tract of *Dermacentor variabilis* (Say). *J. Parasitol.* 67, 737–740, 1981.
83. **Ben-Yakir, D.:** Quantitative studies of host immunoglobulin G in the hemolymph of ticks (Acari). *J. Med. Entomol.* 26, 243–246, 1985.
84. **Minoura, H., Chinzei, Y., Kitamura, S.:** *Ornithodoros moubata:* Host immunoglobulin G in tick hemolymph. *Exp. Parasitol.* 60, 355–363, 1985.
85. **Tracey-Patte, P. D., Kemp, D. H., Johnston, L. A. Y.:** *Boophilus microplus:* passage of bovine immunoglobulins and albumin across the gut of cattle ticks feeding on normal or vaccinated cattle. *Res. Vet. Sci.* 43, 287–290, 1987.
86. **Yunker, C. E.:** Arthropod tissue culture in the study of arboviruses and rickettsiae. A review. In *Current Topics in Microbiology Immunology,* Weiss, E., Ed., Springer-Verlag, New York, Vol. 55, 113–126, 1971.
87. **Řeháček, J., Kozuch, O.:** Comparison of the susceptibility of primary tick and chick embryo cell cultures to small amounts of tick-borne encephalitis virus. *Acta Virol.* 8, 470–471, 1964.
88. **Řeháček, J.:** Cultivation of different viruses in tick tissue cultures. *Acta Virol.* 9, 332–337, 1965b.
89. **Pudney, M., Leake, C. J., Varma, M. G. R.:** Replication of arboviruses in arthropod in vitro systems. *Arctic and Tropical Arboviruses,* Academic Press, New York, 245–262, 1979.
90. **Řeháček, J., Rajčáni, J., Gresikova, M.:** Relationship of Tribec virus to tick cells and tissues. *Acta Virol.* 13, 439–442, 1969.
91. **Libíková, H., Řeháček, J.:** Attempts to prove interference between arboviruses in arthropod tissue cultures. In *Interferon and Interferon inducers,* Foldes, I., Talas, M., Eds., Kiadta a MTA Mikrobiologiai Kutatocsoport, 177–187, 1974.
92. **Libíková, H., Řeháček, J., Rajčáni, J.:** Attempts to prove interference between tick-borne encephalitis and Lipovník viruses in tick tissue cultures. *Čs. Epid. Mikr. Immunol.,* 23, 332–340, 1974.
93. **Řeháček, J., Brezina, R.:** Propagation of *Coxiella burnetii* in tick tissue cultures. *Acta Virol.* 3, 380, 1964.

94. **Řeháček, J., Brezina, R., Majerská, M.:** Multiplication of rickettsiae in tick cells *in vitro*. Acta Virol. 12, 41–43, 1968.
95. **Řeháček, J., Župančičová, M., Brezina, R., Úrvölgyi, J.:** Use of primary tick tissue culture for propagation of rickettsiae of the RMSF-group isolated in Slovakia. Proc. 3rd Int. Colloquium on Invertebrate Tissue Culture, Smolenice near Bratislava, Řeháček, J., Blaškovič, D., and Hink, W. F., Eds., Veda, Bratislave, 445–450, 1973b.
96. **Andreasen, M. P.:** Multiplication of Cowdria ruminantium in monolayer of tick cells. *Acta Pathol. Microbiol. Scand., Sect. B,* 82, 455–456, 1974.
97. **Shatkin, A. A., Beskina, S. R., Medvedeva, G. I., Grokhovskaya, I. M.:** Cultivation of the agent of enzootic abortion of sheep in embryonal cells of Hyalomma genus ticks. *Med. Parazitol. Moscow,* 4, 420–423, 1977 In Russian.
98. **Bhat, U. K. M., Mahoney, D. F., Wright, I. G.:** The invasion and growth of Babesia bovis in tick tissue culture. *Experientia,* 35, 752–753, 1979.
99. **Kurtti, T. J., Munderloh, U. G., Stiller, D.:** The interaction of Babesia caballi kinetes with tick cells. *J. Invert. Pathol.* 42, 334–343, 1983.
100. **Wikel, S. K.:** Resistance to ixodid tick infestation induced by administration of tick-tissue culture cells. *Am. J. Trop. Med. Parasit.* 79, 513–518, 1985.
101. **Varma, M. G. R., Pudney, M., Leake, C. J.:** The establishment of three cell lines from the tick *Rhipicephalus appendiculatus* (Acari: Ixodidae) and their infection with some arboviruses. *J. Med. Ent.* 11, 698–706, 1975.
102. **Kurtti, T. J., Munderloh, U. G., Krych, V.:** Culture of tick and mosquito cells with reduced serum requirements. Presented at Int. Cell Culture Congress, Sendai, Japan, (Abstr.) 1985.
103. **Yunker, C. E., Cory, J., Meibos, H.:** Continuous cell lines from embryonic tissues of ticks (Acari: Ixodidae). *In Vitro,* 17, 139–142, 1981.
104. **Bhat, U. K. M., Yunker, C. E.:** Establishment and characterization of a diploid cell line from the tick, *Dermacentor parumapertus* Neumann (Acarina: Ixodidae). *J. Parasitol.* 63, 1092–1098. 1977.
105. **Holman, P. J.:** Partial characterization of a unique female diploid cell strain from the tick *Boophilus microplus* (Acari: Ixodidae). *J. Med. Entomol.* 18, 84–88, 1981.
106. **Kurtti, T. S., Munderloh, U. G., Samish, M.:** Effect of medium supplements on tick cells in culture. *J. Parasitol.* 68, 930–935, 1982.
107. **Samish, M., Louloudes, S. J., Vaughn, J. L., Kurtti, T. J., Munderloh, U. G.:** Promotion of tick cell growth by proline and fractions from tick eggs. *Int. J. Parasitol.* 15, 21–26, 1985.
108. **Munderloh, U. G., Kurtti, T. J.:** Formulation of medium for tick cell culture. *Exp. Appl. Acarol.* 7, 219–229, 1989.
109. **Nyindo, M. L., Awitti, R. S., Dhadialla, T. S.:** Establishment of embryonic cell lines from the brown ear tick, *Rhipicephalus appendiculatus,* and their immunogenicity in rabbits. In *"Biotechnology in Invertebrate Pathology and Cell Culture",* Maramorosch, K., Ed., Academic Press, New York, 367–374, 1987.
110. **Awitti, L. R. S., Nyindo, M., Dhadialla, T. S., Chimtawi, M.:** Ultrastructure characterization of cultured embryonic cells of *Rhipicephalus appendiculatus* Neumann (Acarina: Ixodidae). *Int. J. Insect Morphol. Embryol.* 16, 369–377, 1987.
111. **Nyindo, M. L., Dhadialla, T. S.:** Expression of adult sex specific proteins, vitellins, in embryonic cell lines of *Rhipicephalus appendiculatus. Int. J. Invert. Reprod. Dev* . 12, 217–222, 1987.
112. **Kurtti, T. J., Munderloh, U.:** The effects of 20-hydroxyecdysone and juvenile hormone III on tick cells. *J. Parasitol.* 69, 1072–1078, 1983.
113. **Brown, S. E., Knudson, D. L.:** Characterization of invertebrate cell lines. IV. Isozyme analysis of Dipteran and Acarine cell line. *In Vitro.* 18, 347–350, 1982.
114. **Pudney, M., Leake, O. J., Buckley, S. M.:** Replication of arboviruses in arthropod *in vitro* systems: an overview. *Invertebrate Cell Culture Applications.* Maramorosch, K. and Mitsuhashi, J., Eds., Academic Press, New York. 159–194, 1982.
115. **Bell-Sakyi, L.:** Continuous cell lines from the tick *Hyalomma anatolicum anatolicum. J. Parasitol.* 77, 1006–1008, 1991.
116. **David-Vest, T. S.:** Propagation and plaquing of Dugbe virus (an ungrouped Nigerian arbovirus) in various mammalian and arthropod cell lines. *Arch. Ges. Virusforsch.* 44, 330–336, 1974.
117. **Banerjee, K., Guru, P. Y., Dhanda, V.:** Growth of arboviruses in cell cultures derived from the tick *Haemaphysalis spinigera. Indian J. Med. Res.* 66, 530–536, 1977.
118. **Pudney, M., Varma, M. G. R., Leake, C. J.:** The growth of some arboviruses in tick cell lines. In *"Tick-borne diseases and their vectors",* Wilde, J. H. K., Ed., Edinburgh, 490–496, 1978.
119. **Munz, E., Reimann, M., Munderloh, U., Settele, U.:** The susceptibility of the tick cell line TTC 243 for Nairobi Sheep Disease virus and some other important species of mammalian RNA and DNA viruses. In *Invertebrate Systems in Vitro,* Kurstak, E., Maramorosch, K., Dübendorfer, A., Eds., Elsevier, North Holland Biomedical Press, 1980, 337–340.

120. **Munz, E., Reimann, M., Jäger, H.**: Qualitative and quantitative detection of Nairobi Sheep Disease virus antigen by immunoperoxidase in BHK-, Sheep Kidney - and tick cell cultures. *Zbl. Vet. Med.* B, 31, 231–239, 1984.
121. **Munz, E., Reimann, M., Mahnel, H.**: Nairobi Sheep Disease virus and Reovirus-like particles in the tick cell line TTC-243 from Rhipicephalus appendiculatus: Experiences with the handling of the tick cells, immunoperoxidase, and ultrahistological studies. In "*Arboviruses in arthropod cells in vitro*", I., Yunker, C. E., Ed., CRC Press Inc., Boca Raton, Florida, 133–147, 1987.
122. **Samish, M., Pipano, E., Hana, B.**: Cultivation of *Anaplasma marginale* from cattle in a Dermacentor cell line. *Am. J. Vet. Res.* 49, 254–256, 1988.
123. **Hidalgo, R. J., Jones, E. W., Brown, J. E., Ainsworth, A. J.**: Anaplasma marginale in tick cell cultures. *Am. J. Vet. Res.* 50, 2028–2032, 1989.
124. **Hidalgo, R. J., Palmer, G. H., Jones, E. W., Brown, J. E., Ainsworth, A. J.**: Infectivity and antigenicity of *Anaplasma marginale* from tick cell cultures. *Am. J. Vet. Res.* 50, 2033–2036, 1989.
125. **Droleskey, R. E., Holman, P., Wagner, G. G.**: *Babesia bovis* infection of a tick cell line studied by electron microscopy. *In Vitro* 67, 209, 1981.
126. **Droleskey, R. E., Holman, P. J., Craig, T. M., Wagner, G. G., Mollenhauer, H. H.**: Ultrastructure of *Babesia bovis* sexual stages as observed in *Boophilus microplus* cell cultures. *Res. Vet. Sci.* 34, 249–251, 1983.
127. **Ronald, N. C., Cruz, D.**: Transmission of *Babesia bovis*, using undifferentiated embryonic cells from *Boophilus microplus* tick eggs. *Am. J. Vet. Res.* 42, 544–545, 1981.
128. **Tully, J. G., Rose, D. L., Yunker, C. E., Cory, J., Whitcomb, R. F., Williamson, D. L.**: Helical mycoplasmas (Spiroplasmas) from Ixodes ticks. *Science*, 212, 1043–1045, 1981.
129. **Kurtti, T. J., Munderloh, U. G., Ahlstrand, G. G., Johnson, R. C.**: *Borrelia burgdorferi* in tick cell cultures: Growth and cellular adherence. *J. Med. Entomol.* 25, 256–261, 1988.
130. **Munderloh, U. G., Krueger, D. E., Ahlstrand, G. G., Kurtti, T. J.**: Ultrastructure of *Borrelia burgdorferi* grown continuously with tick cells. V. Int. Conf. on Lyme borreliosis, Abstract 231, 1992.
131. **Samish, M., Shkap, V., Pipano, E.**: *Besnoitia besnoiti:* Long term cultivation in tick cell lines. *Exp. Parasitol.* 64, 261–263, 1987.
132. **Samish, M., Shkap, V., Bin, H., Pipano, E. M.**: Cultivation of *Besnoitia besnoiti* in four tick cell lines. *Int. J. Parasitol.* 18, 291–296, 1988.
133. **Pudney, M.**: The use of arthropod cell cultures for the in vitro study of filariae (Onchocerca sp.). in *Invertebrate Systems In Vitro*, Kurstak, E., Maramorosch, K., Dubendorfer, A., Eds., Elsevier, Amsterdam, pp. 317–326, 1980.
134. **Pudney, M., Varma, M. G. R.**: Present state of knowledge of in vitro cultivation of filariae. In *The In Vitro Cultivation of the Pathogens of Tropical Disease*, Tropical Disease Research, Series 3, Swabe, Basel, 367–389, 1979.
135. **Litchfield, T. M., Mackenzie, C. D., Varma, M. G. R.**: The conditions required for the maintenance of *Onchocerca lienalis* microfilariae in vitro, *Acta Tropica*, 48, 293–304, 1991.
136. **Nyindo, M., Shatry, A., Awitti, L. S., Chintawi, M., Hendricks, L. D.**: *Leishmania donovani* and *L. major:* Cultivation in vitro in tick embryonic cell lines. *Exp. Parasitol.* 63, 240–242, 1987.
137. **Bhat, U. K. M., Yunker, C. E.**: Susceptibility of a tick cell line (*Dermacentor parumapertus* Neumann) to infection with arboviruses. *Arctic and Tropical Arboviruses*. Kurstak, E., Ed., Academic Press, New York, 263–275, 1979.
138. **Holman, P. J., Ronald, N. C.**: A new tick cell line derived from *Boophilus microplus*. *Res. Vet. Sci.* 29, 383–387, 1980.
139. **Řeháček, J.**: Relation between viruses and tick tissues in vitro. Proc. XII. Int. Contr. Entomol., London, 774–775, 1964.
140. **Řeháček, J., Šutáková, G.**: Virus-like particles in *Dermacentor reticulatus* ticks. *Acta Virol.* 33, 577–581, 1989.
141. **Řeháček, J., Kožuch, O.**: The use of *Hyalomma dromedarii* tick tissue cultures for the isolation of tick-borne encephalitis virus from its natural foci. *Acta. Virol.* 13, 53, 1969.

Chapter 6

LEAFHOPPER CELL CULTURE FOR VIRUS RESEARCH

Toshihiro Omura and Ikuo Kimura

TABLE OF CONTENTS

0-8493-7642-4/94/$0.00+$.50

Toshihiro Omura, Ph.D., is Senior Researcher at National Agriculture Research Center. He received his B.S. in 1971 and his doctorate in Agriculture in 1978 in Plant Virology at Kyushu University, Japan. 1978–1983. Researcher at the Institute for Plant Virus Research. 1983–present. Virus diseases of rice.

I. INTRODUCTION

Insect cell culture has been developed for various kinds of studies such as physiology, genetics, and embryology. The advantages of using cultured cells are their uniform nature, and their capability to multiply in controlled environments including culture media. Their ability to self-propagate and to form monolayers on the surface of culture flasks are their fascinating properties desired for various kinds of studies. One of the research fields benefited by this experimental system is plant virology.

The phytoreoviruses,[1] including wound tumor virus (WTV), rice dwarf virus (RDV), and rice gall dwarf virus (RGDV), and rhabdoviruses, including potato yellow dwarf virus (PYDV)[2] and rice transitory yellowing virus (RTYV),[3] propagate both in plants and leafhopper vectors. The mechanism by which these viruses multiply in plant and insect hosts is of considerable scientific interest. In nature, most leafhopper-borne viruses have no way of dispersal other than being transmitted via vectors. Attempts are being made to interrupt virus infection or multiplication processes in vector leafhoppers, and ultimately to develop control measures for plant virus diseases. Understanding of the fundamental virus multiplication events in host cells and qualified research systems suitable for various research objectives are required for this purpose. The vector cell monolayer has a potential to serve as a host and also provides an excellent virus detection method. The hosts available for the study of leafhopper-borne viruses are plants, vector insects, and monolayer cells of vector insects. Plant protoplast systems for leafhopper-borne viruses have not been developed.

Some assay systems have been developed for plant viruses persistently transmitted by leafhoppers. Infectivity tests of leafhoppers, following an injection of virus suspension into their abdomens,[4-6] or feeding leafhoppers on virus suspensions through a membrane,[7-11] have been used to assay the infectivity of the viruses. Another alternative is the serological titration of virus antigens.[12] The third method is the use of vector cell monolayers. Each method has its own advantages and disadvantages. Hence, depending on the research objectives, we can choose one or another of these methods.

Leafhopper cell culture has a great advantage for analyzing at the molecular level biological events in cells infected with viruses that multiply in leafhopper tissues. This is due to its possibility of obtaining a uniform infection which enables us to trace the infection and multiplication process of viruses and the reactions of hosts to such a process after infection. The vector cell monolayer is a well established system of supporting biochemical analysis and tracing the fate of a virus and of its components. However, an extensive molecular analysis of the pathogen was required before the cell monolayer systems could be used effectively.

One of the current major developments in virus research is the study at the molecular level of the genome and the encoded proteins. Such investigations were also carried out with leafhopper-borne viruses, WTV, RDV, and RGDV.[13] Further progress, using vector cell monolayers, will disclose the biodynamic process of the virus molecules and host origin in infected cells to clarify the molecular entities that occur in hosts after virus infection. These fundamental studies are expected to provide a new approach which will identify weak points of the target virus.

Leafhopper cell lines and details of leafhopper cell culture have been reviewed by Hirumi and Maramorosch,[14] Reddy,[15] and Maramorosch.[16] The reviews of Black[17-19] and Hsu et al.[20] cover most papers concerned with WTV and PYDV in monolayer cells. Current progress in leafhopper cell culture and its application to RDV and RGDV studies were reviewed by Kimura and Omura.[21] As one of the experimental systems for the integrated studies aiming at development of control measures for virus diseases, this chapter will focus on leafhopper cell cultures as a means for virus studies. The methods and applications of leafhopper cell cultures for virus research will be presented in detail, mainly using *Nephotettix* cell cultures and RGDV, supported by the data accumulated during the past three decades.

II. LEAFHOPPER CELL CULTURES

The essential properties of leafhopper cell lines required for virus researches are as follows: the cells should form a monolayer on the surface of culture vessels; the propagation of the cells should be rapid; the damage of cells by subculture should be low; and long-term storage of the cells should be possible. Cell lines that satisfy these requirements and have been used in quantitative assay and multiplication studies of viruses are the AC line series established from *Agallia constricta* for WTV and PYDV,[18] and NC, NN, NV, and RD lines established from *Nephotettix cincticeps, N. nigropictus, N. virescens,* and *Recilia dorsalis,* respectively, used for RDV and RGDV.[21] This section describes the establishment, maintenance, and storage of cells from *N. cincticeps, N. nigropictus, N. virescens,* and *R. dorsalis.*[22,23]

A. ESTABLISHMENT

Leafhopper eggs at the blastokinetic stage[24] are essential for obtaining embryos for tissue cultures. At 25°C, eggs reach the critical stage 7 to 9 days after the initiation of oviposition. Young embryos, removed from eggs, were cut into a few fragments in Tyrode's solution on a hollow slide. The fragments were treated with 0.25% trypsin in Tyrode's solution (pH 7.5) for 10 to 20 min at 30°C, and then transferred to growth medium by coiling the fragments around a fine needle. This coiling is important for the success of cell establishment, i.e., due to the sticky nature of the tissue, it is difficult to recover the tissues from the tube after low speed centrifugation. About 30 fragments were placed together in a drop of 0.3 to 0.5 ml of medium in the center of a small plastic dish placed in a sealable glass dish. Additional medium was placed around the bottom inside rim of the plastic dish to maintain a suitable vapor pressure and thus avoid drying of the medium containing the fragments. After sealing the glass dish with parafilm, the cultures were incubated at 25°C. The growth medium which was derived from Liu and Black's medium[25] was used for culturing NC, NN, NV, and RD cell lines.[22,23] The medium composition was as follows: 500 ml of Schneider's Drosophila medium (revised), 50 ml of medium 199 (10× concentrate) with Hanks' salt and glutamine, without sodium bicarbonate, 25 ml of Medium CMRL 1066 with glutamine, 150 ml of fetal bovine serum heated for 30 min at 56°C prior to use, 500 ml of 0.05 *M* histidine solution, 120,000 units of Penicillin G potassium, 120 mg of Streptomycin sulfate, 6 ml of Neomycin sulfate (10,000 mg/ml), 4 ml of Fungizone (250 mg/ml); pH was adjusted to 6.5 to 6.6 with 2 *N* HCl. The antibiotics used were suitable for the manipulation of NC cells. But it is noteworthy that fungizone used as an antibiotic was reported to have a harmful effect on the growth rate of AC cells.[26] The screening of harmless antibiotics may also be important for attaining stable cell growth. For instance, the recovery rate of frozen cells might be improved by such methods. The role of the antibiotics seems not only to protect the multiplication of possible contaminant microorganisms but also of symbionts included in cultured cells. When Kanamycin (80 mg/l in the total amount of the medium shown above) was used instead of Penicillin and Streptomycin, the cells could not maintain the integrity in about 10 days' culture, probably due to the excess multiplication of symbiont (Kimura, unpublished data). The primary cultures could be maintained for several months by changing the medium once a week. When the cells grew almost confluently to cover the surface of the dish, they were transferred to new dishes. These subcultured cells were then transferred to plastic culture flasks for further subculturing.

Monolayers of cells were usually composed of two types of epithelial cells. The majority were spindle shaped cells and the rest were of circular shape. During the course of subcultures, spindle-shaped cells propagated selectively in this medium. NC, NN, and NV cell lines are similar in shape and size, but RD cell lines are slightly different in shape and smaller in size than the others (Figure 1). The NC, NN, and NV cells grew more rapidly than the RD cells.

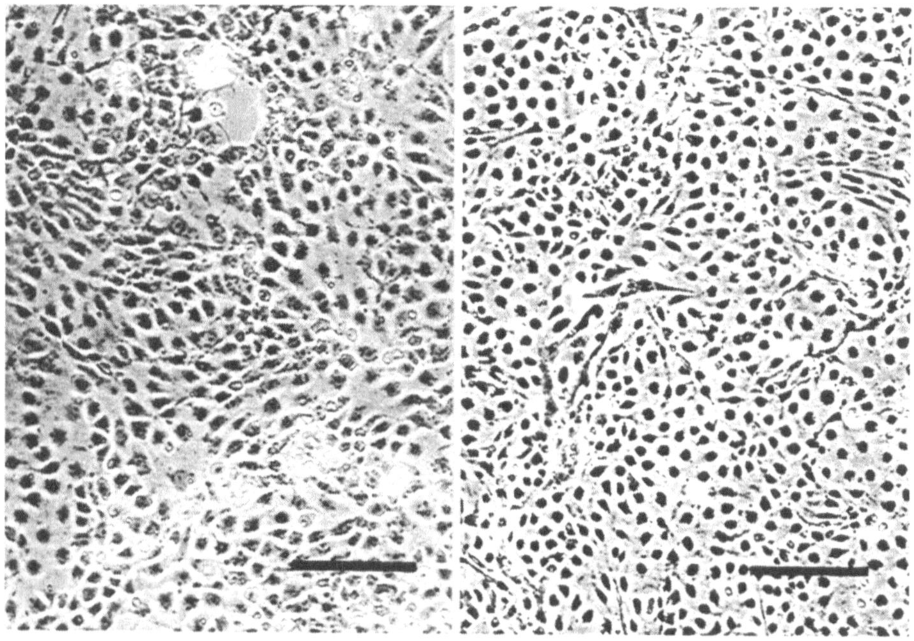

FIGURE 1. Portions of cell monolayers: left, NC-24 cell line established from *N. cincticeps;* right, RD-8 cell line established from *R. dorsalis.* Bar markers represent 100 μm.

B. MAINTENANCE

For subculturing, the monolayers were washed with Tyrode's solution to exclude the effect of medium components on enzymatic activity to trypsin and then treated with 0.05% trypsin in Tyrode's solution for 5 min. Martinez-Lopez[27] used papain instead of trypsin to detach AC cells from a culture flask. The optimal pH of its activity is the same as that for growth of AC cells, resulting in less aggregation of cells, better cell attachment to new plastic, and better cell growth than when trypsin was used. Though there was no apparent difference between the effect of papain and trypsin on cell growth in AC cells, papain might alleviate the damage to cells more delicate than AC cells. In the next step, new medium was added to stop the enzymatic activity, and the cells were detached from the culture flask by pipetting, collected by low speed centrifugation, and then resuspended in a new medium with a vortex blender. The new medium added to stop the enzymatic activity can be substituted for the old medium used for cell culture, thus saving 1/3 of the total medium required. Blending of cells can be done by pipetting. After subculture, the cells in the culture flask are generally divided into three parts. When the cells are diluted more than 1/5, cell growth sometimes is greatly reduced. Optimum culture period depends on cell lines. In NC cells, cultures are transferred every week. When the cells were transferred 2 weeks after each subculture, the growth of the new culture became remarkably slow due to the failure in recovering many cells during transfer. That is, the cell recovery rate is low when the cells are kept long after their confluent growth in the flask. The NC, NN, and RD cells have been maintained for 250 generations by transferring every 10 days (Kimura, unpublished data). The AC cell line has been continued well beyond 300 passages.[19] This suggests that the limited life span which prevents subculturing of some vertebrate cell cultures does not apply to the leafhopper cells. Thus, the leafhopper cell cultures are expected to provide a stable experimental system for years to come, as long as their sensitivity to virus infection does not change.

C. STORAGE

To maintain NC, NN, NV, and RD cell lines without freezing, the cells were incubated at 18°C and the culture medium was changed once a month. The cultured cells almost stopped growing at this temperature. By this method, the cells were maintained for 14 months. The cells grew slowly at 20°C but were damaged if stored at 15°C. The optimum temperature for this type of storage seems to be different for different cell lines. The optimum temperature was 13 to 15°C in AC cultures in CB medium.[18]

NC, NN, NV, and RD cell lines can be frozen and stored in a polypropylene vial for long periods in liquid nitrogen by the same method used for AC cell preservation.[28] Vector cell monolayers at the exponential growth stage were harvested by sedimentation, and the pellet was suspended in the culture medium containing 5% dimethyl sulfoxide. Vials containing suspended cells were incubated overnight at –80°C in a vessel whose wall was thick enough to avoid rapid freezing. Then they were placed in liquid nitrogen at –196°C. Frozen cells were recovered by thawing quickly in a waterbath at 37°C. The liquid-containing cells were diluted up to 10 ml by additions of 1 ml medium per minute, sedimenting by low speed centrifugation, and suspending in 4 ml of new medium. The cells were then dispersed in a 25 cm^2 NUNC flask. About 20% of frozen cells were recovered by this procedure. The NC and NN cell lines have to be defrosted and recovered after 4 years of storage at –196°C to maintain their viability, and NV and RD have to recovered after 2 years. Established cell lines were described by Black,[18] Maramorosch,[16] and Kimura and Omura.[21]

III. VIRUS INOCULATION

Viruses that have been amenable to study by use of leafhopper cell monolayers are phytoreoviruses, WTV, RDV, and RGDV, and rhabdoviruses, PYDV. RTYV may be also a good objective to study using leafhopper cell monolayers because cultured cells of its vectors, *N. cincticeps* and *N. nigropictus,* have already been described in detail by Kimura.[31] The essential procedures used for RDV and RGDV studies are described below.

A. METHODS OF INOCULATION

The cell suspension prepared by the procedure described in Section II.B was layered on coverslips ca. 15 mm in diameter. When the cell monolayers became confluent on coverslips (2 days in NC[2]), they were washed twice with His-Mg, and 50 µl of virus inoculum was layered on the cells. The monolayers were incubated for 2 h at 28°C. Then, the inoculum was removed, cells were washed twice with His-Mg, and 100 to 150 µl of the culture medium was layered. Inoculated cells were incubated for 40 to 48 h at 25°C.

B. SUSCEPTIBILITY OF CELL LINES TO VIRUS INFECTION

Of the NC, NN, NV, and RD cell lines described in the former section, NC and NN were similarly susceptible to RDV,[31] while RD cells were about 10 times less susceptible to infection than NC and NN cells (Kimura, unpublished data). The NV cells from embryos of *N. virescens,* a nonvector of RDV, were not susceptible to RDV infection (Kimura, unpublished data). The AC-20 cells were also not susceptible to RDV infection (Kimura, unpublished data). Differences in susceptibilities of cell lines to WTV or PYDV infections have been described in detail by Black.[18] These cell lines with different susceptibility to virus infection will prove to be an excellent system for studying the mechanism responsible for vector specificity in virus transmission.

C. SUSCEPTIBILITY OF CELL GENERATION TO VIRUS INFECTION

In the NC cell line, no significant loss of susceptibility was observed to RDV infection during 296 passages (Kimura, unpublished data). On the contrary, the AC-20 cell line after about 330 transfers was no longer susceptible to inoculation with PYDV or WTV.[18] At present, it is not certain whether cultured cells lose their susceptibility to virus infection depending on the length of time elapsed after their establishment. Irrespectively of whether or not the susceptibility declines during long-term culturing, storage of the cultures under liquid nitrogen is recommended soon after their establishment to assure the continuity of studies. No detectable loss of infectivity of RDV and RGDV was found in the NC cells recovered after freezing in liquid nitrogen (Kimura, Omura, unpublished data).

D. EFFECTS OF BUFFERS

A 0.1 *M* histidine solution containing 0.01 *M* $MgCl_2$ (His-Mg), devised for the extraction of WTV,[30] was also applicable for RDV.[31] The excellence of this solution for maintaining infectivity of phytoreovirus was confirmed also for RGDV where different buffers were compared for their effects to preserve RGDV infectivity to leafhopper cell monolayers.[32] The fact that this buffer has good effects not only on the cells but also on the virus infectivity was proved by a freezing and thawing experiment. The infectivity of RGDV was damaged in phosphate buffer, while no decline in infectivity was detected in RGDV suspended in His-Mg (Table 1). Purified RDV and RGDV also were found to be stable in His-Mg (Omura, unpublished data). The stability of the three phytoreoviruses in the same buffer provided additional evidence that they belong to the same virus group. Of the virus components, the proteins that are located on the outer part of the outer capsid[33] are thought to receive the direct influence of the solutes in the buffer. In this context, high homology in the primary structure of the outer capsid protein among three phytoreoviruses predicted from the nucleotide sequences[33-35] may explain this phenomenon. Although this may not be so drastic as the effects of buffers, effect of pH on virus infectivity needs to be studied. In His-Mg, the optimal pH value for inoculation of RDV and RGDV is around 6.0[31,21] (Figure 2). The optimum was about 6.6 for WTV.[30]

E. VIRUS PARTICLES AND THEIR ADSORPTION AND PENETRATION INTO VECTOR CELL MONOLAYERS

The optimal inoculation period was expected to differ depending on virus concentration and volume of the inoculum. However, the volume of extracts from RDV-infected leaves, ranging from 0.02 to 0.1 ml used as the inoculum, had no appreciable effect on the number of foci[31] as was also true for WTV.[30] The sufficient adsorption period for infection was affected greatly by virus concentration. The sufficient adsorption period of RDV was about 60, 90, and 120 min when dilution of infected leaf extracts was 10^{-3}, 10^{-4}, and $10^{-4.5}$, respectively.[31]

It was assumed that RDV particles, adsorbed on the surface of vector cell monolayers, enter the cells, probably by phagocytosis. To examine the period required for the penetration of the virus, viral antibody was added at various times after inoculation to inactivate viruses remaining outside of the cells.[31] The results showed that RDV particles penetrated almost completely into the cells 90 to 120 min after the initiation of the incubation.[31] The time required for RGDV penetration into the cells was about the same as that for RDV penetration and was approximately 120 min (Figure 3).

IV. QUANTITATIVE ASSAYS OF VIRUS ACTIVITY

As described by Black,[18] viruses which produce primary local lesions on leaves, such as PYDV, have a greater advantage over the phytoreoviruses, providing means for a quantitative assay. In general, possession of a bioassay system is indispensable for quantitative virus studies. The higher the sensitivity of the assay system, the greater the possibility of tracing virus activity in detail. With the exception of PYDV, there is no local lesion assay for plant

TABLE 1
Effect of Freezing and Thawing on RGDV Infectivity in Rice Extract as Revealed Using *N. cincticeps* Cell Monolayers by Foci Counting

		Freezing[a] and thawing				
Buffer	**Exp. no.**	**Fresh**	**1st**	**2nd**	**3rd**	**4th**
His-Mg[b]	I	385[c]	426	301	320	377
	II	542	476	232	452	395
PB[d]	I	538	75	0	0	0
	II	329	54	0	0	0

[a] Approximately 500 μl of virus materials extracted from infected rice leaves were maintained at –70°C for 1 h and thawed in a water bath controlled at 40°C for 10 min.

[b] 0.1 *M* histidine, 0.01 *M* $MgCl_2$, pH 7.0.

[c] Number of foci per coverslip.

[d] 0.1 *M* Na-phosphate buffer, pH 7.0.

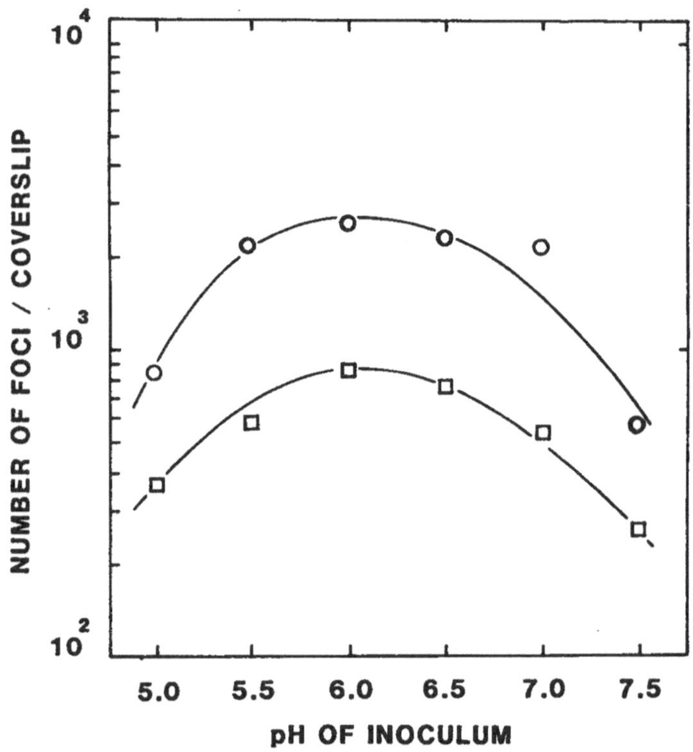

FIGURE 2. Effect of inoculum pH on RGDV infectivity. Infected leaves were homogenized in His-Mg buffer at pH 5.0 to 7.5 and were placed on *N. cincticeps* cell monolayers for inoculation.

viruses that multiply in vectors and are persistently transmitted by leafhoppers. The lack of an appropriate infectivity assay system was one of the major reasons hampering studies of these viruses. The establishment of vector cell monolayers enabled quantitative assay of virus infectivity in a highly sensitive manner, so that these viruses became much easier and attractive to study. This section deals with virus assays and procedures for characterizing the

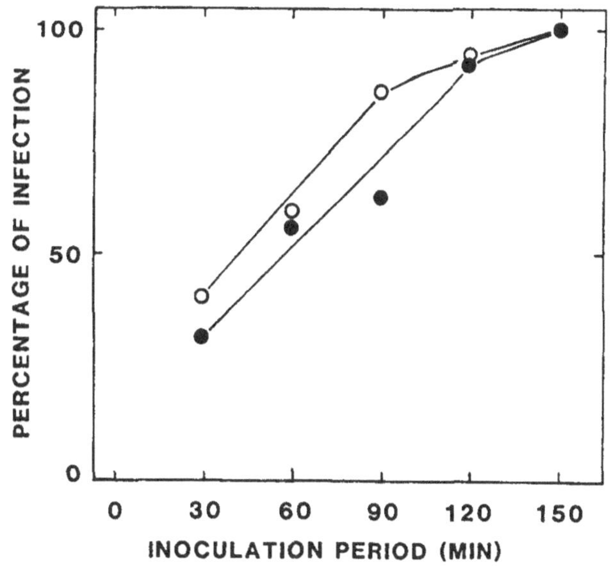

FIGURE 3. Effect of inoculation period on infection of RGDV. Percentage of foci numbers developed on cover slips after 30 to 150 min of inoculation time relative of the number recorded after 150 min inoculation. RGDV-infected rice leaf extracts at 10^{-4} or $10^{-4.5}$ dilutions were used as inocula.

viruses, using vector cell monolayers. The comparative advantages and disadvantages of vector cell monolayers and other assay systems will also be discussed.

A. FLUORESCENT ANTIBODY FOCUS COUNTS

Without fluorescent antibody staining, vector cell monolayers infected with WTV,[18] RDV,[31] RGDV,[32] and PYDV[2] were indistinguishable from uninfected cells by light microscopy. The method was described in detail by Black.[18] The double sandwich method, using FITC-conjugated second antibody to the specific antibody to the virus, is convenient for this study.

Fluorescent cells are counted under a fluorescence microscope at a magnification of ×200 by the focus count method.[30,36] In this method, single infected cells or a cluster of infected cells were designated as infected focus, and one focus was counted as one infection unit. The factor for relating diametral zone counts to counts over whole monolayers on coverslips was determined by dividing the number of foci in the whole coverslip by the average number of foci in horizontal and vertical diametral zones using 10 coverslip monolayers. The average diametral factor with 95% confidence limits is 10.45.[21] When a few infected foci are obtained, such diametral counting is readily correlated with counts of whole coverslip monolayers without losing appreciable accuracy. By this method, quantitative assay of virus concentration becomes possible, as the relationship between virus concentration in the inoculum and number of foci of infected cells is linear in a range of certain dilution.

First sign of infection was detected 12 h after inoculation. The number of infected foci gradually increased up to 40 h after RGDV inoculation using NC cells (Figure 4). No further increase of foci was detected afterward. Therefore, the staining of inoculated cells with FITC is carried out 40 h post-inoculation for RGDV.[32] When the staining of the cells is delayed, the neighboring cells start to fluoresce, and hence the foci counting becomes difficult. This second infection of the neighboring cells can sometimes occur, and this is the reason that we use the term "infected foci" but not "infected cells." The time course appearance of virus-infected foci is similar for WTV[37] and RDV.[23]

When establishment of vector cell lines takes too long or an infectivity assay system is urgently required, the reticule system of Martinez-Lopez and Black[38,39] for counting infected

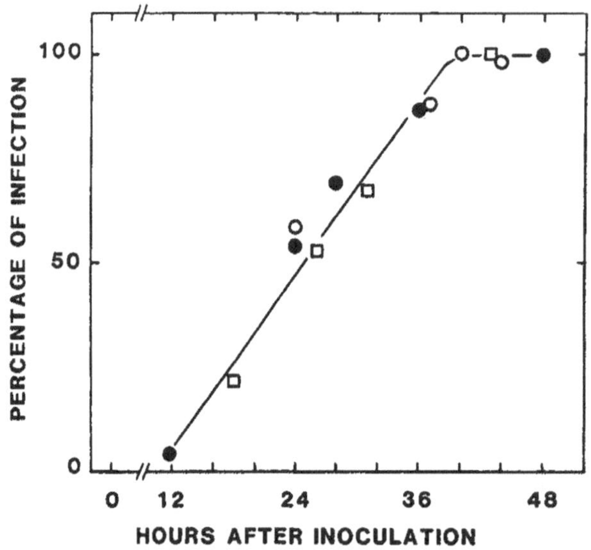

FIGURE 4. Time course of foci appearance in *N. cincticeps* cell monolayers inoculated with RGDV. Foci numbers on each coverslip were counted by immunofluorescent staining at various times after inoculation, indicated as percentage of the number at 48 h.

cells may be applicable as a quantitative assay. Goedigk et al.[40] obtained good results for detecting PYDV essentially following this method.

B. STABILITY OF RGDV IN LEAF SAP

We have to understand the nature of virus infectivity assayed by the vector cell monolayers before proceeding with other virus investigations. Some of the essential results of our study of RGDV are described below.

In parallel with studying the effects of extraction buffers on infectivity, experiments to determine proper specimen dilution, which produce foci proportional to sample dilution, are necessary. In case of RGDV, as shown in Table 2, extracts of virus-infected rice plants at 10^{-6} to $10^{-4.5}$ dilutions in His-Mg are appropriate as the inoculum.

The availability of inocula, which give reproducible infections, is one of the important requirements that increase reliability of experiments. When RGDV-infected plant sap in His-Mg was preserved for different periods at –70°C, no substantial decrease in its infectivity was detected even after 10 years (Table 3). Purified virus preparations can be stored at –80°C for at least a year without detectable loss of WTV[18] infectivity. Thus, inoculum can be stored for a fairly long time and used in other related experiments.

To check the stability of RGDV, virus inocula were subjected to repeated freezing and thawing, and their infectivity was assayed using leafhopper cell monolayers (Table 1). RGDV in leaf extracts in His-Mg was stable following at least four times freezing and thawing. In contrast, a quick decline of the infectivity was detected for RGDV in leaf extracts using phosphate buffer.

Manipulation of specimens before testing their infectivity takes considerable time. Therefore, it is important to know the stability of virus infectivity under different temperatures. As shown in Table 4, RGDV infectivity of leaf extracts is stable for at least 6 h at room temperature in His-Mg. Similar results were obtained also for RDV (Omura, unpublished data). In the case of WTV, after only 2 h at 37°C under the optimal pH, its infectivity was reduced to 55% of that of a comparable control kept at 0°C.[18] In a similar study, RGDV infectivity in His-Mg was retained for 10 min at 50°C but lost at 60°C (Omura, unpublished data). Thus, we can extend further precise studies by understanding the nature of virus infectivity.

TABLE 2
Dilution End Point of RGDV

Buffer	Exp. no.	Dilution						
		10^{-2}	10^{-3}	10^{-4}	10^{-5}	10^{-6}	10^{-7}	10^{-8}
His-Mg[a]	I	>2000[b]	>2000	1327	218	42	2	0
	II	>2000	>2000	>2000	526	70	5	0
PB[c]	I	>2000	>2000	472	28	2	0	0
	II	>2000	>2000	>2000	129	5	0	0

[a] 0.1 *M* histidine, 0.01 *M* $MgCl_2$, pH 7.0.
[b] Number of foci per coverslip.
[c] 0.1 *M* Na-phosphate buffer, pH 7.0.

TABLE 3
Preservation of RGDV in Infected Rice Plants Revealed Using *N. cincticeps* Cell Monolayers by Foci Counting

Buffer	Exp. no.	Duration of freezing at −70°C				
		1 day	7 days	60 days	485 days	10 years
His-Mg[a]	I	406[b]	203	359	372	506
	II	370	212	387	361	736
		Duration of freezing at −70°C				
		1 day	7 days	57 days	200 days	2 years
PB[c]	I	221	211	161	251	179
	II	244	212	143	184	147

[a] 0.1 *M* histidine, 0.01 *M* $MgCl_2$, pH 7.0.
[b] Number of foci per coverslip. The materials other than those maintained for 10 years were stored to preserve at different times and assayed at once.
[c] 0.1 *M* Na-phosphate buffer, pH 7.0.

TABLE 4
Longevity *In Vitro* of RGDV in Leaf Extracts Revealed by *N. cincticeps* Cell Monolayers by Foci Counting

Buffer	Temperature (°C)	No.	Time					
			30 min	6 h	1 day	2 days	5 days	10 days
His-Mg[a]	4	I	346[b]	210	329	345	121	9
		II	322	279	401	244	173	11
	25	I	272	249	51	40	2	0
		II	343	352	62	48	0	0
PB[c]	4	I	231	132	20	1	0	0
		II	335	101	18	0	0	0
	25	I	245	37	0	0	0	0
		II	212	5	0	0	0	0

[a] 0.1 *M* histidine, 0.01 *M* $MgCl_2$, pH 7.0.
[b] Number of foci per coverslip.
[c] 0.1 *M* Na-phosphate buffer, pH 7.0.

C. RELATIVE SENSITIVITY AMONG THE METHODS FOR VIRUS TITRATION

There are three methods available for tracing phytoreovirus activities. The traditional ones are the injection method which consists of injection of pathogen suspensions into the abdomens of vectors, followed by infectivity tests of the vectors[4-6] (Table 5). The second consists of membrane feeding of vectors on pathogen suspensions through a membrane[7-11] and infectivity tests of the vectors. These methods have been based on virus infectivity. However, such methods require longer than 40 days before final results are obtained.[6,11,41] The third method is the serological titration of virus antigens. This method requires only 1 or 2 days to obtain results, and samples can be tested directly without sterilization, but it may not reflect virus numbers nor virus infectivity. Furthermore, large numbers of dilutions should be made to obtain accurate results. The vector cell monolayer technique needs cell culture facilities and it requires rather complicate procedure. But once the system is established, it provides a highly sensitive and reliable method.

The dilution end point of virus-infected leaf extracts for positive reaction by ELISA is 10^{-3} for RGDV.[12] The vector cell monolayer system (Table 2) is 10^3 and 10^4 times more sensitive than the injection method (Table 5) and ELISA, respectively. In PYDV, the number of infections on vector cell monolayers was about $10^{4.5}$ and $10^{3.8}$ times greater than that on the leaves when equal surface area and equal number of cells, respectively, were compared.[42] Another advantage of the vector cell monolayer system is its quantitative nature. Thus, we can titrate virus in different materials more accurately by vector cell monolayers than by the other methods (Table 6). The preciseness of the vector cell monolayer system over ELISA was further confirmed in other experiments on the same materials (Table 7,8).

For detecting virus infectivity, the vector cell monolayer system has an advantage over the other methods in its sensitivity. However, each method has respective advantages and disadvantages depending on the research objectives (Table 9). Hence, choice of one among these methods can be made according to the purpose of the study.

Quantitative assay of virus activity using vector cell monolayers enabled a detailed analysis of virus multiplication in infected rice plants, vector insects, and inoculated monolayer cells[21] (Table 8). The first sign of virus reproduction was detected at 5 days, 2 days, and 8 h after inoculation of rice plants, exposure of vector leafhoppers to the virus, and inoculation of vector cell monolayers, respectively.[21] Such a quick detection of virus after infection of vector cell monolayers has never been obtained by the other methods. In fact, the first detection of virus by ELISA was at 15 days, 8 days (Omura, unpublished data), and 22 h (Table 7) after inoculation in rice plants, vector insects, and vector cell monolayers, respectively.

V. MULTIPLICATION OF VIRUSES IN VECTOR CELL MONOLAYERS

A. GROWTH CURVE OF VIRUSES

To study virus multiplication, it is desirable to obtain an inoculum concentration that results in 100% infection of cells and attains synchronous infection of vector cell monolayers. Such an inoculum can be prepared by diluting RDV-infected leaf extracts $10^{-2.5}$ (Kimura, unpublished data). In the case of RGDV, infection reached a level of 100% with the following: $10^{-3.5}$ from infected leaves, $10^{-4.5}$ from viruliferous insects, $10^{-5.5}$ from infected monolayer cells, and 10^{-5} from purified preparations with an absorbance at A_{260}.[32] Exposure of AC-20 cells to a fresh concentrated lysate prepared from cells persistently infected with WTV resulted in a synchronous infection of more than 90% of the cultured cells.[43]

Measurable increase in RGDV infectivity, assayed by the monolayer cells of extracts of vector cell monolayers, inoculated with an inoculum which causes 100% infection was

TABLE 5
Infectivity Test of RGDV in Rice Extracts by the Injection Method

	Dilution of infected plant				Healthy Control
Number of insects that	10^{-2}	10^{-3}	10^{-4}	10^{-5}	10^{-2}
Injected	50	50	50	50	90
Survived[a] 12 days	9	13	16	23	22
Transmitted virus	5	4	2	0	0

[a] *N. nigropictus* at 2 to 3 instar nymph was injected with approximately 10 μl of RGDV-infected leaf extracts in 0.1 *M* phosphate buffer, pH 7.0.

TABLE 6
Relative Concentration of RGDV in Cultured Cells, Viruliferous Insects, and Rice Plants Revealed Using *N. cincticeps* Cell Monolayers by Foci Counting

	Number of foci/coverslip at dilutions of			
Host	$10^{-6.5}$	10^{-6}	$10^{-5.5}$	10^{-5}
Cultured cells	745	2638	>3000	>3000
Viruliferous insects	25	125	680	>3000
Infected plants	0	48	238	1068

TABLE 7
Detection of RGDV Antigen by ELISA in Cultured Monolayer Cells of *N. cincticeps* at Various Times after Inoculation

	Hours after inoculation						
Exp.	2	8	16	22	28	34	40
I	<1[a]	—	<1	1	8	64	64
II	—	<1	—	4	—	—	128

[a] Numbers in the table indicate the reciprocals of dilution end point for positive reactions, multiplied to 10^3.

detected about 8 h after inoculation[32] (Table 8). Infectivity of the extracts increased sharply between 8 and 40 h after inoculation and reached a plateau that was maintained for at least the following 24 h. The increase in virus infectivity in the cultured cells during 8 to 40 h spanned 5 log units; the doubling time for the increase of RGDV infectivity during this period was about 2 h. RDV infectivity of the inoculated vector cell monolayers was detected for the first time at 12 h after inoculation, and it reached the plateau 20 h after inoculation, with a doubling time of 96 min.[31] In WTV, the doubling time of the virus at 30°C occurred in less than 58 min, and the virus infectivity reached the plateau 28 h after inoculation.[18] The question whether these differences in virus multiplication rates are due to the principal propagation

TABLE 8
Infectivity of RGDV in Cultured Monolayer Cells of *N. cincticeps* at Various Times after Inoculation Detected by Foci Counting Using Cultured Monolayer Cells of *N. cincticeps*

	Hours after inoculation						
Exp.	2	8	16	22	28	34	40
I	23×10^{2}[a]	426×10^{2}	499×10^{4}	163×10^{6}	891×10^{6}	109×10^{7}	474×10^{7}
II			227×10^{4}	85×10^{6}			161×10^{7}

[a] Number of foci per coverslip when extracts of the cells were used as inocula.

TABLE 9
Efficacy of Vector Cell Monolayer Method, Vector Injection Method, and ELISA in RDV or RGDV Studies

	Methods		
Characteristics	Vector cell monolayer	Vector injection	ELISA
Sensitivity	10^{-7}	10^{-4}	10^{-3}
Days required for obtaining results	2	40–60	1–2
Time required for the testing of 100 specimen	2 days	15 days	2 h
Suitable for biochemical studies	+	–	–
Assay of pathogenicity on plants	–	+	–
Assay of infectivity	+	+	–

ability of each virus or due to the culture conditions, such as temperature, as discussed by Black,[18] needs to be clarified. The synchronous and rapid RDV or RGDV virus multiplication suggests that vector cell monolayers are excellent experimental systems for detailed studies of biological events, as indicated earlier for WTV.[44]

B. VIRUS PERSISTENCE

The infection of leafhopper monolayer cells AC-20 by WTV is noncytopathic and it persists during repeated subculturing.[17,18] The cells appear normal when observed by the light microscope and they grow at a similar rate as uninfected cells,[37] even thought the majority of the cells is infected with the virus.[45] NC cells systemically infected with RDV also showed vigorous multiplication like those of uninfected cells (Kimura, unpublished data). Furthermore, the infected cells can be maintained for several months at 17°C when culture medium was changed monthly (Kimura, unpublished data). Such cells can provide an excellent system for studying virus activity in cells as described in the following section.

C. ELECTRON MICROSCOPY

The finding of viroplasms by electron microscopy came as a surprise, indicating that there were entities which support some step of virus propagation that can be visualized.[46] Now we can use molecular probes, as stated by Nuss and Dall.[13] The electron microscopy combined with bio-probes will visualize the biodynamic processes of virus multiplication in vector cell monolayers having the advantages mentioned above. One of the open questions in reovirus research, the sorting and packaging mechanism of the segmented genome into one virus particle,[13] may also be clarified by using these systems.

VI. VIRUS-MONOLAYER CELL INTERACTION

When extracts of RDV-infected plants were treated with CCl_4, they lost their infectivity to vector cell monolayers.[47] Not only crude plant extracts, but also virus particles purified by the routine method using CCl_4 could no longer infect vector cell monolayers. However, both the CCl_4-treated extracts and the purified virus showed infectivity when injected into the vector, *N. cincticeps.* The possible explanation of these phenomena was that CCl_4-treated RDV particles lost some essential materials required for adsorption of particles to the cell surface. On the other hand, the virus particles were introduced directly into the insect cells when cells became injured by fine glass needles used for injection. Recently, RDV was purified without CCl_4 treatment (Omura, unpublished data). This purified virus showed infectivity to the vector cell monolayer and to vector insects after injection. Thus, the adsorption of RDV to the surface of cultured leafhopper cells seems to be regulated by a molecular entity which is removed from virus particles by CCl_4 treatment.

The interaction between WTV and leafhopper cell cultures was studied at the molecular level after the penetration of virus into the cells.[43-45] Cells were used shortly after inoculation (acute phase: through day 5), and the virus-infected cells were subcultured for a long period (persistent phase). Virus-specific polypeptide synthesis and viral genome RNA accumulation increased to the maximum level during the acute phase and then decreased to 5 to 20% of the maximum level at the acute phase after 10 to 15 passages. All the cells at this passage remained infected. Based on comparative studies of the amounts of viral proteins, genome RNA, transcripts, and the products from the transcript in the cell-free translation system, the authors concluded that the contributing factor to the decline in viral polypeptide synthesis appears to be a change in the functional activity of the pool of viral transcripts.[43-45]

VII. CONCLUSION

Among phytoreoviruses, RDV ordinary strain,[48] severe strain[48] and nonvectorial strain,[49] and RGDV[50] represent promising materials for comparative studies of biological divergence and components. During the course of biochemical analysis of these viruses, we have encountered difficulties in further analysis due to limitations of the technique and virus multiplication systems. Therefore, we concentrated our efforts on analysis of virus components at the molecular level, using molecular tools in probing biological processes in infected hosts.[13] Besides, we also worked to develop host systems which support synchronous virus infection and multiplication. One of the possible systems for such purposes is the use of the protoplast system which proved to be excellent in studying plant viruses.[51,52] However, our attempts to infect protoplasts from barley (Kano, personal communication) or rice callus cells (Nishiguchi, personal communication) with RDV or RGDV failed, and we were unable to obtain anticipated results.

As described in this chapter, leafhopper cell monolayers represent an excellent system for studying viruses that multiply in plants and leafhopper vectors. The notable advantages of the vector cell monolayer system is the uniform 100% infection of cells with virus using diluted inoculum, which enables us to follow synchronous virus multiplication. The vector cell monolayer also provides a very sensitive bioassay system for tracing the fate of virus infectivity in different hosts and in different conditions. The self-propagative nature of leafhopper cells under favorable conditions ensures a stable supply of an assay system with a stable sensitivity to virus infection. Thus, the vector cell monolayer systems convert the leafhopper-borne viruses from one of the most difficult viruses to work with to those that permit most rapid and precise infectivity measurements, supplying a system for synchronous virus development.[18] We believe that the vector cell monolayer systems and molecular probes, along with reliable traditional methods or systems, will facilitate studies on the interactions between hosts and viruses at molecular as well as individual levels. New ideas and techniques to controlling virus diseases will be devised during the course of such fundamental studies.

ACKNOWLEDGMENT

We thank Dr. H. Kano for his continued interest and valuable suggestions and we thank Dr. H. Hibino for his encouragement of our work and for reviewing this paper.

REFERENCES

1. **Boccardo, G. and Milne, R. G.** Plant reovirus group. *CMI/AAB Descriptions Plant Virus* No. 294, 1984.
2. **Black, L. M.** Potato yellow dwarf virus. *CMI/AAB Descriptions Plant Virus* No. 35, 1970.
3. **Shikata, E.,** Rice transitory yellowing virus. *CMI/AAB Descriptions Plant Viruses* No. 100, 1972.
4. **Storey, H. H.** Investigations of the mechanism of the transmission of plant viruses by insect vectors. *I Proc. R. Soc. (London)* B113, 463, 1933.
5. **Black, L. M.** Further evidence for multiplication of the aster-yellows virus in the aster leafhopper. *Phytopathology* 31, 120, 1941.
6. **Maramorosch, K.** Mechanical transmission of curly top virus to its insect vector by needle injection. *Virology* 1, 286, 1955.
7. **Kimura, I. and Fukushi, T.** Studies of rice dwarf virus. *Ann. Phytopathol. Soc. Jpn.* 25, 131, 1960.
8. **Carter, W.** A technic for use with Homopterous vectors of plant disease with special reference to the sugar beet leaf hopper *Eutellix tenellus* (Baker). *J. Agric. Res.* 34, 449, 1927.
9. **Severin, H. H. P. and Swezy, O.** Filtration experiments on curly top of sugar beets. *Phytopathology* 18, 681, 1928.
10. **Chiu, R. J., Liu, H. Y., Macleod, R., and Black, L. M.** Potato yellow dwarf virus in leafhopper cell culture. *Virology* 40, 387, 1970.
11. **Kimura, I.** Acquisition of rice dwarf virus through an artificial membrane by a vector insect, *Nephotettix cincticeps. Ann. Phytopathol. Soc. Jpn.* 45, 563 (abstract in Japanese), 1979.
12. **Takahashi Y., Omura, T., Shohara, K., and Tsuchizaki, T.** Comparison of four serological methods for practical detection of ten viruses of rice in plants and insects. *Plant Dis.* 75, 458, 1991.
13. **Nuss, D. L. and Dall, D. J.** Structural and functional properties of plant reovirus genomes. *Adv. Virus Res.* 38, 249, 1990.
14. **Hirumi, H. and Maramorosch, K.** Cell culture of Hemiptera. In *Invertebrate Tissue Culture,* Vol. 1, Vago, C. Ed., Academic Press, New York, 1971, 307.
15. **Reddy, D. V. R.** Techniques of invertebrate tissue culture for the study of plant viruses. *Meth. Virol.* 6, 393, 1977.
16. **Maramorosch, K.** Leafhopper tissue culture. In *Leafhopper Vectors and Plant Disease Agents,* Maramorosch, K. and Harris, K. F., Eds., Academic Press, New York, 1979, 485.
17. **Black, L. M.** Insect tissue cultures as tools in plant virus research. *Ann. Rev. Phytopathol.* 7, 73, 1969.
18. **Black, L. M.** Vector cell monolayers and plant viruses. *Adv. Virus Res.* 25, 191, 1979.
19. **Black, L. M.** The controversy regarding multiplication of some plant viruses in their insect vectors. *Curr. Top. Vector Res.* 2, 1, 1984.
20. **Hsu, H. T., Nuss, D. L., and Adam, G.** Utilization of insect tissue culture in the study of the molecular biology of plant viruses. *Curr. Top. Vector Res.* 1, 189, 1982.
21. **Kimura, I. and Omura, T.** Leafhopper cell cultures as a means for phytoreovirus research. *Adv. Dis. Vector Res.* 5, 111, 1988.
22. **Kimura, I.** Establishment of new cell lines from leafhopper vector and inoculation of its cell monolayers with rice dwarf virus. *Proc. Jpn. Acad.* 60(B), 198, 1984.
23. **Kimura, I.** Quick detection of rice dwarf virus infection employing its vector cell monolayers. *Jpn. Agric. Res. Q.* 19, 109, 1985.
24. **Hirumi, H. and Maramorosch, K.** Insect tissue culture: use of blastokinetic stage of leafhopper embryo. *Science* 144, 1465, 1964.
25. **Liu, H. Y. and Black, L. M.** Improvements in establishing and growing leafhopper cell cultures. *Proc. Am. Phytopathol. Soc.* 3, 234 (Abstract), 1976.
26. **Martinez-Lopez, G. and Black, L. M.** Development of a new medium for the culture of Agallian leafhopper cells. *In Vitro* 13, 777, 1977.
27. **Martinez-Lopez, G.** Development of a Medium for the Culture of *Agallian* Leafhopper Cells. Ph.D. Thesis, University of Illinois, Urbana, IL, 1973.

28. **Hsu, H. T. and Black, L. M.** Preservation of viable leafhopper vector cell cultures under liquid nitrogen. *Phytopathology* 59, 1032 (Abstract), 1969.
29. **Chiu, R. J. and Black, L. M.** Assay of wound tumor virus by the fluorescent cell counting technique. *Virology* 37, 667, 1969.
30. **Kimura, I. and Black, L. M.** Some factors affecting infectivity assay of wound-tumor virus on cell monolayers from an insect vector. *Virology* 46, 266, 1971.
31. **Kimura, I.** A study of rice dwarf virus in vector cell monolayers by fluorescent antibody focus counting. *J. Gen. Virology* 67, 2119, 1986.
32. **Omura, T., Kimura, I., Tsuchizaki, T., and Saito, Y.** Infection by rice gall dwarf virus of cultured monolayers of leafhopper cells. *J. Gen. Virol.* 69, 429, 1988.
33. **Omura, T., Ishikawa, K., Hirano, H., Ugaki, M., Minobe, Y., Tsuchizaki, T., and Kato, H.** The outer capsid protein of rice dwarf virus is encoded by genome segment S8. *J. Gen. Virol.* 70, 2759, 1989.
34. **Noda, H., Ishikawa, K., Hibino, H., Kato, H., and Omura, T.** Nucleotide sequences of genome segment S8, encoding a capsid protein, and S10, encoding a 36K protein, of rice gall dwarf virus. *J. Gen. Virol.* 72, 2837, 1991.
35. **Xu, Z., Anzola, J. V., and Nuss, D. L.** The 3′-terminal sequence of a wound tumor virus transcript can influence conformational and functional properties associated with 5′-terminus. *Virology* 170, 511, 1989.
36. **Spendlove, R. S.** Microscopic techniques. *Meth. Virol.* 3, 475, 1967.
37. **Chiu, R. J. and Black, L. M.** Monolayer cultures of insect cell lines and their inoculation with a plant virus. *Nature (London)* 215, 1076, 1967.
38. **Martinez-Lopez, G. and Black, L. M.** A simple and accurate method for the measurement of the growth of cell monolayers. *In Vitro* 9, 1, 1973.
39. **Martinez-Lopez, G. and Black, L. M.** Measurement of the growth of cell monolayers in situ. *Methods Cell Biol.* 15, 407, 1977.
40. **Goedigk, K., Adam, G., and Mundry, K. W.** The spike protein of potato yellow dwarf virus and its functional role in the infection of insect vector cells. *J. Gen. Virol.* 67, 2763, 1986.
41. **Omura, T., Morinaka, T., Inoue, H., and Saito, Y.** Purification and some properties of rice gall dwarf virus, a new phytoreovirus. *Phytopathology* 72, 1246, 1982.
42. **Hsu, H. T. and Black, L. M.** Comparative efficiencies of assays of a plant virus by lesions on leaves and on vector cell monolayers. *Virology* 52, 284, 1973.
43. **Nuss, D. L. and Peterson, A. J.** Expression of wound tumor virus gene products in vivo and in vitro. *J. Virol.* 34, 532, 1980.
44. **Peterson, A. J. and Nuss, D. L.** Regulation of expression of the wound tumor virus genome in persistently infected vector cells is related to change in translational activity of viral transcripts. *J. Virol.* 59, 195, 1986.
45. **Peterson, A. J. and Nuss, D. L.** Wound tumor virus polypeptide synthesis in productive noncytopathic infection of cultured insect vector cells. *J. Virol.* 56, 620, 1985.
46. **Shikata, E. and Maramorosch, K.** Electron microscopy of wound tumor virus assembly sites in insect vectors and plants. *Virology* 32, 363, 1967.
47. **Kimura, I., Omura, T., and Minobe, Y.** Mechanism of the transmission of rice dwarf virus by *Nephotettix cincticeps* (NC) — adsorption of RDV particles on vector cells. *Ann. Phytopathol. Soc. Jpn.* 50, 430, 1984.
48. **Kimura, I., Minobe, Y., and Omura, T.** Changes in a nucleic acid and protein component of rice dwarf virus particles associated with an increase in symptom severity. *J. Gen. Virol.* 68, 3211, 1987.
49. **Kimura, I.** Loss of vector-transmissibility of an isolate of rice dwarf virus. *Ann. Phytopathol. Soc. Jpn.* 42, 322, 1976.
50. **Omura, T. and Inoue, H.** Rice gall dwarf virus. *CMI/AAB Descriptions Plant Viruses* No. 296, 1985.
51. **Takebe, I. and Otsuki, Y.** Infection of tobacco mesophyll protoplasts by tobacco mosaic virus. *Proc. Natl. Acad. Sci. U.S.A.* 64, 843, 1969.
52. **Takebe, I.** Protoplasts in the study of plant virus replication. *Compr. Virol.* 11, 237, 1977.

Chapter 7

INVERTEBRATE CELLS FOR MAMMALIAN VIRUS RESEARCH

Hans Koblet

TABLE OF CONTENTS

0-8493-7642-4/94/$0.00+$.50

Hans R. Koblet, M.D., born 1928, is a full Professor of Biochemistry, specializing in Molecular Biology, at the Institute of Medical Microbiology of the University of Berne, Switzerland. He received his M.D. degree in 1959. He then worked in Pathology (Berne), in Biochemistry (Berne, Tufts University, Boston), in Molecular Biology (Geneva) and in Virology (Zurich). Since 1974 he has been Professor at the above mentioned institute. 1987, Sabbatical in Japan. Gesellschaft für Virologie; American Soc. for Virology; TCA; Swiss Soc. Biochem., Mol. Biol., Cell Biol. Alpha virus infection of mosquito cells in culture; persistent infection, post-translational modifications of viral proteins, fusogenic effects.

I. SCOPE OF THIS CHAPTER

This review will focus on the molecular biology, biochemistry, and cell biology of arthropod cells in culture and their mammalian viruses. There is no evidence for any differences in the nucleotide (nt) or amino acid sequences (i.e., of the primary products of a virus) in organisms or cells which are far apart phylogenetically. If viruses change their genomes, they do so by intrinsic mechanisms, and the infected cells or organisms may play the role of the selecting arbiter. However, coexistence of viruses and cells leads to an unstable dynamic equilibrium with mutual influences. Products of the viruses may be modified by the biochemical characteristics of the infected cell type; the cell may provide factors favoring or inhibiting replication; the viruses may damage the cells in a more or less specific way, and mutant viruses may appear which survive in one host but not in another. Therefore, a few remarks must be made not only on the immediate fate of a given virus and its host cell, but also on the evolution of viruses. According to the definition of the World Health Organization, arboviruses (*ar*thropod *bo*rne viruses) are defined as viruses which are maintained in nature through biological transmissions between susceptible vertebrate hosts by hematophagous arthropods. They multiply and produce viremia in the vertebrates, multiply in the tissues of arthropods, and are passed on to new vertebrates by the bites of arthropods after a period of extrinsic incubation. The definition is appropriate for the daily diagnostic and epidemiological work with viral (v) species, but it is not concordant with the host ranges of taxonomic families and/or genera.

Therefore, a mere description of arboviruses according to the definition would be insufficient. This would not do justice to the paramount role of arthropods in epidemiology and virus evolution. The relation of arthropods to virus taxonomy is complicated: in some families, most or all members are arthropod-borne and are transmitted by these hosts; in other families, certain genera or species only behave correspondingly. Thus, a brief background describing taxonomy, the vectors, the cells, the viral relationships, and the role of the arthropods in this biological network will be given. Most arthropod-borne viruses infecting vertebrates have been studied in cell cultures of mammalian origin. Reviewing only the data from invertebrate cell culture would not give a coherent picture. Therefore, each part of this chapter presenting the interaction of a given virus family or genus with arthropod cells will start with a brief presentation of the basic data available. Taxonomic information was taken from the most recent "Classification and Nomenclature of viruses".[1] Much information can be gained from *Virology*.[2]

Arthropod (insect) viruses proper and recombinant applications of Baculoviruses will only be considered marginally.

II. THE ROLE OF ARTHROPODS IN VIROLOGY

There are three lines of arguments demonstrating a role of arthropods in virology which seems to be more than fortuitous: (1) the role of arthropods in epidemiology of humans, animals, and plants (2) the large number of viral families with members which either infect arthropods, or which infect arthropods and are transmitted to animals or plants (Table 1), and (3) the fact that viral superfamilies can be created, the families of which have related genomes, whereby some families or genera are transmitted by arthropods or infect arthropods and other families do not. One could claim that the host range of the superfamilies has to be considered rather than the host range of the families, genera, and species in order to assess the host range potential. In most (all?) superfamilies, invertebrates play their role as hosts and/or vectors. One could argue that arthropods were the mixing vessels and mailing stations of viruses during evolution and that in view of the rapid evolution some families, genera, or species lost their capacity to infect arthropods after having been firmly established in "new" hosts such as mammals. This notion is not new.[3] Arboviruses have a strong tendency to establish chronic

TABLE 1
Host Ranges of Viral Families/Genera[a]

Basic criteria	Families/genera	Vertebrates (V)	Invertebrates (I)	Plants (P)	Remarks
ds DNA	Pox	x	x		
envelope	Herpes	x			
	Hepadna (Pararetro)	x			
	Baculo		x		
	Polydna (multiple)		x		One particle
ds DNA	Adeno	x			
no envelope	Irido	x	x		
	Papova	x			
	Caulimo (Pararetro)		(x)	x	
ss DNA	Parvo	x	x		
no envelope	Gemini (1–2)		x	x	Separate packaging
ds RNA	Reo (10–12)				VIP
no envelope					1 particle
	Orthoreo (10)	x			
	Rota (11)	x			
	Aquareo (11)	x	x		
	Orbi (10)	x	x		Arbovirus
	Colti (12)	x	x		
	Cypo (10)		x		
	Phyto and Fiji (12/10)		x	x	
	Birna (2)	x	x		1 particle
	Crypto (2)			x	Intracellular
ss RNA	Corona (1)	x			
Envelope, + sense	Flavi (1)				
	Pesti	x			
	Hepatitis C	x			
	Flavi	x	x		Arbovirus
	Cell fusing agent (CFA)		x		
	Toga (1)				
	Alpha	x	x		Arbovirus
	Rubi	x			
	Retro (2)	x			1 particle
No envelope, + sense	Calici (1)	x			
	Picorna (1)	x	x		
	Tetra (1–2)		x		
	Noda (2)	x	x	x	VIP 1 particle
	ca. 27 groups of plant viruses (1,2,3)		x	x	Separate packaging
Envelope, – sense	Filo (1)	x			
	Paramyxo (1)	x			
	Rhabdo (1)				VIP, Arbovirus?
	Vesiculo	x	x		
	Lyssa	x	x		
	Plant Rhabdo		x	x	
	Arena (2)	x			1 particle

TABLE 1 (Continued)
Host Ranges of Viral Families/Genera[a]

Basic criteria	Families/genera	Vertebrates (V)	Invertebrates (I)	Plants (P)	Remarks
	Bunya (3)				VIP, 1 particle
	Bunyamwera	x	x		Arbovirus
	Phlebo	x	x		Arbovirus
	Nairo	x	x		Arbovirus
	Hanta	x			
	Tospo		x	x	
	Orthomyxo				
	A (8)	x			
	B (8)	x			
	C (7)	x			
	D (6–7)	x	x		
No envelope, – sense	Tenui (4)		x	x	Separate packaging

[a] The terms "viridae" for families and "virus" for groups or genera are not given. Numbers indicate numbers of genomic segments; +, positive polarity; –, negative polarity. The host ranges cannot be taken as host ranges of species. Separate packaging indicates individual packaging of genomic segments; 1 particle indicates copackaging.

(persistent) infections in arthropods, and the persistent (circulative and propagative), semi-persistent (forgut-borne), or nonpersistent (stylet-borne) transmission pathways to plants are well known.

A. THE ROLE OF ARTHROPODS IN EPIDEMIOLOGY AND EVOLUTION IN THE WILD

It makes no sense to recapitulate the abundant knowledge on the role of arthropods in human, veterinarian, and agricultural epidemiology. These fields have been covered.[1,2,3a,4-11] It becomes quite clear now that there are specialized and intricate relationships between a given virus (species), its "vectors", and the "hosts" in a dynamic and evolving ecological niche. These relationships have been discussed in many papers.[1,2,7,11,12-34]

One aspect should be mentioned here. It is the notion of "new" diseases and the emergence of "new" viruses. It is intriguing that in recent times, several, sometimes spectacular, invasions of viruses into new niches have taken place, provoking the uneasy feeling that the worldwide spread of viruses has accelerated since the 1960s. RNA viruses represent the overwhelming majority of these invaders. A large proportion of recent outbreaks is due to viruses with segmented genomes (influenza, reo-, bunya-, and arenaviruses). Most of these so-called new or emerging RNA viruses are zoonotic and are transmitted by vectors and/or mechanically. However, there are no new viruses in the sense that the nt sequence of the genome has been recently and freshly composed; there are only evolving viruses. The recent outbreaks can be explained by viral traffic, and (man-made) disturbances of original ecological niches, sometimes combined with viral evolution. Vectors can travel themselves. The number of papers dealing with this alarming field is rapidly increasing.[35-45] As arthropod borne viruses lead mostly to a persistent infection in arthropods *in vivo* and *in vitro,* there must be a vast potential for viral evolution. Therefore, the position of arthropods in the taxonomy of viruses reflects a transient contemporary picture of viral evolution.

B. THE POSITION OF ARTHROPODS IN THE TAXONOMY OF VIRUSES

The taxonomy of viruses is mainly still based on classical criteria, such as the symmetry (icosahedral, bacilliform, rod-like, helical, complex), the lipid bilayer envelope surrounding the inner parts of the virion (the genome containing nucleocapsid), and the type of the genomic nucleic acid (double-stranded (ds) DNA, single-stranded (ss) DNA, ds RNA, and ss RNA). In addition, viruses can be grouped according to the host kingdoms they parasitize. The most recent classification scheme lists some 2430 viruses belonging to 73 families or groups.[1] There are no ss DNA viruses with an envelope, and ds RNA viruses with an envelope occur in bacteria only. All other combinations of nucleic acid type and envelope/no envelope can be found in vertebrates (V) and in invertebrates (I). The genomic ss RNA can either be positive-stranded (+) (acting as a messenger RNA) or negative-stranded (–) (no messenger function). The positive-stranded viruses without an envelope represent the overwhelming majority of all v families, reflecting the many families or groups which infect plants (P). In many cases, the plant viruses are transmitted by arthropods. There are families with segmented genomes. In the case of the parvoviruses (ss DNA, no envelope) infecting V and I, the bipartite geminiviruses, related to the monopartite geminiviruses, may be the correlate in plants. The ds RNA bisegmented birnaviruses have a host range encompassing V and I. The correlate in plants may be the ds RNA bisegmented cryptoviruses (Table 1).

The Reoviridae have 10 to 12 segments of linear ds RNA and a host spectrum ranging from V and I to P. This is a typical case demonstrating the *relativity of the host range.* The genera Orthoreovirus and Rotavirus are specialized for V, and its members are not transmitted by arthropods. They are considered as true vertebrate viruses. The genera Orbivirus and Coltivirus infect mammals and I; they are transmitted by arthropod vectors and are considered to be true arboviruses. The genus Cypovirus is considered to contain true arthropod viruses. The genera Phyto- and Fijivirus represent plant viruses which are transmitted in a persistent propagative manner by arthropods. However, it would be unreasonable to elevate these genera to the status of families merely on the basis of different host ranges and transmission pathways; the relationships (genome type and replication strategy) are too strong and do not allow such a separation. The bunyaviruses (enveloped, tripartite negative, or ambisense ss RNA genome) are considered to be true arboviruses as far as they circle between vertebrates and arthropods. However, there is again a genus, the Hantavirus, which is not transmitted by arthropods; the genus Tospovirus consists of plant viruses several of which are transmitted by thrips.

Not all families of the negative-ss monopartite or multipartite genome viruses infect arthropods. On the other hand, comparisons of the gene orders and the replication strategies show definite relationships, so that a superfamily of minus-stranded viruses can be created. This shows the relativity of another taxonomic criterion, *the number of genome segments.*

There is an even more complicated situation in the case of the positive-ss RNA viruses; some families are confined to vertebrates, some are true arboviruses (the alphatoga- and the flaviviruses). But the genus Rubitogavirus (Rubella) is not transmitted by arthropods. All plus-ss RNA viruses of animals are monopartite with the exception of the bipartite Nodaviridae which are considered to be insect viruses; however, many plus-ss RNA plant viruses are multi (bi- and tri-) partite. Most virions of plus-ss RNA families of animals have an envelope, whereas such viruses of plants never possess an envelope. Yet all the plus-sense viruses are more or less distantly related and can be collected into superfamilies irrespective of the host range, number of genome segments, and envelope. This shows the relativity of a third taxonomic criterion, *the presence or the lack of an envelope.*

In effect, several classical taxonomic criteria are useful for diagnostic purposes but irrelevant with respect to evolutionary and host range considerations. The configuration of the genome (polarity, gene order, replication strategy, gene functions) is important. Is it not extraordinary that all these sometimes strange setups (ss DNA, ds RNA, ss RNA of negative and positive polarity) are found throughout all kingdoms and that they can be grouped into superfamilies with host ranges always also comprising arthropods? The minor criteria such as host range, virion structure, segmentation of the genomes, individual (plants) or common (animals) encapsidation of the segments, and envelope seem to be a result of the v evolution, the mode of transmission between organisms and within an organism (plants!) and the organization of the host cells (animal plasma membranes [PM] vs. plant cell walls). These aspects have been reviewed.[46,47]

Those readers interested in invertebrate viruses (Baculoviridae, Polydnaviridae, Entomopoxvirinae, invertebrate iridoviruses, Nodaviridae, Nudaureliaviridae (Tetraviridae), insect picornaviruses, Birnaviridae, parvoviruses of invertebrates) are referred to Adams and Bonami, and Kurstak.[48,49]

C. VIRAL EVOLUTION AND THE ROLE OF ARTHROPODS

The discussion above seems to show a systematic involvement of invertebrates in the host ranges of viral superfamilies. The moving and piercing arthropods might well be the places of transshipment and the mailing stations for viruses during their evolution. However, a mere glance at Table 1, showing the host ranges of the viral families, might be misleading, since it might cover up a more or less accidental, nonsystematic role of the arthropods.

In principle, viral nt sequences, gene orders, and segment composition (of a partite genome) change by inherent stochastic mechanisms. Whereas DNA and DNA viruses seem to be quite stable as a result of repair and editing systems, the evidence points to a substitution rate of RNA viruses that exceeds that of the mammalian host by a factor of six orders of magnitude.[39,50-54] Due to the notorious infidelity of viral RNA transcriptases/replicases based on the lack of a 3′ → 5′ exonucleolytic correction activity, the substitution frequency at a single defined site approximates 10^{-3} to 10^{-4} *in vivo* and *in vitro*.[53,55]

The important point is that single substitutions at critical positions may lead to drastic phenotype changes, such as extended or reduced host ranges, altered cell tropisms, increased or attenuated virulence, escape from immune selection, thermosensitivity (ts) (growth in culture at below the normal temperature only), and small plaque size (sp). This heterogeneity has been designated with various conceptional terms.[54] One is "quasispecies", describing a distribution of substituted RNAs around a master sequence. This type of evolution (genetic drift) might be considered as a rapid, short-term evolution leading to diversity within families, genera, or groups. It is clear that arthropods living with changing temperatures and allowing persistent infections are breeding fields for this type of evolution. There are also transient changes and synergistic viral interactions, e.g., protein mixing in progeny virions without exchange of genetic material after coinfection with heterologous viruses (phenotypic coat mixing, mixed type particles). This phenomenon can temporarily expand the tissue tropism.

Virtually irreversible changes can be introduced by reassortment and (RNA-RNA) recombination. These changes might be considered as abrupt but lasting long-term evolutionary changes.[51,52,54] Reassortment (pseudorecombination) takes place when two strains of related viruses with segmented genomes coinfect one cell. Progeny virions may then contain a complete genomic set with genes of either parents. The mechanism is common in Influenza A viruses (genetic shift); however, reassortment does occur in all (animal) viral families with segmented genomes. It has been described in the case of Reoviridae,[56-60] Bunyaviridae,[61-65] Arenaviridae, and has been shown in plant viruses with segmented genomes.[66] RNA-RNA recombination represents another long-term evolutionary event.[67-69] It has been described for picornaviruses, coronaviruses, alphatogaviruses, and even plant tricornaviruses. With respect to taxonomy, RNA-RNA recombination has led to new species within a genus. However, there is also a convincing example of an RNA-RNA rearrangement creating a new genus. It concerns the only representative of the Rubitogaviruses, Rubella.[70] Thus, point substitutions, recombination, and reassortment may combine in divergence and gradually evolve into new genera and even viral families.

There is no evidence of overall unequal substitution rates between the various RNA virus families. On the other hand, there are families which evolve slowly. Such discrepancies between substitution and evolution rates can most probably be explained by low multiplicity transmission and multiple selective pressures.[53] This seems to be the case for the monopartite flavi- and alphaviruses, always transmitted in tiny volumes of tick and mosquito saliva. In addition, these viruses have a multiple host specificity which exerts additional selection pressures. Alphaviruses kept continuously in the same cell type in culture rapidly develop mutant viruses (see below). These problems with respect to the New World alphaviruses have been discussed recently.[71]

The inherent noisiness of all RNA-dependent RNA and DNA polymerases may endanger the stability and replicability of too long an RNA. RNA genomes are often divided into several pieces. In the case of ss positive-sense RNA viruses, this seems to be correlated with the biology of the host and the mode of transmission. Thus many ss positive-sense RNA viruses of plants have segments which are individually packaged; in contrast, the ss positive-sense RNA viruses of animals are monopartite. The (only) exception, the Nodaviridae, package their two segments into one particle. However, these viruses exhibit relationships and can be linked into supergroups. It is therefore most interesting that a bipartite alphatogavirus could be produced in the laboratory[72] which corroborates the notion that viral RNA genomes can divide into modules while conserving replication and translation strategies and functions.[51]

The assumption of a modular evolution lends credence to the concept of superfamilies or supergroups, collections of families related by similar genome organizations, translation strategies, and sequence motifs mainly in mandatory parts of transcription/replication enzymes. This concept of superfamilies is obviously tantamount to declaring that the taxonomic criteria segmentation, encapsidation, envelope, virion structure, and host range of species and genera are of relative importance in evolution.[46,47] It places emphasis on the role of arthropods in viral evolution. More information on the Picorna-like (Poty-like) supergroup comprising ss positive-sense RNA viruses such as calici- and picornaviruses and several plant virus groups has been given by King et al.;[73] on the alpha- or sindbis-like (Tobamo-like) and the flavivirus-like supergroups encompassing ss-positive RNA viruses such as flavi-, pesti-, Hepatitis C, and plant viruses by Goldbach et al.;[74] on the relations to ds RNA viruses by Bruenn;[75] Pereira;[76] on the minus-ss RNA virus supergroup encompassing the monopartite Paramyxoviridae, the Rhabdoviridae, and the Filoviridae and the multipartite Orthomyxoviridae, Bunyaviridae, Arenaviridae, and Tenuiviruses of plants by Peters.[77] The retroviruses which are able to transcribe their RNA genome into DNA by reverse transcription have their correlates in form of hepadnaviruses (Hepatitis B), the caulimoviruses of plants, and widely distributed transposable elements. The most retrovirus-like transposon is typified by the Gypsy element of *Drosophila,* close relatives of which are also found in plants and fungi.[78] The whole field has been recently reviewed again.[79,80]

III. BIOCHEMISTRY OF ARTHROPODS, CELL CULTURE, AND MEDIA

Differences between invertebrate and vertebrate viral cycles can be found wherever the cell provides original contributions such as host factors for transcription/replication/translation, post-translational modifications, compartmentalization of v products and transport functions, maturation sites, and regulatory functions. Concentrations of critical metabolites and building blocks may be different. Last but not least, the reaction of the cells to the trauma of a viral infection may vary. Therefore, attention is drawn to literature covering some aspects of fundamental biochemistry. The general biochemistry of protein glycosylation[81-83] and of protein acylation[84-86] have been reviewed. More specifically, sialic acid,[87] insect glycolipids,[88] fatty acids,[89-92] lipids and lipoproteins,[93] sterols,[94,95] invertebrate lectins,[96] and glycoprotein biosynthesis[97] have been treated. The *Drosophila* cytoskeleton had a chapter of its own.[98] The biology of mosquitoes was treated in a recent book,[99] and the genetics[100] and biochemistry of mosquito cells in culture[101] have been reviewed. Some notes on post-translational modifications as far as they have general validity are given here.

Glycosylation of secreted or membrane proteins, whether host- or virus-coded, is a cotranslational event catalyzed by the host cell. Depending on the host cell, chemically different virions may arise even if the genome is identical.

Briefly, the events can be summarized as follows. A dolichyl-P-P-glycan is formed on the cytoplasmic side of the rough endoplasmic reticulum (RER) and transferred to the luminal side.[83] This step can be inhibited by tunicamycin. In principle, the glycan consists of Glc_3 Man_9 $(GlcNAc)_2$ in higher and lower eukaryotic cells. This glycan is covalently bound by block transfer to asparagine residues ("N-glycans") in the amino acid sequence context -asn-X-ser/thr, as soon as the acceptor site appears on the luminal side during biosynthetic translocation of the protein through the RER membrane. The trimming of the original glycan to a high mannose type is started in the RER by glucosidases I and II, whereby the terminal glucoses are removed; then some mannose residues are cleaved by ER mannosidase. Golgi-α-mannosidases I and II cleave additional mannose residues. The result is a low mannose-type glycan. After that, Golgi transferases of vertebrates form a complex-type glycan, containing mannose (Man), *N*-acetyl-glucosamine (GlcNAc), galactose (Gal), sialic acid (Sial), and fucose (Fuc).[81,82] It was generally assumed that insects cannot form complex glycans, so that degraded chains remain in a maximal trimming state ($Man_3(GlcNAc)_2$).[102-104]

Complex glycan formation starts by addition of GlcNAc in the medial Golgi cisternae (rabbit liver, LLC-RK cells), whereas Gal and Sial are added in the trans-Golgi cisternae.[105] The endo-β-*N*- acetylglucosaminidases (Endo-D, -F, -H) cleave certain glycan types within the di-*N*-acetyl-chitobiose stem: Endo-F cleaves all the glycans mentioned, Endo-H the high mannose types down to $Man_5(GlcNAc)_2$-N..., and Endo-D unsubstituted low mannose types ($Man_3(GlcNAc)_2$-N...). This pathway represents not only a series of consecutive biochemical reactions but also a series of intracellular compartments through which secreted and house-keeping, such as PM, membrane proteins have to pass in order to reach their destinations in the cell (exocytotic pathway). With the help of the endoglycosidases, the position of a transported protein can be analyzed. In addition, the sugar composition of the glycan chains can be manipulated by means of trimming inhibitors.[106,107]

Important readings on arthropod cell culture per se are found in References 108 to 113. More specialized literature concerning individual cell lines mentioned in this work will be provided in Tables 2 and 3.

Several papers and books concerning the growth characteristics of "arboviruses" in various arthropod systems in culture are recommended.[5,9,22,23,113-129]

It may be mentioned here that the *A. albopictus* C6/36 clone supports the growth of a large range of viruses far beyond the arboviruses proper.[130] Reoviruses showed good replication; poliovirus, enterovirus 69, and some coxsackieviruses (all picornaviruses) persisted over several passages; mumps (Paramyxoviridae) and vaccinia (Chordopoxvirinae) viruses also grew in C6/36 cells.[131] This shows the potential of mosquito cells to support viral growth; it also demonstrates how easily such cells could be contaminated in the laboratory or could carry contaminating and hidden viruses at the time of preparation.[132,133]

Finally, literature concerning media is listed.[112,115,127,134-142] Mitsuhashi has compiled the formulas, for example, for the Mitsuhashi-Maramorosch (MM) medium.[143]

IV. ARTHROPOD CELL CULTURE FOR MAMMALIAN VIRUSES

A. ALPHATOGAVIRUSES*

1. Introduction

The Togaviridae represent an RNA virus family of enveloped virions (60 to 70 nm diameter). They contain 1 molecule of positive-sense ss monopartite linear RNA, which is 5′-capped and 3′-polyadenylated. The gene order is 5′-nsP1-nsP2-nsP3-nsP4-C-E_3-E_2-6K-E_1-3′. nsP designates the nonstructural proteins which together form the transcriptase-replicase

* For further readings, see References 2,5,7,9,11,22,23,109,115–117,119–122,125–127,129,144–166.

TABLE 2
Susceptibilities to Viruses of Vertebrate Cell Lines As Mentioned in the Text

Cell line	Origin, full length name	Alphatoga-virus	Flavi-viridae	Bunya-viridae	Reo-viridae	Rhabdo-viridae
A6	Xenopus, kidney[203]	X				
AGMK	African green monkey kidney	X	X	X	X	X
Balb 3T3 A3-1-1	Mouse fibroblast 3T3	X				
BB	Brown bullhead fish trunk[203]	X				
BF-2	Bluegill fish trunk[203]	X				
BHFTE	Bighorn sheep fetal tongue					
BHK(-21)	Baby hamster kidney	X	X	X	X	X
--	Bone marrow, human, immature		X			
--	Brain, mouse, primary	X				
BSC (-1, 40)	African green monkey kidney	X	X	X		X
BSR	A clone of BHK-21					X
C-1008	Vero cells, polarized			X		
CEF	Chick embryo fibroblast, primary	X	X	X	X	X
CER	Chick embryo related fibroblast					X
CHO(K1)(ATCC CCL 61)	Chinese hamster ovary	X			X	X
CHO Lec-1	Chinese hamster ovary, lectin-resistant					X
COS (-1) (ATCC CCL 1650)	SV40 transformed AGMK	X		X		X
CPAE	Bovine pulmonary artery, endothelial				X	
CSL (-503)	Ovine lung, fetal, fibroblast				X	
CV (-1) (CCL 70)	A monkey kidney line	X				
DE	Bejing duck embryo, primary		X			X
DRG	Dorsal root ganglia, rat embryonic					X
--	Endothelial cells, umbilical or brain, mouse[711]	X	X			
EPC	Carp, epithelioma papulosum Cyprini					X
GL	Gekko lung					X
H 35, H 4S	Rat hepatoma (Reuben) and a clone thereof					X
HEL	Human embryonic lung		X	X		X
HeLa (S3)	Human cervical epitheloid carcinoma	X	X	X	X	X
Hep	Human hepatoma	X				
--	Hepatocytes, primary			X		
HEp-2 (ATCC CCL 23)	Human laryngeal epithelial carcinoma	X	X		X	X
HL (-60)			X			
IMR (-32) (ATCC CCL 127)	Human neuroblastoma	X				X
K (-562)	Human leukemia	X	X			
KB			X		X	
L 929 (CCL 1)	Mouse continuous fibroblast	X	X	X	X	X
L 6	Rat myoblast (Jaffe)	X				
L 8, L 8 Cl 3 U	Rat skeletal muscle and a clone thereof					X
--	Rat embryonic muscle, primary					X
L-A 9	Mouse fibroblast			X		
LK	Lamb kidney				X	
LLC-MK2	A rhesus monkey kidney line	X	X	X	X	
LLC-RK1	A rabbit kidney line					
--	Lymphoblastoid cells (EL 4 $H-2^b$, RDM4 $H-2^k$)	X				
--	Human lymphocytes					X

TABLE 2 (Continued)
Susceptibilities to Viruses of Vertebrate Cell Lines As Mentioned in the Text

Cell line	Origin, full length name	Alphatoga-virus	Flavi-viridae	Bunya-viridae	Reo-viridae	Rhabdo-viridae
MA (-104)	Rhesus monkey, fetal kidney continuous		X			
MA (-111)	Rabbit, fetal kidney, continuous		X			
MDBK	Madin Darby bovine kidney	X			X	
MDCK	Madin Darby canine kidney	X	X			
MRC-5	Human fibroblasts		X			
MS	Monkey, stable				X	
--	Mouse neuroblastoma, primary		X			X
NA (-1300) (C-1300)	Mouse neuroblastoma, cloned					X
NBL (-6)	Equine dermis					
Nil (-2)	Hamster fibroblast	X				X
Nil (-8)	Hamster fibroblast	X				X
NRK	normal rat kidney	X				
P 388 D1			X			
PK (E)	Porcine kidney (embryo)	X	X			
PS	Porcine kidney, stable		X	X	X	X
RTG (-2)	Rainbow trout gonad (for Birnaviruses)					X
SK	Sheep kidney			X		
SW (-13)	Human adenocarcinoma	X	X	X		
3T3	A mouse fibroblast line	X				X
TK (-143)	A human line	X				
U	A human amnion line		X			X
U (-937)	A human mononuclear line		X	X		
UT (-1)	A mutant chinese hamster line					X
VERO (ATCC-CCL 81)	African green monkey kidney	X	X	X	X	X
VSW	Viper spleen[203]	X				X
WIT (-38)(CCL 75)	Human lung fibroblasts					X
XL-2	Xenopus laevis line[712,713]	X		X	X	
XTC-2	Xenopus laevis[712]		X			
YAC	A natural killer cell target cell	X				

complex; C is an abbreviation for (nucleo)capsid protein, E for envelope protein. C and the E proteins correspond to the structural proteins. The partial or complete sequences of many species have been elucidated. The references can be found in Table 4. E_1 and E_2 form heterodimers, which are clustered into heterohexamers $((E_1\ E_2)_3)$. In some species, E_3 is lost; in some species, it is an external spike protein $((E_1E_2E_3)_3)$. Most neutralizing antibodies react to E_2. There are about seven antigenic complexes.

The Togaviridae consist of the genera (1) Alphavirus (formerly Arbovirus group A, prototype Sin; and (2) Rubivirus (prototype and single member, Rubellavirus). The alphaviruses with about 26 members are arthropod-borne and transmitted to a wide range of vertebrates by mosquitoes; Rubivirus has only man as a host. It will not be discussed. There are various reasons why alphaviruses have been intensively studied. They provide the most rewarding part of this chapter, and much can be learned also with respect to the other mammalian viruses growing in invertebrate cells.

The nature of the host cell "receptors" is still controversial. The problem will be discussed below. After binding, virions are taken up by receptor-mediated endocytosis. In the acidic compartment of the endosomes, at the critical pH of about 6, a conformational change of the E_1 protein is triggered in such a way that the virion membrane fuses with the endosomal

TABLE 3
Susceptible Invertebrate Cell Lines As Mentioned in the Text

Cell line	Ref.
1. Invertebrate lines from anopheline and culicine mosquitoes (Diptera)	
Aedes (A.) aegypti (Mos. 20) (ATCC CCL 125)	135, 169, 234, 359, 502
clone C-17 (Mos. 20 A)	503
A. albopictus (Singh, ATCC CCL 126)	169, 227
clones (cloned once) LT-C (4,7,10,19,22)	138
C7–10 (clonal derivative)	138
u4.4	141
C6/36 (cloned twice)	130
A. malayensis (Mos. 60)	137
A. pseudoscutellaris (AP61, Mos. 61)	137
A. triseriatus	323, 585, 586
Anopheles stephensi (Mos. 43)	714
Culex (Cx.) quinquefasciatus	715
Cx. theileri	526
Haemagogus equinus GML-HE12	716
Toxorhynchites amboinensis TRA 171,284	525
TA9	587
2. Invertebrate lines from phlebotomines and culicoides (biting midges) (Diptera)	
Culicoides variipennis CuVa (ABAD 1087)	617
Lutzomyia longipalpis sandfly LL–S	717
Phlebotomus papatasi sandfly PP9	624
3. Invertebrate lines from ticks (hard, ixodid; soft, argasid) (Acarina)	
Boophilus microplus	113, 120
Dermacentor (D.) andersoni	620
D. mitens ANE 58	718
D. parumapertus RML 16	719
D. variabilis RML 19, 20	719
Hyalomma dromedarii	720
Rhipicephalus (R.) appendiculatus (RAE 25)	120, 139, 471, 721
R. sanguineus RML 21, RSE 8	139, 722
4. Invertebrate lines from others	
Antheraea eucalypti moth (Lepidoptera)	687, 723
Drosophila (D.) melanogaster 77 OM3 (Diptera)	663
D. melanogaster Blood III	598

membrane. Thereby the nucleocapsid containing the genomic RNA is translocated to the cytoplasmic surface of the endosomal membrane.[167] The RNA and the C-proteins dissociate by an interaction of the nucleocapsids with ribosomes.[168] The ns proteins are the product of a first round of protein biosynthesis, in which only the ns genes of the incoming genomic RNA are translated. They first transcribe full-length minus-strands. These in turn serve as a template either for production of full-length progeny plus-strands (49S, often called 42S RNA), or for production of a 26S subgenomic mRNA by internal initiation (promoter) at the junction region. This junction region delineates in the full-length minus-strands the complementary ns genes (3′ two thirds) from the structural genes (5′ one third of the minus-strand). The two plus-strands are capped, methylated, and polyadenylated. Thus, four ns proteins perform three RNA initiations, the elongation, the capping, the methylation of the cap, and the proteolytic processing of the ns polyprotein. This type of RNA production leads to double-stranded (ds) intermediary RNA forms, such as replicative intermediates and replicative forms. One of these is the 22S ds RNA equivalent to a full-length plus and minus complex. Minus-strand synthesis is shut down after about 4 h post-infection (hpi) in vertebrates, while plus-sense RNA synthesis continues. The 26S RNA serves as a messenger (m) for the structural proteins. They are formed as a polyprotein. The individual proteins are products of virus-coded cotranslational

TABLE 4
Alphatogavirus: Vectors, Sequences

Genetic cluster	Vector	Sequence information
1. Semliki subgroup		
Semliki Forest virus (SFV; old world)	Mosquitoes	151, 325, 430, 431
Ross River virus (RRV; old world)	*A. aegypti, A. camptorhynchus, A. polynesiensis, A. vigilax; Cx. annulirostris, Mansonia uniformis*	432, 433, 434
O'Nyong-nyong (ONN; old world)	*Anopheles funestus, gambiae*	435, 436
Chikungunya (CHIK; old world)	*Aedes* (sylvatic), *A. albopictus, A. aegypti; Mansonia africana* (Bedbugs?, ticks?)	
Getah/Sagiyama (old world)	*A. vexans nipponii, Cx. tritaeniorhynchus*	
2. Western equine subgroup		
Western equine encephalitis virus (WEEV) (new world)	*Cx. tarsalis*	437, 438
Sindbis virus (Sin; old world)	*Aedes, Cx. pipiens*, and *univittatus Culiseta*	439, 440, 441, 442
Aura (new world)		
Mayaro, Una (new world)	*Haemagogus/Psorophora* (mosquitoes)	
Ockelboo (old world)		443
3. Eastern equine subgroup		
Eastern equine encephalitis virus (EEEV) (new world)	*A. sollicitans; Culiseta melanura*	444, 445
Venezuelan equine encephalitis virus (VEEV; new world)	*A. aegypti, A. taeniorhynchus, Culex (Melanoconion), Mansonia dubitans*	444, 446, 447
4. Other		
Barmah Forest (BF; old world)		437
Highlands virus (new world)		444
Middelburg virus (old world)		432, 437, 441, 442

and host-coded co- and post-translational endoproteolytic cleavage events. Therefore, the gene order of the 26S mRNA could also be written as 5′-C-p62-6K-E_1-3′. Traditionally, the Sin precursor for E_3 and E_2 is named pE2. The first cotranslational cleavage is an autoendoproteolytic event taking place between the C protein and the E_3 part of p62 due to a serine-type proteinase activity present in the carboxy(C-)terminal domain of C. However, free C protein cannot be detected. It seems to bind transiently to ribosomes before entering the nucleocapsids. Only the 49S plus-strands can bind C protein to form nucleocapsids on basis of a packaging signal in the 5′-region. This arrangement does not provide the 26S RNA with a packaging signal. After cleavage of the C protein in the cytoplasmic compartment, the E proteins are translocated through the membranes of the RER. E_3 contains the signal for the precursor p62, E_2 for 6K, 6K for E_1. The cleavages between p62 and 6K, and 6K and E_1 occur cotranslationally by signalase(s). E_3 and E_2 are separated post-translationally, probably by a furin-like host endoproteinase. In addition, the host contributes a series of post-translational modifications such as glycosylation of the envelope proteins, acylations in the transmembrane and cytoplasmic domains of the E proteins, and acetylation of the aminoterminal serine residue of p62. Thus, p62 as well as E_1 represent type I transmembrane proteins, the C termini facing the cytoplasm, the amino(N-) termini being located in the lumen of the ER, and one transmembrane domain connecting the compartments. Soon after biosynthesis, the envelope protein configuration becomes p62-E_1 and (p62-E_1)$_3$ so that an immature spike is preformed in the ER (heterooligomerization). This structure migrates through the Golgi complex to the plasma membrane (PM) together with the small 6K peptide which is integrated into the membrane. The position of the transport signals is unclear. However, the E_1 protein expressed alone cannot migrate out of the ER-Golgi region, and p62 seems to be the vehicle of the whole complex, but the cytoplasmic domain of p62 (E_2) does not contain a localization signal. The E_3 part of p62 is responsible for oligomerization. In vertebrates, p62 is cleaved in the trans-Golgi network (TGN) or in the PM; however, cleavage and viral morphogenesis are not tightly coupled. Uncleaved p62 can be integrated into virions in vertebrate cells. The transport time in vertebrate cells from the end of biosynthesis to the appearance of the spikes in the PM is about 25 min. In vertebrate cells, assembly and budding occur mainly at the PM.

Packaging signals of the genomic RNA interact with the N terminal part of the C protein; and the cytoplasmic domain of E_2 is able to bind (noncovalently) the C-terminal part of the C protein, providing the matrix protein function of negative-stranded viruses and linking the nucleocapsid to the envelope. The morphopoiesis seems to be quite a forceful event as soon as the relative concentrations of the proteins in the maturation compartment are high enough. Small amounts of 6K are integrated into the virions; the bulk seems to stay back in the PM. In the case of some alphaviruses, the E_3 protein is lost into the medium (Sin). Progeny virions are detected within 3 to 4 h post infection (hpi) at 37°C. Extensive cytopathic effects (CPE) and cell death develop in vertebrate cells within 12 hpi.

2. Vectors and Cell Cultures

For reviews see References 4,6,7,11,16,22,30,160

For infection patterns with the life vectors, see, for example, Weaver et al. or Grimstad.[17,71,447,725]

A compilation of vectors and sequence information is presented in Table 4. The contents are not exhaustive.

3. Selected Topics

a. Analysis of the Virions

With the advent of mosquito cell culture and invertebrate media, it became possible to investigate whether a given protein composition influences the lipids of a bilayer.[134-136,169] No significant differences with respect to the lipid classes of the virion and the PM, respectively (SFV, BHK-21), were detected.[170] A similar conclusion was drawn from an analysis of the phospholipids of SFV grown in *A. albopictus* (Singh). However, two thirds of the 16,000 phospholipid molecules present in a virion had different polar groups as compared to their counterparts grown in BHK cells, demonstrating that an appreciable amount of these molecules is not specified by the viral surface proteins but by the composition of the host PM and that virions are not completely self-determined.[171] The differences in lipid composition of the same virus grown in different cells (Sin, SFV in BHK-21, *A. albopictus* [Singh]) were reflected by analysis of the microviscosity of the virion membrane.[172]

Sugar components of the glycan moieties of the alphavirion glycoproteins were also analyzed. GlcN, Man, Gal, Fuc, and Sial were detected in vertebrate cells; however, differences in Sial content were found depending on whether the virions were grown in bird or mammalian cells.[173-176]

An N-glycosidic type of oligosaccharides similar to that of serum glycoproteins (SFV, BHK-21) was proposed.[177,178] Sin grown in *A. albopictus* (Singh) cells lacks Sial, since invertebrate cells in general lack Sial (but see below).[87] However, this absence had no effect on the antigenic properties and the ratios of particle/hemagglutination units and particle/plaque-forming units (PFU) as compared to those of virions grown in chick or BHK cells.[102] It was also shown that switching of the hosts from BHK-21 to *A. albopictus* (Singh) cells led to a loss of Gal and Sial from the virion glycans and of most of the cholesterol from the virion membrane, whereas the composition and the content of SFV proteins were not affected.[179] The trimming inhibitors[106,107] deoxynojirimycin (dNM), deoxymannojirimycin (dMM), and swainsonine prevented the formation of Endo-H resistant structures in C6/36 cells but had little effect on virus formation and on the biological activities of the virus. The intracellular and extracellular particle count and the PFU remained essentially unchanged.[180]

b. Binding to the Cell Surface

The binding structures on the virion are the spikes. Anti-E_2 antibodies can be neutralizing if not protective. The domain E_2 aa 173-220 (Sin) seems to be a major antigenic and attachment domain.[181] Upon binding, a first conformational alteration of the spikes (prior to entry and not triggered by low pH) seems to occur.[164,182] The glycans of the envelope proteins seem to play a minor role in binding, as their chemical composition can be modulated without consequences for infectivity (see below).[180] However, the complementary binding structures on the cell surface are not clearly known.

A large body of conflicting data has been collected, using insect as well as vertebrate cells. One of the problems is the number of cell types in culture which can be infected (Table 2), reflecting the host range. Another problem is the fact that in mosquito cells (C6/36) binding and endocytosis cannot easily be separated; endocytosis is going on even at 4°C.[183,184] On the virion side, even single amino acid changes in the envelope proteins can alter the affinity,[185,186] and, on the cellular side, receptor concentrations may be regulated according to developmental stages or cell culture conditions.[187] Consensus may be possible for the following data and conclusions. Alphaviruses bind specifically (saturation, competition) according to a spatial saturation model with multivalent binding; BHK cells possess at least 10^6 receptors per cell;[188] chicken cells 10^5 for Sin;[189] neuronal cells 1.3×10^6 for a neurovirulent Sin strain, but only 5×10^4 for an avirulent variant;[185,186] mesenteronal epithelial cells of brush border fragments of the susceptible *Cx. tarsalis* $1.8 - 3.5 \times 10^6$ for WEEV. The affinity constant was calculated to be $2.2 \times 10^{11}\ M^{-1}$.

Corresponding cells of the refractory *Cx. pipiens* were not able to bind the virus.[190] Binding is dependent on the ionic strength of the medium (Sin and chicken cells,[191] SFV and Vero cells[184]). However, there is no easy agreement with respect to the nature of the receptors. Binding to various mammalian cells was sensitive to proteolytic treatment, but mosquito cells were insensitive to proteases, phospholipases, and neuraminidase.[192] On the other hand, Sin bound to liposomes composed of phospholipids and cholesterol.[193] Negatively charged phospholipids or polyanions were proposed as receptor components in several systems.[194-196] Other authors implicated carbohydrates such as neuraminic acid (SFV, Vero cells[184]) as binding structures (Sin[197]). Such proposals give rise to some fundamental questions. Are there common components, e.g., in glycoproteins or glycolipids, so that a single cell type possesses more than one protein moiety as receptor? What kind of sugar residues would mimick sterically the neuraminic acid in the case of mosquito cells? Glucuronic acid? What positively charged groups would be carried by the virion? We have clear evidence that Ca^{2+} provided by the virion is needed (SFV, Vero cells[184]). However, the individual viral envelope proteins do not show classical Ca^{2+} binding sites.[198] Alternatively, do highly conserved proteins exist in organisms phylogenetically widely apart? Examples in favor of all these ideas have been published. Histocompatibility antigens for SFV (Reference 199; but see also Reference 200), chicken fetal antigen of an apparent molecular weight ($M_{r_{app}}$) of 48 kDa for Sin,[197] a lymphoblastoid cell line surface protein (90 kDa) for Sin, and a chicken cell protein (63 kDa) precipitated by a putative antireceptor antiserum for Sin, have been proposed as receptors.[201,202] The latter study gives a good example for the possibility of several receptor types: an antibody escape mutant of the virus was taken up despite the presence of the antireceptor antiserum, and this antiserum did not recognize the binding counterpart of BHK cells. On the other hand, the high affinity laminin receptor (67 kDa), a highly conserved protein, has recently been identified as also serving as a receptor for Sin. Antilaminin receptor antibody partially blocked Sin binding to C6/36 cells also, implying that this protein is conserved in mosquitoes.[203] Another example demonstrated that the number of receptor molecules can differ between cell types and can be modulated by the cell culture medium. With an anti-idiotypic antibody approach, neuroblastoma cells were found to express a 110- and a 74-kDa receptor which may be regulated after birth.[187]

Obviously, studies with negative controls are not available. It might be interesting to investigate cell types *not* able to bind alphaviruses, such as *Spodoptera frugiperda* cells or human erythrocytes.

c. Uptake by Endocytosis

Many reviews have appeared dealing with this topic.[204] Alphaviruses are taken up by "adsorptive", receptor-mediated endocytosis in higher eukaryotes and in mosquito cells in culture.[183,205] However, the lysosomotropic weak base chloroquine (0.1 m*M*) which elevates the endosomal pH over the critical level for fusion to occur had no effect on Sin production in *A. albopictus* cells.[206,207] However, this concentration is low and the resistance of mosquito cells (C6/36) towards chloroquine is high; SFV was still produced in presence of 10 m*M* chloroquine.[156] I explain this by assuming that the mosquito cells possess high energy stores and correspondingly strong pump/exchanger actions. Certainly, the fusion from within (FFWI) taking place from 7 hpi onwards at an acidic pH of the medium only is fully compatible with uptake by endocytosis.[159,208,243] Endocytosis of SFV by C6/36 cells was shown by electron microscopic investigations to take place at pH 7 even at the low temperature of 4°C into coated and noncoated pits and vesicles.[183]

An interesting point is the fact that insects are cholesterol auxotrophs obtaining cholesterol from dietary sources.[95,209] In contrast to mammalian cells, insect cells can be partially depleted of sterols by culture in media devoid of low density lipoproteins (LDL).[210] However, it had been shown that the fusion of SFV with liposomes (occurring only at low pH) required cholesterol in the liposomal membrane (about 1 cholesterol molecule per 4 phospholipid molecules); that the 3β-hydroxyl group is a critical part of the cholesterol molecule, and that the purified E_1 ectodomain of the virus required both low pH and 3β-hydroxysterol in order to trigger its acid-induced conformational change.[211-213] In fact, cholesterol-depleted C6/36 cells were able to bind, internalize, and acidify SFV virions, but they were blocked in SFV fusion and subsequent virus replication. However, the depleted cells could be infected by the cholesterol-independent vesicular stomatitis virus which also enters by endocytosis and a low pH-mediated fusion.[214] It may well be that the localization and concentration of cholesterol in mosquitoes in the wild after the blood meal may contribute to the persistent infection with alphaviruses.[214]

d. Electron Microscopy Studies

For reviews see References 129,144,215,216.

A. albopictus (Singh) cells were investigated after infection with SFV.[217] The virus titer approached the maximum 12 hpi; however, electron-dense cytoplasmic inclusions, containing vesicles and virions, and cytoplasmic vacuoles similar to the type-1 cytopathic vacuoles (CPV) as described in vertebrate cells started to appear at 24 hpi.[218,219] In the same cell line, RR established a persistent noncytopathic infection with no significant change of the cell division rate. The virions matured within large, electron-dense cytoplasmic inclusions and also at the PM. CPVs as seen in the case of SFV were absent. At the time of the transition to a persistent infection, 40 to 60 hpi, the cytoplasmic inclusions were transformed into vacuoles, and virus disappeared from these structures.[220] In another study, using Sin and *A. albopictus* cells, nucleocapsids and virions could be detected in complex vesicular structures which sometimes fused with the PM releasing the mature virions. Complementary findings demonstrated that free nucleocapsids were rarely seen in the cytoplasm, and budding of virions from the cell surface was very infrequent.[221] There was no CPE and no noticeable reduction in the rate of the cell growth. It is noteworthy that there were multinucleocapsid containing enveloped structures in the medium also; however, it is still unclear as to how they are formed, since budding into the interior of these structures was not observed. A remark of caution with respect to CPE is called for: clear statements on cytopathogenic and noncytopathogenic infections can only be made with regard to cloned cells. Dual sites of virus maturation both within the cell and at the PM of mosquito cells will often be encountered in high producer clones and may lead to CPE. In three subcloned *A. albopictus* lines infected with Sin, clear differences were visible: C7-10 cells, highest titers, severest CPE, budding particles at the cell surface and at membranes of internal vesicles and exocytosis, electron-dense inclusions with nucleocapsids; C6/36, similar to C7-10 cells; line u 4.4, no CPE and no obvious morphological changes of the cells, no budding at the cell surface. This study also provides evidence that clones mostly represent different tissues in the larval or adult insect.[222] In the case of Mayaro virus, another alphavirus grown in C6/36 cells, the maximum titer was reached after 50 h (3×10^9 PFU/ml). Membrane-coated inclusion bodies and cytoplasmic vesicles containing lamellar structures, nucleocapsids, and mature virions appeared. The vesicles fused with the PM. However, budding into vacuoles was not observed at any time; no enveloped virions were detected in the cytoplasm, and cytoplasmic nucleocapsids were scattered and rare; budding through the PM was rare.[223] In a similar study, alphavirus infected *A. pseudoscutellaris, A. aegypti* and *Anopheles stephensi* cells were investigated. CPV-1 were seen, and 4 hpi, electron-dense bodies developed and reached their maximum after 6 to 8 h; at a later time, these bodies contained nucleocapsids and "myelin-like" structures.[224]

e. Acute Infection of Mosquito Cells, Time Course, and Cytopathic Effects

As compared to vertebrate cells, the period of acute infection is prolonged in mosquito cells. Whereas vertebrate cells are invariably killed by wild-type (wt) alphaviruses, mosquito cells can develop a state of everlasting persistent infection. In the case of vertebrate cells in culture, CPE precede cell death; they can be defined as the morphological changes of cells and cell cultures, such as rounding or starlike appearance of the cells, detachment of the cells from the culture vessel, aggregation of the cells, and holes in the monolayer. In CEF infected with wt Sin, newly synthesized v envelope proteins were detected at the cell surface after 2 h of infection and virions were released after 3 h.[225] CPE followed and cells were killed after 10 to 14 h. Shutoff of host protein synthesis developed between 2 to 4 hpi.[226] Earlier studies with mixed type mosquito cell cultures started immediately after the establishment of the cell lines and mainly concerned acute infection by arboviruses.[135,169,227-233] It was soon recognized that multiplication of the viruses depended upon the cell lines, that a CPE could not be detected in a regular fashion, and that there was evidence for a chronic infection.[135,228,234-238]

The picture becomes clearer when cloned cells are analyzed. LT-C-7 cells exhibited extensive CPE after infection with Sin, EEEV, and vesicular stomatitis virus (VSV). However, the virus yield as compared to a CPE-resistant clone was not significantly different. CPE and v RNA synthesis were enhanced at 34°C relative to 28°C in contrast to the resistant clone.[138] Tooker and Kennedy[239] made 115 clones of *A. albopictus* (Singh) cells and infected them with SFV. About 70% of the clones gave low yields (8×10^6 to 2×10^8 PFU/ml) of virus without CPE; 30% gave high yields (10^9 to 10^{10} PFU/ml) and suffered a CPE. In the low producer lines, the negative-strand RNA synthesis and all subsequent virus-specific events were markedly reduced. When CPE-positive and CPE-negative clones of *A. albopictus* cells were fused, the CPE-positive phenotype always proved to be dominant. The Sin-provoked CPE was associated with a high level of v RNA synthesis and a rapid production of virions during the early periods of infection.[240] Similar findings apply for the high-producer *A. albopictus* clone C6/36. Mild to extensive CPEs were visible after infection with the alphaviruses SFV, Sin, and CHIK and the flavivirus Dengue (types 1 to 4). All cells were infected as indicated by the immunofluorescent assay. Cell surface projections contacting neighboring cells were intensively stained with fluorescent antibodies reflecting viral budding from the PM.[130] In fact, SFV, WEEV, and Sin are formed within the C6/36 cells *and* at their PM.[156,222,241,242] In C6/36 cells infected with SFV (MOI:10), the appearance of viral products and the CPE were clearly correlated. At 3 hpi the 49S and the 26S mRNAs were detectable; intracellular viral proteins were visible at 6 h and prominent in fluorograms at 8 hpi. At 8 to 10 hpi, the CPE (starlike appearance of the cells contacting neighbors) started to be visible. At 16 hpi, it was complete and host protein synthesis was shut off. The data demonstrated a nonsynchronized expression of v products at the PM between 7 to 16 hpi in a cloned cell line.[148,156,159,241,243]

The absence of CPEs explains that uncloned mosquito cells are not very useful for plaque tests. Of nine alphatogaviruses, only one (CHIK) formed plaques in *A. albopictus* (Singh) cells.[244]

f. Cytopathic Effects and Cell-Cell Fusion (FFWI)

I distinguish the CPE from FFWI quite strictly. (1) CPE is a sign of expression of v envelope proteins in the PM and reflects the tendency of v spikes to bind to neighboring PM. The infected PM mimicks the virion. Since there is intracellular alphavirus formation in mosquito cells, concomitant transport of envelope proteins into the PM and budding of virions at the PM may be the hallmark of high producer clones. High virus production seems to be correlated with shutoff of host protein synthesis, and this in turn may increase the damage at the PM as described in vertebrate cells.[245-247] Obviously, the original uncloned cell lines are inhomogeneous, and the coexistence of low producer and high producer cells in one culture may favor the transgression into a persistently infected state by selection.[222,239] In view of the data of Tooker and Kennedy[239] demonstrating a bimodal distribution of the virus yields of the clones, it may be speculated that the different clones possess different concentrations of (a) host component(s) needed for cooperation with the v RNA polymerase (see n. and o. of this section).[248] The high producer clone C6/36 does not only restrict virus production and enter a persistent infection; a high percentage of the cells, irrespective of the growth stage, perished during acute infection with SFV. Surviving cells became persistently infected. Thus, for unknown reasons (defective, interfering RNA? see below) the cloned cells may become inhomogeneous during infection, the majority being high producers, showing CPE and perishing, the minority restricting the production, dividing and surviving and becoming permanently infected.[249] Coevolution phenomena of cells and virus might be possible. Evidence for such developments has been provided with other systems.[250] (2) Fusion from within (cell-cell fusion, FFWI) is also explained by the presence of v spikes in the PM. However, it is an artificial phenomenon triggered by intentionally lowering the pH of the medium to 6, thus enforcing a conformational change of the E_1 protein needed for fusion to occur.[156,159,208,241,251] Generally, this phenomenon can only take place in a quasi-invisible manner when virions are taken up into the acidic endosomal compartment. Some cell lines in culture lower the pH of the medium by metabolic activity and allow FFWI to occur spontaneously.[148] It is probable that such events have been taken to be a CPE. The biochemistry of fusion (of C6/36 cells) after infection with alphaviruses has been treated in depth and will not be reviewed here.[156,159,164,165,204,208,252-254]

g. RNA Biosynthesis

Extraction of RNA from Sin infected *A. albopictus* (Singh) cells yielded the mature v 49S RNA, the 26S RNA, a 32S RNA, and a ribonuclease-resistant 20 to 22S replicative form RNA.[255] In the course of such studies it was found that the larger ribosomal RNA of mosquitoes — the 28S rRNA — is inherently unstable at higher temperatures (due to an inbuilt nick),[256] a finding later confirmed in *A. aegypti* cells.[257] Polysome-associated v RNA in SFV-infected *A. albopictus* (Singh) cells consisted of the four ss species "42"S, 38S, 33S, and 26S. Nucleocapsids of 140S containing the genomic type RNA were also detected, and protein was rapidly integrated into nucleocapsids. This part of the biosynthetic machinery is very similar to that of infected vertebrate cells.[258-260] It is clear now that there are only two plus-stranded alphavirus RNAs with respect to sequence: the genomic 49S and the subgenomic 26S mRNA. The 38S and 33S RNAs may be conformational variants of the two main species.[261] They are regularly found in variable amounts in mosquito and vertebrate cells. In a cloned CPE-positive cell line of *A. albopictus*, infection with Sin led to a significant decrease of host nuclear RNA synthesis at 12 hpi.[138]

Thus, the multiplication strategy in the different hosts is the same; virus-specified events per se involve the same components and the same ss and ds RNAs occur. This extends even to the defective interfering (DI) RNAs and the (nontranslated) termini.[149,239,262-266]

h. Protein Biosynthesis

For reviews see References 153,159,267.

The statements concerning the quasiconstancy of the v RNA are obviously also correct with respect to the amino acid sequences of the proteins. Thus, information on protein biosynthesis in virus-infected mosquito cells is relevant only with respect to the time course and the reactions of the infected cells. Again, the remarks on cloned vs. uncloned cells apply. For example, mainly the C protein was visible in gels after a latent period of 4 to 6 h in *A. albopictus* (Singh) cells infected with Sin (strain MRM 39); a large host protein background hampered the analysis. There was no switch off of host protein biosynthesis, and even the differential inhibition of host protein synthesis by use of hypertonic media was impossible.[268,269] After infection of the same cell line with RRV or SFV, no shutoff of host protein biosynthesis could be observed. Viral polypeptide synthesis was maximal between 10 and 24 hpi and became undetectable by 72 hpi.[238]

On the other hand, a general shutoff is not rare in high-producer clones. In C6/36 cells at 16 hpi with SFV, essentially only v proteins are formed. As in vertebrate cells infected with alphaviruses, the phenomenon is not sudden, but gradually and regularly augments over hours. We adopt therefore Wengler's view that, as in vertebrate cells, the inhibition of host protein biosynthesis and the increasing biosynthesis of virus-specified proteins can be explained by the substitution of host mRNAs by v mRNAs in the polyribosomes.[226,267] This assumption is corroborated by the observation that infected C6/36 cells at pH 7 can keep their ionic "milieu intérieur" constant.[159] Eventually, migration of excess C protein into nuclei/nucleoli might play a role, for example by reducing host transcription.[270-272] A recombinant C-protein expressed by a baculovirus recombinant in *Spodoptera frugiperda* cells also was karyophilic and associated with nucleolus-like structures.[272a] Transient enhanced synthesis of cellular proteins has also been described. A high producer clone from Singh's line (AI CL 9) was infected with Sin or CHIK. Up to 2×10^{10} PFU/ml were produced after 24 h. After 2 days, the infected cells synthesized a large amount of a nuclear protein (43 kDa) for a period of 2 to 3 days; this proved to represent a host heterogeneous nuclear RNA ribonucleoprotein. Minor proteins (P80, P71, P41, and P26) were also overexpressed.[273] It is unclear whether these reactions are to be explained in terms of a stress response as described in chick or BHK cells.[274] In such cells, despite the inhibition of host protein synthesis, several stress proteins (hsp 89, hsp 73, hsp 31, and hsp 27) were induced by infection with Sin.

At the height of the v production before the onset of the persistent infection (C6/36, SFV, 5×10^9 PFU/ml), 24 to 48 hpi, important regulations take place which are not observed in the original *A. albopictus* (Singh) or the low-producer *A. pseudoscutellaris* (10^6 to 10^7 PFU/ml) lines. A secreted host protein (54 kDa) was no longer secreted, irrespective of whether the C6/36 cells were grown in MM or medium 199; later on, it appeared again in the medium. Another secreted host protein (62 kDa) disappeared and reappeared only if the cells were kept in medium 199. The two proteins are monomeric and nonrelated as assayed by peptide mapping; they are secreted in all the cell lines mentioned.[275] This regulation may be due to the transient high virus burst, to the CPE, and/or to the shutoff.

Sometimes, v proteins of unusual M_r can be detected in SFV infected C6/36 cells. Immunoprecipitation and peptide mapping showed a truncated C protein (26 kDa), which was also found in ribosomal subunits, but not in virions. Similarly, a smaller variant of E_1 could be visualized. These polypeptides are most probably degradation products. However, no systematic investigation in relation to the establishment of the persistently infected state has been performed.[276]

i. ***Post-Translational Modifications: Glycosylation***

In alphaviruses, there are several interesting glycan heterogeneities. (1) The potential glycosylation sites of the different species are only partially conserved. (2) Intersite heterogeneity exists in a given protein with respect to glycan composition. (3) Intrasite heterogeneity may also be detected. For example, Sin, grown in hamster and chicken cells, exhibits the following positions: E_3, 14, E_2, 196 (complex) and 318 (high mannose), E_1, 139 (complex) and 245 (complex and high mannose); SFV: E_3, 13 and 60, E_2, 200 and 262, E_1, 141 only; ONN: E_3, 12, E_2, 263 and 345; E_1, 141 only.[277,278] This list provides examples of all heterogeneities mentioned.

The E_1 glycan of SFV grown in C6/36 is also heterogeneous with a low mannose type (about 80%, Man_3-$(GlcNAc)_2$) and a hybrid type (about 20%, GlcNAc $Man_3(GlcNAc)_2$, Endo-D *and* Endo-H resistant).[279]

Experiments with the glycosylation inhibitor tunicamycin and the trimming inhibitors dNM and *N*-methyl-1-deoxynojirimycin (MdN) (glucosidase inhibitors, resulting in N-glycans of the type Glc_3Man_{7-9} $(GlcNAc)_2$) as well as dMM (inhibitor of the Golgi mannosidase I, resulting in N-glycans of the type Man_{7-9} $(GlcNAc)_2$) and swainsonine (inhibitor of the Golgi mannosidase II, resulting in N-glycans of the type Man_5 $(GlcNAc)_2$) in vertebrate and mosquito cells led to several important conclusions:

1. In vertebrate cells, glycosylation per se is needed for protein transport, precursor cleavage, and virus formation to take place. In the chain of trimming reactions, glucose removal is a critical step. However, unfavorable effects of the trimming inhibitors such as reduction of virus yield and inhibition of pE2 cleavage were reversed at 30°C. The formation of complex glycans is much less critical.[280-282]
2. Virions formed in vertebrate cells and mosquito-grown virions did not exhibit antigenic differences.[102,283]
3. Glycan processing in the exocytotic pathway is determined primarily by the accessibility of the oligosaccharides within the three-dimensional structure of the target protein to the trimming enzymes, so that the glycan type and its complexity at a given site is mainly a function of the protein configuration and largely independent of the host cells;[277,284,285] however, the host cell can play a certain role.[278] This would explain why complex sites (in vertebrate cells) correspond to low mannose sites (in mosquito cells) when complex types cannot be formed due to lack of host enzymes.[104] The glycans of the alphavirus envelope proteins seemed to correspond to the glycans of the proteins of the host cell.[97,103,286]
4. Glycosylation per se is also needed for viral maturation to occur in C6/36 cells infected with SFV. However, the trimming inhibitors dNM, dMM, and swainsonine, which are biochemically active in C6/36 cells, allowed the production of fusogenic infectious virions in normal amounts without any shift in the ratio of particles to infectious units. These results demonstrate that "correct" glycosylation is not a prerequisite for the biological activity of alphaviruses, not even for transport and p62 cleavage in C6/36 cells (at 28°C).[180,279]

Recently, an additional chapter was opened. There are not only intrasite-, intersite-, and cell-dependent heterogeneities, but also time-dependent changes. Some insects seem to be able to transfer fucosyl- and terminal GlcNAc residues.[279,287,288] The α-1,3-fucosylation of the asn-bound GlcNAc residue may create an antigenic determinant since mammalian glycoproteins lack this modification.[289] *Spodoptera frugiperda* (Sf9, Lepidoptera) cells were competent to trim N-glycans of the influenza hemagglutinin expressed by a Baculovirus vector to trimannosyl cores and to further process these by the addition of fucose into the form $Man_3[Fuc](GlcNAc)_2$.[290] A similar finding was reported for a chimeric respiratory syncytial virus (Paramyxoviridae) protein expressed in Sf9 cells.[291] In an intriguing series of papers, it was demonstrated that a recombinant human plasminogen (one glycan chain) expressed in *Spodoptera* had about 40% of a complex bisialobiantennary type identical in structure with that of the human plasma protein;[292] that the cells contain the glycosyltransferase genes required for assembly of N-linked complex oligosaccharides and thought to be missing in insects; that, after the introduction of the baculovector, either the enzymes or the genes were activated, so that the infection altered the normal glycosylation process in a time-dependent manner;[293] and that the activation enhanced an α-D-mannosidase activity that catalyzed the trimming of $Man_6(GlcNAc)_2$ to $Man_5(GlcNAc)_2$, the preferred substrate of GlcNAc-transferase I.[294] Most significantly, *Mamestra brassicae* and *Manduca sexta* (Lepidoptera) cells infected with the vector mentioned above also showed the ability to assemble complex-type glycans, which was brought about by the infection.[295]

j. Post-Translational Modifications: Oligomerization, Acylation, p62 Cleavage, and Protein Transport

An interesting series of other post-translational modifications occurs in the exocytotic transport pathway. Oligomerization, acylation, and p62 cleavage seem to be connected somehow, but the exact interdependencies are unclear. Thus, the three topics are examined together without claiming that they are necessarily linked.

Oligomerization (for reviews see References 296,297) is a prerequisite for spike formation, transport of the proteins from the ER to the maturation site, and virion maturation itself. It takes place very soon after biosynthesis in the RER. If correct folding of the SFV p62 and E_1 proteins in COS or HeLa cells is inhibited by the glycosylation inhibitor tunicamycin, misfolded aggregates and aberrant interchain disulfide bonds will form within less than 1 min after the biosynthesis of the proteins. They aggregate nonspecifically at random as a mixed type, which can be small or comprise several million Da and involve proteins not from single polysome products only. Chaperonines such as BiP (for a review see Reference 298) are noncovalently associated. These aberrant complexes are retained in the ER and most probably reflect their poor solubility.[299] In the case of alphaviruses, the proper sequence of oligomerization is $p62 + E_1 \rightarrow (p62\ E_1) \rightarrow (p62\ E_1)_3$. It occurs in the RER within 5 min after biosynthesis in higher eukaryotes and C6/36 cells, so that an immature spike complex is preformed.[300-308] As indicated in the introduction, p62 is the vehicle for the transport of E_1. The E_3 part of p62 seems to play a major role in oligomerization. The envelope proteins of SFV are arrested in the RER of BHK-21 cells at 39°C if Cys 58 of E_3 is mutated to tyr. However, the block is released at 28°C.[309] Similarly, if E_3 of SFV (BHK-21 cells) is replaced by a cleavable artificial signal sequence, preserving the topology, heterodimerization is abolished, E_1 is retained in the ER, and transport of E_2 is inhibited; however, E_1 can be complemented with wt p62 provided by a separate coding unit.[310] Properly folded and oligomerized complexes will then emigrate and take the exocytotic transport route through the Golgi complex to the PM as described for SFV in BHK-21 cells.[311] This is only partially the case in SFV infected C6/36 cells (see below).[308]

Acylation (for reviews see References 84–86) is a rather puzzling process, because the function of this modification remains unclear. As a rule, palmitate is incorporated as a thioester of cysteine (cleavable with hydroxylamine) in the transmembrane/cytoplasmic domains of the v envelope proteins in vertebrate cells and in C6/36 cells.[312-317] It is clearly a post-translational event taking place 10 to 20 min after blockade of protein biosynthesis (Sin, CEF; SFV, C6/36 cells), so that palmitoylated envelope proteins in mature extracellular virions appeared 10 to 20 min before the [^{35}S]methionine-labeled virions in a double-labeling, pulse-chase experiment (Sin, CEF).[312] In higher eukaryotes, acylation occurs before p62 cleavage. Palmitoyl-CoA represents the donor lipid (SFV, BHK).[316] Palmitate is strongly preferred as compared to other fatty acids.[315] In C6/36 cells, it is either integrated into phospholipids or proteins, but it is metabolized neither in the free nor in the protein-bound state.[317a] In SFV E_1, one residue is bound to the single cys 433 of the transmembrane domain.[318] In the case of p62/E_2, the sites and the number of fatty acid residues incorporated are less clear. In the extracellular virions grown in C6/36 cells, exhibiting an equimolar relation of E_1 and E_2, there is a [^{3}H]palmitate radioactivity ratio of about 2:5, so that we assume that there are 3 residues in E_2. The candidate cysteines of E_2 (SFV) are cys 385 (transmembrane domain) and cys 395, 415, 416 (cytoplasmic domain). At least in certain cases (VSV, Sin in CEF) acylation seems to be needed for virus assembly and budding.[313] It should be added here that the 6K peptide also becomes acylated as shown in vertebrate cells; this may play a role in assembly and budding.[319-322] Several studies clearly showed that in vertebrate cells (CEF, BHK) the primary acceptor proteins of Sin or SFV are pE2 and p62, respectively, and E_1;[312,316] p62 appears to be fully acylated at the same intracellular site as E_1, so that there is a clear precursor-product relationship between p62 and E_2 as is the case in pulse-chase experiments of [^{35}S]methionine-labeled material.

This situation is different in C6/36 cells. Our data (Schärer and Koblet; and Vallan and Koblet, in preparation) are compatible with the following interpretation.[308,317,317a,354] E_1 and p62 form a whole spectrum of complexes of different sizes, in which mainly E_1 is acylated. Parts of these complexes are retained for long times and p62 always remains Endo-H sensitive (see below). When the correct conformation and oligomerization is reached, they feed the transport chain, whereby p62 becomes monoacylated in the transmembrane domain. Soon thereafter, p62 is cleaved and E_2 becomes fully acylated. It has been shown that in C6/36 cells, p62 cleavage occurs early, in a pre-Golgi compartment.[308] Thus a significant pool of acylated p62 cannot be established at any time. The experiments do not prove a strict coupling between cleavage and full acylation of E_2; this may be a topographical coincidence in a pre-Golgi compartment. However, by compacting oligomerization, acylation, and p62 cleavage into an early compartment of the transport pathway, the stage would be set in C6/36 cells for an intracellular virus formation. If all this is true, one might speculate that the underacylated p62 of C6/36 cells is not able to make contact with the nucleocapsids, whereas the fully acylated p62 of vertebrate cells is competent to provide this binding. This would explain the observation (discussed below) that p62 can be integrated into virions in vertebrate cells, while p62 cleavage is mandatory for virus formation in mosquito cells.[323]

p62 (pE2) cleavage: The cleavage of the precursor p62 (SFV) into E_2 and E_3, most probably by host endoproteinases, occurring next to an arginine pair, is an important event in the luminal part of the trans-Golgi network of BHK cells or in the PM.[324] The cleavage is then followed by removal of the C-terminal arg of E_3.[325] The cleavage activates the fusion potency of the E_1 protein and thus infectivity.[251,326,327] It should be noted here that in BHK cells infected with Sin, E_2 was also visible in an Endo-H sensitive form (after immunoprecipitation) at the end of a pulse of 5 min duration only.[328] However, it seems to be clear by now that the event is not a generally mandatory morphopoietic cleavage always coupled to virus formation. There are several reports demonstrating that uncleaved p62 can be integrated into virions grown in vertebrate cells; this integration is not possible in mosquito cells, and an unproved explanation has been given above for this discrepancy. One example concerns a cleavage mutant with a new glycosylation site, which hinders the cleavage. The resulting pE2 was integrated into virions grown in vertebrate cells (BHK) but the virions were attenuated.[329] In *A. triseriatus*, u 4.4, C6/36, and C7-10 cells the mutant pE2 was transported to the cell surface, but 1000x less virus was made and released from C6/36 cells than from BHK cells. Spontaneous revertants restored the ability to grow, and pE2 cleavage did take place.[323] Similarly, moderate doses of the ionophore monensin reduced the titer of Sin in BHK cells and inhibited the cleavage of pE2. The protein migrated to the cell surface and replaced E_2 in virions (75%).[330] Another example concerns mutant CHO-K1 cells resistant to *Pseudomonas* exotoxin A. These cells failed to cleave pE2 of Sin. However, this protein migrated to the PM and was released in noninfectious virions. It contained complex glycans. Trypsin treatment of the particles restored the infectivity.[331] As stated above, p62 cleavage is a pre-Golgi event in C6/36 cells. E_2 already appears after a 5 min pulse with [^{35}S]methionine, and p62 always remains fully sensitive to Endo-H.[308] This may explain why p62 is cleaved in presence of tunicamycin in C6/36 cells. In vertebrate cells, tunicamycin leads to retention of the v oligomers in a pre-Golgi compartment and to inhibition of p62 cleavage.[104,299]

Transport: The transport of the envelope proteins of Sin in BHK-21 cells from the ER to the PM at 37°C took about 30 to 40 min.[332] Similarly, the passage of the SFV proteins in BHK-21 cells required about 45 min with 15 min in the ER and 15 min in the Golgi stacks.[311] In C6/36 cells infected with SFV, the first labeled virions started to appear in the medium after 70 to 90 min.[308] However, at that time nonacylated proteins in aggregates and p62 with high mannose glycans exclusively (about 50%) were still present in the cells.[317,317a] This dispersion in space of a cohort of commonly labeled proteins which heterooligomerize together is unusual. Certainly, different individual protein populations may have widely different transport rates, exhibiting half-times from 14 to 140 min for the ER-Golgi track, while the migration from the Golgi to the medium seems to be equally rapid ($t_{1/2}$ about 15 min).[333]

Since E_1, E_2, and p62 are also transported as complexes in C6/36 cells, E_1 can be used as a positional marker for the transport and processing of E_2 and p62. The oligosaccharide of the E_1 protein was modified first to an Endo-D-sensitive (half-life 35 min) and then to an Endo-H-resistant structure (half-life 55 min), whereas the oligosaccharides of p62 remained sensitive to Endo-H throughout.[308] p62 never reached a state where it became (partially) sensitive to Endo-D. It took at least 2 h of chase for more than 95% of the E_1 population migrating together with E_2 to acquire Endo-H resistance. All this suggests that the cleavage products of p62 are slowly and asynchronously released, whereas the precursor seems to be retained in a pre-Golgi compartment. The prolonged survival of heavy complexes showed that a pool of briefly pulse-labeled envelope proteins became heterogeneous with respect to their oligomerization state. The v envelope proteins seem to heterooligomerize into the correct form in C6/36 cells at different rates, so that they reach the cleavage compartment at different times. A rate-limiting step seems to be built into the transport pathway. We postulate that in C6/36 cells oligomerization, acylation, and cleavage of p62 occur in the compartment, where the rate-limiting step leads to an asynchronous release of the preformed spike. This compartment may be the ER or a transitional compartment. Here, intracellular virus formation might occur.[317,317a]

k. Maturation

For reviews see References 129,144,152,153,159,215.

Maturation of enveloped virions can be defined as the binding of the inner parts of the virion to the transmembrane roots of the spikes, so that the nucleocapsids become coated with a host membrane studded with v envelope proteins. Alphaviruses bud at the PM of vertebrate cells into the outside world. They have a T = 4 envelope surrounding a T = 3 nucleocapsid.[306,307,334,335] Thus 80 spikes consisting of $(E_1E_2E_3)_3$ (SFV) or $(E_1E_2)_3$ (Sindbis) are opposed to 180 molecules of C protein in the nucleocapsid. Recently, this concept of virion organization has been questioned again. Paredes et al.[336,336a] assert that the nucleocapsid also contains 240 copies of the C protein. The question is not trivial. All the structural proteins are products of a polyprotein coded by a polycistronic messenger; they are made in equimolar amounts. In a T = 4/T = 3 configuration, an excess of C proteins without a home is synthesized. What happens to these "homeless" molecules?

The interaction of the C protein with the genomic RNA is specific. The segment required for binding *in vitro* and for packaging *in vivo* lies between the nt 746 and 1226 of the genomic RNA in the coding region of ns P1.[337] The extended and flexible N terminal arm (residues 1 to 103) of the C protein with its basic domains binds to the RNA.[338] The interaction of the C proteins with the cytoplasmic domain of the E_2 protein then drives the budding.[335,339-342] Three possible pathways to alphavirus maturation in mosquito cells can be distinguished:

1. Nucleocapsids are formed in the cytoplasm, migrate to the PM on unknown routes and bud in a process similar to that described in vertebrate cells.[343]
2. Virus maturation takes place in membrane-rich cytoplasmic vesicles ("virus factories"), where nucleocapsids are formed and enveloped and finally released as regular virions after a fusion event between the vesicle and the PM.[221,223]
3. Mixed type, dual maturation also occurs.[129,156,220,222,241,242] Virions are found in vesicles and nucleocapsids in the cytoplasm. Release of virions would occur simultaneously by budding (explaining FFWI) and by vesicle fusion with the PM.

The compartmentalization might protect the mosquito cell from the CPE.[221,344,345] In fact, cells in which nucleocapsids are observed in the cytoplasm also seem to be those which display a CPE.[343] These phenomena as described in mosquito cell culture bear resemblance to those taking place *in vivo*. *Cx. tarsalis* and *A. dorsalis* were infected with WEEV and the mesenteron was investigated. Accumulation of nucleocapsids and maturation occurred along basolateral membranes 22 to 24 h after ingestion.[346]

The site of maturation is not completely fixed. For example, in CEF infected with SFV, it was possible to enforce intracellular maturation into dilated vesicles of the Golgi area with the ionophore monensin. A retention signal is not necessarily needed to enrich v proteins in a given compartment, and an increased spike concentration may be sufficient to trigger envelopment. Correspondingly, by cell fractionation of SFV infected C6/36 cells, we found a highly infectious fraction exhibiting E_1 still in an Endo-H sensitive state but not containing p62. This may reflect a maturation site in the smooth ER or a transitional compartment (Naim and Koblet, unpublished).

I. *Influence of Temperature on SFV-Infected C6/36 Cells*

For reviews see References 347, 348.

In vertebrate cells, membrane and membrane protein movement can be blocked at two sites by lowering the temperature. At 15°C, the traffic between ER and Golgi is arrested in a transitional compartment.[349] At 20°C, the proteins reach the trans-Golgi stack but not the PM.[349-351] Again, the behavior of mosquito cells is different. The incubation times until Rift Valley fever virus (Bunyaviridae) could be transmitted by living *A. fowleri* were 11 days at 28°C and 35 days at 17°C, respectively.[352] The relative number of infected cells was much lower at 13°C (38%) than at 26°C (75%) in *Cx. pipiens*.[353]

In C6/36 cells infected with SFV, there is a single clear transport block at 12°C. At higher temperatures below 28°C, there was a prolongation of the eclipse; between 14 and 19°C this prolongation amounted to 3 h per degree centigrade. The maximum titers were reached at 24 hpi (28°C, 4.2×10^9 PFU/ml), at 42 hpi (16 to 18°C, 4×10^8 to 6×10^8 PFU/ml), and at 48 hpi (14°C, 2×10^7 PFU/ml). No virions were produced at 12°C, a temperature still allowing protein- and RNA biosynthesis. At 12°C, the v proteins remained in a fully Endo-H sensitive state (high mannose types) and p62 was not cleaved. However, the E_1 protein was acylated even at 8°C. As soon as the temperature was raised to 14 to 16°C, p62 was cleaved and E_2 became acylated.[317a,354] The temperature block did not lead to a relative increase of intracellular virions. Thus, we assume that the viral proteins are retained in the RER, on the cis-side of p62 cleavage and virion maturation. The spread of the transport rates was very pronounced. After a labeling period of 5 min only, the first wave of E_1 (40%) became resistant towards Endo-H after 2 h and the last wave (100%) after 5 h at 24°C (Vallan and Koblet, in preparation).[354] The data above 12°C are compatible with the assumption that the rate of protein synthesis may determine the rate of oligomerization and this, in turn, the rate of transport.[355]

m. Persistent Infection and Interference; Small Plaque and Thermosensitive Mutants and Defective Interfering Particles/RNA

For reviews see References 23,149,154.

Regularly, alphavirus-infected cloned or uncloned mosquito cells develop a persistent infection. Davey and Dalgarno[237] demonstrated that in *A. albopictus* (Singh) cells SFV-specific RNA synthesis was restricted already 12 hpi and that the percentage of the apparently infected cells sharply declined from 100 to 2% by 48 hpi. Subculturing such cells always promoted a small burst of virus. At 8 to 10 dpi, small plaque (sp) variants appeared and replaced the wt virus completely after 16 days. No interferon-like activities and no antiviral agents in infected cells or in the medium were detected. Superinfection with SFV gave no increase in yields (homologous interference), confirming earlier data.[234,356-358] Such cells divided and virus was continuously released; they could be infected with heterologous viruses (no heterologous interference).[235,359] *A. aegypti* ("line 59") cells adapted to grow at 15°C produced high titers of Sindbis virus at 8 to 10 dpi and virus was shed over a period of 130 days; the large plaque (lp) size did not change. The conclusion was that homologous interference is not a prerequisite for the persistent infection.[234,360] After infection of *A. albopictus* (Singh) cells with Sin, heterologous superinfection with related alphaviruses (SFV, CHICK) 3 days after the initial infection was not feasible; however, superinfection with a bunyavirus was possible. Thus, the cells restricted the replication of both homologous and heterologous viruses from the same genus. Restriction started 1 h after infection with the first virus and lasted indefinitely with respect to the homologous virus but only for 8 days in the case of heterologous alphaviruses.[361]

Curing of persistently infected cell cultures is possible. Riedel and Brown[345] showed that in *A. albopictus* (Singh) cultures in the persistent phase of Sin infection nearly 100% of the cells appeared to be virus-negative. However, homologous superinfection was impossible. When intrinsic reinfection was prevented by eliminating the virions in the culture fluid with antibodies, curing occurred and the cultures could be reinfected. Cloning experiments with *A. albopictus* (Singh) cells infected with Sin for 6 months or more demonstrated the coexistence of both virus-producing and non virus-producing cells. This behavior could be reproduced by recloning virus producing clones. Curing with anti-Sindbis virus antiserum was feasible.[362] This datum suggests that there are always cells in a persistently infected culture which manage to become virus-free but will be reinfected by virus-producing neighbors, so that the persistent infection is maintained indefinitely. This may explain the maintenance of the chronic infection, but it does not explain its establishment, the restriction of the virus production. Several explanations have been offered for this phenomenon.

Interferon-like substances were invoked,[363] but also (a) low M_r substance(s) released from *A. albopictus* (Singh) cells infected with Sin for a period of 6 months. Three days after infection this substance started to appear and acted, in contrast to interferon, in a virus- *and* cell-specific manner (Reference 364; see also Reference 365). Three cell lines — C6/36, u 4.4, and C7-10 — differed in expression of this type of interfering activity.[366]

Mutant virus was also incriminated. Indeed, sp variants of Sin interfered with the growth of the wt (but not EEEV) in BHK-21, KB-, and *A. aegypti* (Peleg) cells (homologous, nonhost-restricted interference).[367] These sp variants were not defective as they were able to replicate autonomously in the various cell types. In many cases, the sp variants obtained from mosquito cells were ts (RNA^+ or RNA^-). The ts variants could also be generated in vertebrate cells at low temperature.[368] ts mutants that fail to synthesize RNA at the nonpermissive temperature (RNA^-) were ordered by complementation analysis into 4 RNA^- groups. These mutants contain ts lesions in the viral ns proteins. The gene order of the complementation groups is NH_2–B–(A/G)–G–F–COOH.[369] The complementation groups C, D, and E are RNA^+ and make v RNA at the nonpermissive temperature. Group C mutants exhibit lesions in the C protein, D mutants in E_1, and E mutants in the E_2 protein.[370] Such a sp ts RNA^+ Sin mutant from persistently infected *A. aegypti* cells interfered with the replication of the wt at permissive *and* nonpermissive temperatures. Again, this mutant was not defective.[371] In *A. albopictus* cells, ts mutants of Sin gave rise to only 2 complementing combinations.[344] Thus, nondefective but mutant viruses provoked nonhost-restricted homologous interference, allowing heterologous replication. However, sp ts mutants appear after the establishment of the persistent state. They seem to be a reflection of the quasi-species nature of RNA viruses, tolerated in cells kept at low temperature.[54] Therefore, this explanation for the establishment of the persistent infection is not satisfactory.

Finally, the question concerning defective-interfering particles (DI) can be asked.[358,372,373] DI of SFV, containing deletions, generated by serial high multiplicity passage on BHK cells interfered with v RNA formation in vertebrate cells only but not in *A. albopictus* cells.[374-376] In another study, DI particles of Sin were generated by serial undiluted passages in *A. albopictus* cells. In these cells, this virus stock interfered with the replication of the wt and provoked the synthesis of abnormally small intracellular (deleted) v RNAs. It did not interfere with replication in chick cells, and DIs of chick cells did not interfere in mosquito cells.[377,378] Intracellular v RNAs were analyzed by blot hybridization early (1 to 6 dpi) and late (150 to 600 days) after infection of C6/36 cells with SFV. There were two relevant points. (1) Several defective v RNAs were detected already after 24 h, before, during, and after the establishment of the persistent infection. They resembled the defective interfering RNAs of vertebrate cells generated by serial undiluted passages of standard SFV. (2) However, these defective v RNAs were rarely released in virions from the C6/36 cells.[379] Eaton[380] found small amounts of 49S, 26S, and 22S ds v RNA together with new species of small ss RNAs (M_r 0.65×10^6 to 1×10^6) in *A. albopictus* (Singh) cells. Logan[381] described not only defective intracellular RNAs but also DI particles after serial undiluted passage of SFV in C6/36 cells adapted to vertebrate medium. The virions contained RNAs of M_r 0.61×10^6 to 0.78×10^6 and provoked homologous interference.

The complicated situation concerning the establishment and maintenance of the persistently infected state can be summarized as follows:

1. One hour pi the infected mosquito cells become resistant to superinfection with homologous and heterologous alphaviruses.[344,361]
2. After 12 to 24 h, defective v RNAs appear in the infected cells.[379]
3. After 2 to 3 days, the proportion of virus-producing cells decreases to less than 10%.[237,361] At this time, a virus- and cell-specific factor may be shed into the medium, restricting virus production.[364]
4. At 8 dpi, 8 to 10% of the cells allow the replication of other species of the genus but remain resistant to homologous virus.[361]
5. sp variants start to appear after 8 to 16 d;[237,367,382] and ts mutants can be detected 38 dpi.[344,368,371,382] After 48 days, Eaton had evidence for DI particles.[380] It is clear that this time course of appearance of sp, ts, and DI virions cannot be the cause of the initial restriction of the acute infection. Rather, it is the early occurrence of intracellular defective RNAs and/or the shedding of a restriction factor.
6. However, permanently infected cells can be cured or cure themselves;[345,362] it can be assumed that the chronic infection is sustained by reinfection. The occurrence of variant viruses of all sorts must be taken as a normal evolution of a quasi-species not suffering the selective pressures of the wild (change of hosts, small transmission volumes).[44,45] Reinfection may then be dampened by the presence of intracellular defective RNA and extracellular sp and ts virions competing with residual wt virus. In addition, at 28°C, ts mutants can arise and be tolerated.

A final word of caution: mutant viral clones, restricted in *A. albopictus* LT-C-7 cells, produced normal yields of virus in intact mosquitoes. This shows how difficult it may be to extrapolate data from cultured cells to the levels in nature.[383]

*n. **Mutant Alphaviruses***

For reviews see References 150,166.

It would lead too far to review all the known ts mutants and the complementation groups (see above), but some mutants relevant to mosquito cell culture will be discussed. Morphological variants of Sin grown in *A. albopictus* (Singh) cells have been described.[384] Multiploid virions containing a variable number of nucleocapsids in a common envelope are known in case of Sin, CHIK, VEEV, and WEEV.[385] One of these mutants is Sin ts 103.[386,387]

Quasi-species formation started in *A. albopictus* cells infected with WEEV with single-site changes (RNA^- or RNA^+, ts); late cultures then engendered double and triple mutants (RNA^-, sp, and ts phenotypes coupled).[388] The late passage ts mutants grew much better in C6/36 cells than the wt WEEV. Coinfection with a late passage mutant and wt WEEV yielded virions of both parental types and phenotypically mixed particles (no homologous interference). Some variants induced CPE, others exhibited no CPE.[242,389]

There are several examples of Sin host range mutants replicating normally in mosquito cells, being restricted in vertebrate cells, and vice versa. For example, the mutant $SV_{AP15/21}$ (sp ts, RNA^+, complementation group E) generated in C7-10 cells was restricted in chick or BHK cells, but replicated normally in the mosquito cells at 34.5°C.[390] The defect disturbed budding through the chick cell PM. E_2 and pE2 exhibited an abnormally high $M_{r_{app}}$. The standard strain possessed 3 glycosylation sites (positions 196, 318, and 232) and $SV_{AP\,15/21}$ had even a 4th (position 275) in the E_2 protein.[391] Revertants lost all the phenotypic changes. The explanation for this may be that the extra glycan is complex, bulky, and polar in vertebrates.

Another mutant was depressed in vertebrate cells but grew as well as the wt WEEV in C6/36 cells. It was a sp ts variant from persistently infected C6/36 cells (16 months pi) with a smaller pE2, and a smaller, nonglycosylated E_3 released into the culture medium. E_2 appeared to be unharmed.[392] Other examples of altered structural proteins in sp ts alphaviruses generated in persistently infected mosquito cells have been described.[393]

On the other hand, there are mutants restricted in mosquito cells, as for example $Sindbis_{CL35}$ and $Sindbis_{CL58}$, RNA^- in mosquito cells.[394] The first mutant had two amino acid changes, and the latter mutant one, at position 142, in nsP4 (the RNA polymerase).[395] A silent substitution at nt 5751 following the opal codon near the end of the nsP3 coding sequence of the Sin genome was also restricted in mosquito cells at 34.5°C. A possible explanation for such host ranges provoked by silent substitutions could be the involvement of host factors regulating v RNA biosynthesis.

The glycosylation mutant S12 of Sin restricted in C6/36 cells has been mentioned above.[323,329,330] The mutant showed clearly for the first time that the cleavage of pE2 is not mandatory for maturation in vertebrates, but that the defect leads to an attenuated phenotype. This is reminiscent of the myxo- and paramyxoviruses where inability to cleave the precursor is correlated with loss of virulence.

Another mutant had a single amino acid change of the E_2 protein of VEEV (E_2 ile 207 $\rightarrow$ phe) in a region involved in binding to the cell surface. It decreased the viral growth in C6/36 cells and was restricted in its ability to infect and disseminate from the midgut after oral infection. However, it replicated as efficiently as the parental virus when inoculated into *A. aegypti* mosquitoes.[396]

Sindbis virus mutated in the cytoplasmic domain of the E_2 protein and in the 6K protein grew also slower than the wt virus in mosquito cells in a host-restricted manner.[730]

There is a series of natural and artificial substitutions in interesting regions of the genomic RNA. They concern the "junction" region and the untranslated ends of the RNA as well as a highly conserved region in the nsP1 gene.

The junction presents the promoter region (in the minus strand) for the synthesis of the 26S mRNA.[397] In Sin, it encompasses the nt –19 to +5, the 26S RNA starting nt being number 1 at position 7598 counted from the 5′ end of the genomic RNA. The SV_{CP} (clear plaque) has a substitution (U → C) at position –6 (7592) which is silent. In infected mosquito C7-10 cells, it led to a decreased ratio of 26/49S RNA and to an increased CPE; this was not the case in CEF cells. Thus the efficiency of the promoter was lowered in mosquito cells only, and the question of a host factor must again be posed.

The 5′ nontranslated region of the genomic RNA of alphaviruses seems to be the complementary equivalent of the minus-strand promoter at which the synthesis of full-length plus-strands starts. The first 44 nt at this 5′ end are important for replication. Discrete changes of this region of Sin genomic RNA had different effects in C6/36 cells as compared to CEF cells; either initiation of translation of the ns proteins can be modulated here or cellular factors are needed for the replication of the minus-strand template.[398] Similar conclusions concerning putative complexes between host or virus proteins acting at the 5′ nontranslated region to promote RNA synthesis were drawn by constructing RRV-Sin chimeras.[399]

There were 42 site-specific substitutions in the 3′ nontranslated region (alphaviruses: 121 to 524 nt in the plus-strand before the poly(A)) introduced into the genomic RNA of Sin, primarily in the 3′ terminal 19 nt containing conserved region. This element may play the role of the promoter for the initiation of minus-strands. Some substitutions led to a delay in growth, others led to lowered virus yields; some were more severely impaired in C7-10 or C6/36 cells than in chick cells or vice versa. There were different effects in different hosts suggesting interaction with host proteins of this untranslated part of the RNA. Again, the wt 3′ nontranslated region seems, like the 5′ region, to represent a compromise between best replication in mosquitoes and the best replication in vertebrates.[400]

Indeed, host factors were found in Vero cells for the minus-strand 3′ end of Sin, EEEV, and Rubella virus, whereby one of the proteins also bound to the 3′ end of the plus-strand RNA (common protein for both elements);[401] binding assays to oligo nt probes representing the 3′ terminal sequence of the minus-strand identified proteins from chick cell extracts.[402]

Finally, there is a highly conserved domain of 51 nt in the region encoding ns P1 (nt 155 to 205 in Sin). Here, 25 site-specific substitutions were constructed; 21 were silent but changed the stability of possible hairpin structures. All silent substitutions were viable; 19 were severely impaired for growth in both CEF and C6/36 mosquito cells, and 2 grew better than the wt, again suggesting a compromising wt sequence.[403] An essential conclusion of all these studies is that effects of such substitutions/deletions must be investigated separately in mammalian, avian, and mosquito cells.[404]

Especially interesting are host-specific ts mutants. In normal Sin genomic RNA, there is an in-frame opal termination codon separating the ns P3/ns P4 coding regions. Readthrough by an unknown mechanism allows ns P4 to be translated. The opal codon was replaced by either serine-, tryptophan-, or arginine-sense codons, or the amber or ochre codons, and virions were gained by transfection of CEF with the RNA. All mutants grew in bird and mammalian cells at 37°C and in C7-10 mosquito cells at 30°C. However, the amber and ochre substitutions did not form plaques and were ts in C7-10 cells at 34.5°C.[405]

According to Stollar, there are five low M_r substances which are substrates for alphavirus encoded enzymes: ATP, CTP, GTP, and UTP for the polymerase, GTP for the RNA-guanylyltransferase and capping, and S-adenosyl-methionine (adomet) for an RNA-methyltransferase.[406] The levels of the compounds can be changed by depriving (mosquito) cells of methionine,[407] by lowering the GTP concentration with mycophenolic acid or ribavirin,[408] or by addition of adenosine which decreases the concentrations of CTP and UTP, or by addition of guanosine, which decreases the concentrations of ATP, CTP, and UTP.[409] Some of these effects are drastic in mosquito cells only and negatively influence the standard virus production.[410] Viral RNA (Sin) was not produced in the infected *A. albopictus* cells in lack of methionine. However, the mutant SV_{LM21} produced high virus yields in methionine-free medium in mosquito cells.[411] The "lesion" was localized at 2 adjacent nonconservative amino acid changes at positions 87 and 88 of ns P1, leading most probably to an increased affinity of the adomet binding site.[412-415] Similarly, the mutant SV_{MPA} was resistant to mycophenolic acid and, again, the changes mapped into the ns P1 gene.[416,417] These elegant identifications of enzymatic functions and their location were only possible by performing comparative virology in distant hosts. Thereby the relations between viruses of the Sindbis-like (Tobamo-like) supergroup were further substantiated.[418] It has been demonstrated that the putative methyltransferase domain located near the N terminus of nsP1 is conserved throughout the supergroup, but is not present in the other supergroups.[80,419,420]

o. Host Factors

Upon infection of vertebrate cells, host protein synthesis is gradually and increasingly shut down and the CPE develops concomitantly.[226,267] A similar phasing takes place in C6/36 cells, extended from 10 to 16 hpi.[159] Therefore, constant host macromolecular synthesis is not needed for viral replication. Neither enucleation nor a brief treatment with actinomycin D before infection inhibited production of Sin or SFV in vertebrate cells.[421,422] However, Sin production was inhibited in vertebrate cells after prolonged application of actinomycin D.[423,424] When inhibitors of protein biosynthesis were added after formation of the ns proteins and the minus-strand RNA, the production of v RNA in a vertebrate system was not drastically reduced.[425]

On the other hand, evidence was provided with several *A. albopictus* cell lines that host components were involved in at least two distinct steps of Sin production: for v RNA and for v structural protein synthesis.[141,421,422] Whereas vertebrate cells continued to make v RNA, utilizing ns proteins synthesized before addition of cycloheximide,[425] v RNA synthesis was blocked when cycloheximide was added to mosquito cells after commencement of the synthesis of the ns proteins.[141] It is clear by now that such factors exist. Additional information is available throughout this review (CPE and cell-cell fusion; mutant alphavir–uses).[239,395,398,399,401,402]

p. Mosquito Cells As a Tool

For reviews see References 122,426.

It is evident that mosquito cells such as the C6/36 clone can be used for diagnostic, epidemiological, and productive purposes. For example, mutant Getah (alphaviruses) and Japanese encephalitis (Flaviviridae) viruses were detected in C6/36 cells after transfer from *Cx. tritaeniorhynchus* caught in the field; at 28°C, they had a 1000-fold higher plating efficiency than in BHK-21 cells (37°C), because they had characteristics of host-dependent ts mutants.[427] This does not only demonstrate the usefulness of these cells for daily work, but also the possibly hidden contaminants due to viral *in vivo* mutants, when new invertebrate cell lines are established.[132,389] An example has been given as to how the noncontaminated C6/36 cells can be used to detect virus in contaminated (resistant by unknown homologous interference) mosquito cell lines.[130,428] Practical aspects of the work with mosquito cells have been reviewed.[122,426] Methods for mass production of C6/36 cells in suspension have been presented.[429]

B. FLAVIVIRIDAE

Relevant papers and books are References 1,2,4–7,9,11,13,16,22,23,109,116,118,120, 125–127,145,147,148,161,162,216,448–460.

1. Introduction

The virions of the Flaviviridae are enveloped, spherical particles with a diameter of 40 to 60 nm. They contain one molecule of positive-sense (infectious), single-stranded, monopartite, 5′ capped RNA ("44S RNA"), usually without 3′ Poly(A), in an isometric ribonucleoprotein core which is about 35 nm in diameter. A single long open reading frame codes for a polyprotein. Two or three membrane-associated proteins (E,M) and a core protein (C) constitute the protein components of the virion. Recent sequence information shows quite unequivocally that the Flaviviridae are composed of the genera Flavivirus (formerly Arbovirus group B) with about 70 species; Pestivirus (mucosal disease virus group); and hepatitis C virus group.

Pestivirus (bovine viral diarrhea or mucosal disease virus, hog cholera, or european swine fever, border disease of sheep) has no known (arthropod) vectors and a limited host range confined to mammals. Hepatitis C virus has no other natural host besides man. It is transmitted by injection (blood transfusion, intravenous drug abuse).

The prototype member of the genus Flavivirus is the YF virus. Eight subgroups or antigenic complexes were established within the genus on the basis of cross-neutralization. Taking amino acid sequence homologies, sero-complexes, and vectors into account, Table 5 might be tentatively constructed.[1,454,457,458,461]

The host range of the family is extremely interesting. On the one hand, the CFA may be a true insect virus and represent a "missing link";[461] on the other hand, there are two antigenic complexes within the genus Flavivirus and two genera within the family without vectors. The pestiviruses and the Hepatitis C viruses will not be discussed here. It is interesting to note that the flaviviruses take an intermediary position between the Tobamo-like (Sindbis-like) and the Poty-like (Picorna-like) supergroups of the ss, positive-sense RNA viruses.[75,80,462]

Biosynthesis can be described briefly as follows. At the 5′ terminus of the genomic RNA is located a m^7GpppAmp cap. In YF, the start codon is 118 nt downstream of the 5′ terminus. Then 10′233 nt correspond to the open frame and 511 nt remain untranslated at the 3′ end. The gene order is 5′–C–prM–E–NS1–NS2A–NS2B–NS3–NS4A–NS4B–NS5,NS indicating the nonstructural proteins. Thus, in contrast to the togaviruses, the structural genes occupy the 5′ part of the genome. This basic arrangement is also found in the Picornavirus family.

Attachment and entry have not been intensively studied; the receptors have not been identified with certainty (see below). Recently, a 110-kDA protein of PKE cells (part of fibronectin?) has been incriminated as a receptor for the TBE-E protein.[729] The incoming RNA is the template for the formation of minus strands which in turn serve for the production of progeny plus strands. No subgenomic messengers can be detected. After uncoating the infecting virion, RNA is translated on the basis of a single initiation site, so that a formal polyprotein is synthesized. Three structural and seven NS proteins are formed: C (V2), prM (NV2,GP20), E (V3), NS1 (NV3, GP44), NS2A (NV2, P19), NS2B (NV1, P10), NS3 (NV4, P71), NS 4A, NS4B (NV2$^1/_2$, P21), and NS5 (NV5, P98) (older nomenclature for KUN in brackets).[463] The proteins are generated by co- and post-translational processing events in the following way: anchored C/prM/E/NS1 by signalase(s); NS2A/NS2B/NS3/NS4A/NS4B/NS5 by NS3 in partial cooperation with NS2B.

Signalase(s) produce a precursor to the C protein (residues 1 to 123) and the proteins prM and E; each protein contains the signal for the following protein in its C-terminal domain. Therefore, the primordial C protein remains fixed to RER membranes at its C-terminal sequence. NS3 may then cleave the amino acids 106 to 123 away prior to or during assembly of the intracellular particles. Such early particles may consist of C (residues 1 to 105), prM (residues 124 to 290), and E (residues 291 to 787)(WN). The final cleavage of the other precursor, prM, between amino acids Arg-Arg/Lys-Arg 215 and 216, concomitantly with the release of the virions from the cells, would be due to a host enzyme in the later parts of the exocytotic transport pathway.[464] The M protein corresponds to the C-terminal part of the prM protein, possessing the membrane anchor sequence.

The N-terminal segment (about 180 residues) of the peripheral membrane protein NS3 contains the active site of a trypsin-like serine proteinase (His 53, Asp 77, Ser 138 for YF) cleaving the NS proteins as shown above, under the condition that the polyprotein is incorporated into RER membranes (WN, DEN).[465] For each of the NS proteins NS2A, NS2B, NS3, and NS4B, the carboxy end sequence is Lys-Arg.

TABLE 5
Flaviviridae: Vectors, Cells, Sequences

Genus, antigenic complex	"Species"	Vector	Invertebrate cells	Sequence information
A. Flavivirus				
Dengue complex	Dengue 1–4 (DEN)	*A. aegypti* *A. albopictus* *A. scutellaris*	*A. albopictus*,[120,227] *A. aegypti*,[120,448,502] C6/36, *A. malayensis*,[137] *A. pseudoscutellaris*[120,137] (AP61), *Toxorhynchites amboinensis* (TRA 284)[525]	527, 528, 529, 530
Japanese encephalitis complex	Japanese encephalitis (JE)	Many species of *Aedes*, *Anopheles*, and *Culex* (Cx.) *Cx. tritaeniorhynchus*	*A. albopictus* (C6/36) *A. malayensis*, *A. pseudoscutellaris* (AP61), *Toxorhynchites amboinensis* (TRA 284)	531
	Murray valley encephalitis (MVE)	*Cx. annulirostris* *Cx. pipiens* *Cx. tritaeniorhynchus*	*A. aegypti*, *A. albopictus*	532
	St. Louis encephalitis (SLE)	*Cx. pipiens* *Cx. quinquefasciatus* *Cx. tarsalis*	*A. albopictus* (*D. variabilis*)	533
	West Nile fever (WN)	*Cx. pipiens* *Cx. tritaeniorhynchus* *Cx. univittatus*	*Cx. Theileri*[526] *A. aegypti*[120,227] *A. albopictus* *A. malayensis* *A. pseudoscutellaris* (AP61) *Drosophila*, tick cell lines	534
	Kunjin (KUN), Kokobera and others	*Cx. annulirostris*	*A. albopictus* (C6/36)	535
Unassigned complex	Yellow fever (YF), Wesselsbron	*A. aegypti* *A. africanus* *A. albopictus* *A. furcifer* *A. luteocephalus* *A. pseudoscutellaris* *Haemagogus mosquito*	*A. aegypti*,[502] clone C17[503] *A. albopictus* (C6/36) *A. pseudoscutellaris* (AP61) (*D. parumapertus*)	536
Two tick-borne complexes (TBE)	European and far eastern TBE, Langat, Omsk hemorrhagic fever, Louping ill and others	*Ixodes ricinus* *Ixodes persulcatus* *Dermacentor (reticulatus)* *Haemaphysalis*	Tick cell lines as *R. appendiculatus* *Boophilus microplus* *D. parumapertus* *D. variabilis*	490, 537, 538
Cell fusing agent (CFA)	CFA	No vertebrate hosts; insect virus?	*A. albopictus*	461
Rio Bravo and Modoc complexes		No vectors		
B. Pestivirus		No vectors Only mammalian hosts		
C. Hepatitis C group		Only man as host (?)		

The NS proteins form the transcriptase-replicase complex, but apart from NS3 (helicase?, endoproteinase) and NS5 (polymerase) functions have not been assigned. NS1 is found in an intracellular and in a secreted form (DEN, JE, YF). It dimerizes in vertebrate (BHK) and mosquito cells (DEN, SLE, Powassan), thereby increasing in hydrophobicity.[466] It is released slowly into the medium and may facilitate the intracellular assembly of the virions.

The E protein and in many cases the prM and the NS1 proteins are N-glycosylated.

Maturation of flaviviruses can occur on intracellular membranes of the ER or the Golgi apparatus, and two types of viral particles, the intracellular and the extracellular types, have been described. The extracellular virions contain C (V2), the membrane-associated nonglycosylated M (V1), and the transmembrane protein hemagglutinin (E, V3). The cysteine residues of the (WN) membrane proteins are all intramolecularly disulfide-bonded.[467] In intracellular virions, prM-E heterodimers are present; when the N-terminal, glycosylated region of the prM protein is lost from the virus after the cleavage, the E protein trimerizes.[468,469] This situation resembles that of the alphavirus p62 cleavage.

Thus, in summary, the basic principles of the gene order and translation are reminiscent of the picornaviruses, and the cleavage of a heterooligomerized spike precursor of the alphaviruses. NS5 exhibits regions of homology with NS genes of double- and positive-stranded RNA viruses of animals and plants,[470] and may be the RNA-dependent viral RNA polymerase.

2. Vectors and Cell Cultures

For a review see Reference 458.

Papers have been presented concerning arboviruses in tick cells.[114,118,120,140,471-473] The overview of Pudney et al.[120] on the growth of arboviruses in many mosquito- and tick cell cultures is recommended. A compilation of vectors and cell culture types can be found in Table 5. The contents are not exhaustive.

3. Selected Topics

a. Entry and Uncoating

Several studies have been performed to define the entry pathway of flaviviruses, but there is no consensus with regard to a generally valid mode of uptake. It is thought that the trimerized E proteins form the ligands for the PM receptors. Amino acids 52, 364, and 367 of the E protein of JEV affect early virus-cell interaction in Vero cells and virulence in mice.[474] The receptors are unknown. WN virus enters under physiological conditions by receptor-mediated endocytosis; at an acidic pH of the medium, the virus penetrated and uncoated directly at the surface by membrane-membrane fusion (Fusion from without, virus-cell-fusion, FFWO).[475] This is the typical sign but no proof for an endosomal uptake (for a review see Reference 159). The conformational change needed for fusion to occur took place in virions in a pH-dependent manner with an upper threshold of pH 7 and a maximum conversion at pH 6.4 and below. In parallel, free or surface-bound virions lost infectivity.[476] Therefore, it was suggested that there are two entry modes: (1) by endocytosis and (2) by direct fusion at the PM, as is the case in paramyxoviruses.

On the other hand, TBE, Langat, and JE viruses assayed for FFWO on C6/36 cells exhibited the fusion phenomenon only at an acidic pH with an optimum at pH 6, similar to SFV. Monoclonal antibody escape mutants with single mutations in the E protein revealed either a reduced fusion activity or a lower pH threshold (e.g., E Lys 171 $\rightarrow$ Glu).[477] Correspondingly, at high MOIs of DEN 1, 2, 3, 4, JE, and YF, FFWO with C6/36 cells was found at pH 6 and 39°C.[478] However, JE (SA_{14}) and DEN-2 (PR-159) entered C6/36 cells and human blood monocytes directly into the cytoplasm. Virions attached at physiological pH; the PM dissolved at the binding site and disintegrated. Coated pits did not form and there were no virion-containing vesicles.[479] Thus, receptor-mediated endocytosis seems to be ruled out in the entry process of some JE and DEN-2 strains into C6/36 cells. Similar findings were reported for JE entry into permissive mouse neuroblastoma cells.[480] These findings were confirmed when comparing the modes of entry into C6/36 cells by SFV and JE.[183] SF virions appeared in coated and uncoated cytoplasmic vesicles, while JE virions penetrated at the surface and disintegrated at nearby sites. Viral endocytosis was still possible at 4°C in C6/36 cells.

Thus, there is some confusion. WN,[475] YF,[481] and KUN viruses[482] seem to penetrate by receptor-mediated endocytosis or by a dual pathway, while JE and DEN 2 may enter by direct penetration.[479] More investigations are needed to systematize the pH ranges of the conformational changes, the virus strains and sources, the virus-producing cells, and the target cells. The difficulties do not become smaller when analyzing FFWI (see below).

b. RNA Biosynthesis

Obviously, the v RNAs formed in all types of cell cultures must be quasi-identical. The genomic 42-44S RNA was predominant in WN-infected BHK-21 and *A. albopictus* cells. It was capped ($m^7GpppAmpN_1$ or $m^7GpppAmpN_1mpN_2$).[483] A 20S RNA is a ribonuclease-resistant ds replicative form. Heterodisperse partially ribonuclease-resistant 26S RNA represents a replicative intermediate.[484] Transcription/replication seems to take place in the perinuclear region of the host cells; early protein synthesis proceeds in the perinuclear RER.[485-488] JE-specific RNAs were synthesized in subcellular fractions from infected C6/36 cells. A strong synthetic activity was found in the nuclear fraction, probably because NS3 and NS5 are membrane-associated proteins.[489]

An interesting difference seems to exist between mosquito-borne and tick-borne flavivirus genomic RNAs. Whereas a poly(A) tail is lacking in these RNAs of mosquito-borne flaviviruses, the TBE strain Neudoerfl exhibits a short 3′noncoding region followed by poly(A). However, even after introduction of the TBE genomic RNA into C6/36 cells by transfection, there was no detectable level of viral growth.[490]

c. Protein Biosynthesis and Transport

A. albopictus (Singh) cells were infected with KUN, WN, JE, or DEN. KUN yielded 10^7 to 10^8 PFU/ml at 28° and 37°C after a latent period of 10 to 12 h. By 24 h, 90% of the cells were infected. After pulse-chase labeling at 24 hpi, at least 5 to 6 NS proteins and the structural proteins could be discriminated. Conversion of anchored C to mature C seemed to proceed very slowly as compared to the rapid conversion in Vero cells. The glycosylated proteins migrated more rapidly in gels when extracted from mosquito cells than when extracted from Vero cells. There was no switch off of host protein synthesis in the mosquito cells.[269] Glycan heterogeneities were also noticed in case of the DEN-2 glycoproteins synthesized in C6/36 cells.[486] Interesting differences between Vero cells on the one hand and C6/36 cells, TRA 284 cells, and AP61 cells on the other hand after infection with JE (at 24 hpi, no CPE) have been described.[491] The E and the two NS proteins NS1 and NS1′ (products of alternative cleavages at the carboxy end) were processed differently, not only with respect to the glycans. E was released slowly from Vero cells ($t^1/_2 > 6$ h) and even more slowly from mosquito cells ($t^1/_2 > 8$ h); NS1 and NS1′ were retained in undegraded form in mosquito cells and contained high mannose glycans for more than 8 h after synthesis. The co-retention of all these proteins in the secretory pathway provides evidence for a morphopoietic role of NS proteins. Phenotypic changes were described when KUN was either grown in *A. albopictus* (Singh) or C6/36 cells. The virions were very labile, poor in hemagglutinating activity, and E exhibited an apparently smaller size when grown in the Singh's cells.[492] Monoclonal antibodies were found which neutralized YF vaccine virus grown in C6/36 cells only. It may be that the well-known differences in glycosylation influence the architecture of surface epitopes in the case of flaviviruses.[493]

d. Electron Microscopy; Assembly

Flaviviruses are usually located in cytoplasmic vacuoles in close association with the RER, together with smooth convoluted membranes. Within the vesicles, "thread-like" structures are often present and have been postulated to be progeny viral RNA.[487,494,495] After infection of Vero cells with KUN, microtubule paracrystals were observed, associated with the NS3 protein.[488,495] The vesicles appeared at around 9 to 10 hpi, the paracrystals at about 12 hpi in Vero cells. However, paracrystals were not seen in C6/36 cells. In mosquito cells, the morphological changes provoked by the infection with KUN appeared later than in vertebrate cells. After 1 day, intracellular RER and SER membranes proliferated and some of the vacuoles contained whorls of fibers. Between 24 and 30 h, clusters of virions were enclosed within cytoplasmic vacuoles; viral particles were rare in the cytoplasm. By 48 h, large numbers of mature particles were associated with the RER, with the vesicles and the convoluted smooth membranes.[487] After infection of C6/36 cells with JE, 500 to 2250 virions per cell were released into the medium at 24 and 72 hpi, but budding processes at the PM could not be observed. Particles at various maturation stages appeared within the cisternae of the RER at 24 hpi, and they seemed to be carried to the surface within secretory vesicles for release.[496] No nucleocapsids were visible in the cytosol. Differences from that picture were noted in case of DEN-2 infected C6/36 cells. Maturing virions appeared within cytoplasmic vacuoles and on the surface of the cells from 24 hpi onwards; nucleocapsids seemed to assemble in the cytoplasm, then budding at the vacuoles and the PM. Virions that budded into the vacuolar lumens were released by exocytosis.[497] In C6/36 cells, DEN-2 reached titers of more than 10^8 PFU/ml after 5 days. Morphogenesis took place in close relationship with RER membranes; mature virions were observed mostly in the corresponding cisternae. Budding at the PM was rare. In the RER cisternae, a fine granular substance and a network of vesiculotubular structures appeared, as described in *Cx. pipiens* salivary glands after infection with SLE.[498,499] An excessive proliferation of vacuoles and ER was also described in YF-infected Vero cells; the nascent virions appeared to be released by exocytotic fusion events of vesicles with the PM.[481] Based on all these findings, Hase et al.[496] postulated two types of maturation: a cis-type (e.g., DEN-2), where nucleocapsids form in the cytoplasm and bud either into the lumen of vacuoles or into the outside world at the PM; and a trans-type (e.g., JE), where the whole assembly is supposed to take place in the RER. In this case, virions would be carried within transport and secretory vesicles. In the first case, the RER tends to fragment; in the second type, the RER exhibits hypertrophy and dilatation of the cisternae. The maturation type can be tested in C6/36 cells. DEN-2 (strain PR-159) was so far the only flavivirus undergoing the cis-type maturation (for a review see Reference 455). Some flaviviruses, such as YF or DEN-4 viruses may combine the two types, whereby the virions are mainly released by cytolysis (of vertebrate cells). There may even be differences between strains. For example, the DEN-2 NGC strain exhibited a clear-cut trans-type maturation.

There are many unanswered questions. How do virions enter the vesicles if there is no budding into the lumen of the RER cisternae? What is the biochemical link between infection and ER expansion? For virus formation to occur, contacts must be established between the prM-E complex and the C protein; therefore whether the anchored C protein is rapidly cleaved and detached from the RER membrane or whether it remains fixed to the membrane due to a late cleavage may be very important for the maturation type.

e. Cytopathic Effects and Fusion from Within (FFWI)

Syncytium formation in *A. albopictus* (Singh) cells was documented for DEN-2.[500] WN, JE, DEN, and SLE induced a cytopathic response in *A. albopictus* cells.[227,233] WN, JE, YF, DEN 1,2, and 4 produced plaques.[230,501] In *A. malayensis*, WN infection resulted in the formation of syncytia 3 dpi. The syncytia detached from the vessel surface and were replaced by cells of normal appearance. Infection with JE also led to syncytia. In *A. pseudoscutellaris*, WN, JE, and DEN-2 provoked syncytia as early as 3 dpi.[137] No cytopathic effect could be detected in *A. aegypti* cells infected with MVE.[494] However, in the high-producer clone C17 of *A. aegypti* infected with YF, a clear CPE (vesiculation, clumping of the cells, no syncytia) was visible.[502] Antiviral substances could not be detected.[503]

As discussed in Section IV.A.3.f., I distinguish a CPE without FFWI at physiological pH from a fusion phenomenon occurring after a conformational change of spike proteins at mildly acidic pH. According to this view, FFWO (high multiplicity binding of virions to the PM, exposure to low pH), FFWI at a low pH of the medium, and endosomal fusion during infection by enveloped viruses taken up by endocytosis are intimately related (for a review see Reference 159). Obviously, there are enveloped viruses entering by fusion at pH 7 such as paramyxoviruses (e.g., Sendai virus); accordingly, they provoke a FFWI at physiological pH. In the case of the flaviviruses, there is conflicting evidence for these relations as reviewed in Section IV.B.3.a.

Unfortunately, the pH of the medium at the time of spontaneous FFWI has not been specified in many reports. FFWI occurred in *A. albopictus* cells which can lower the pH of the culture medium by metabolic activity. As the CPE was dependent in many cases upon flask types and media, I think that the syncytium formation was triggered whenever the cells were under conditions appropriate to lower the pH of the medium spontaneously.[159] C6/36 cells which do not exhibit such an activity, infected with JE (strain Ja O Ar S 982), fused only from 30 hpi onwards if the pH of the medium was lowered to 6.3 or below.[504] The same was true in the case of infection of the *A. albopictus* clone C7-10 with DEN-2 and SLE.[505] Syncytium formation was maximal 36 to 48 hpi at pH 5 to 6.5 for DEN- and 5.0 to 5.5 for SLE-infected cells. Syncytia were not observed in Vero or BHK cells under these conditions. Fusion was shown to be ATP-dependent (as is the case in alphavirus-infected cells[159]). Lysosomotropic amines inhibited the replication of SLE virus in both mosquito and vertebrate cells. All these findings are consistent with the idea that low pH-induced fusion is necessary for virus entry (JE, DEN-2, SLE).[504,505]

The conflicting findings can be reconciled. (1) Some flaviviruses such as JE, WN, or YF may have a very broad range of pH at which the conformational change of the E protein can take place. An entry of such viruses at the PM might then be observed. (2) FFWI needs the expression of spike proteins at the PM. Release of viruses by exocytotic events may circumvent this route. However, one type of maturation or the other does not seem to be specifically related to a given virus taxon.[496,497] A virus may exhibit one or the other type of maturation based on passage history and adaptation to a host cell system.[496,497] This question deserves more investigation.

Indeed, strain differences leading to differences in fusogenicity have been described for JE after infection of AP 61 mosquito cells.[506] A possible explanation for the lack of cytopathogenicity by the YF 17D vaccine strain was recently proposed; the E protein of this strain was highly susceptible to cleavage.[507]

There is another interesting facet to the problem. Treatment of flavivirus-infected mammalian and mosquito cells with acidotropic amines at late stages of the cycle inhibited the proteolytic processing of the prM protein to M. Virions generated under such conditions contained prM rather than M; they had a lower specific infectivity than virions containing the M protein. This suggests that prM is cleaved in an acidic late compartment of the exocytotic pathway, protecting the E protein from the acid-induced, fusogenic, conformational change to the last moment, when virions are released into the medium of pH 7. This situation is reminiscent of the alphaviruses (p62 cleavage, see above).[508] In fact, FFWO between TBE or JE on the one hand and C6/36 cells on the other could only be induced at low pH, but virions grown in presence of ammonium chloride (TBE, Langat) and containing the precursor prM did not cause FFWO.[477]

Finally, CFA has to be mentioned; it stems from the *A. aegypti* (Peleg) cell line. It led to marked syncytium formation of *A. albopictus* (Singh) cells. It failed to replicate in or cause fusion of BHK, KB, or Vero cells.[23,509,510] The nucleotide sequence is now known; CFA probably diverged from an ancestral flavivirus and may represent an insect virus.[461]

f. ***Persistent Infection***

For a review see References 119,154.

Persistent infections of mosquito cells with MVE and JE were described at the very beginning of the mosquito cell culture period; even curing with specific antibodies was demonstrated. Superinfection with KUN was possible. After the 20th cell passage, the pathogenicity of the viruses for mice decreased.[229,511] Similar findings concerned WN and *A. albopictus* cells.[356] *A. albopictus* (Singh) cells became persistently infected with DEN, produced ts viruses, and were resistant to (homologous) superinfection with all DEN types, but were not resistant to superinfection with CHIK.[512]

DEN infection also led to persistent infection in the cell line TRA-171 from the nonvector mosquito *T. amboinensis* and all serotypes became ts or sp (DEN-2);[513] defective DEN particles were not found.[514]

Persistent infections with SLE were established in cell lines of *A. albopictus, A. dorsalis,* and *Cx. tarsalis.* A high proportion of the cells remained infected. Viral titers decreased over time, but relatively large quantities of viral antigens continued to be produced (host-restricted, defective mutants?). Homologous, but not heterologous interference, could be demonstrated. In this study, only *Cx. tarsalis* cell cultures could be cured with anti-SLE antibodies. Therefore, the authors speculated that *Cx. tarsalis* cells can cure themselves and become reinfected, whereas *Aedes* cells passage the virus by intracellular transfer during cell division.[515] ts mutants were present mainly in *A. dorsalis* and *Cx. tarsalis.* sp mutants predominated in *Cx. tarsalis* after establishment of the persistent state. Growth restrictions of the viruses could be shown after transfer to other cell lines. This was the first report demonstrating host-restricted mutants in flavivirus-infected mosquito cells. In addition, a degradation or cleavage product of the E protein (p27) appeared. Truncated v RNAs were shown in Northern blots, but no DI particles were detected.[516] Similar findings were published by Ng and Westaway.[450]

In summary, the evolution of flavivirus-infected mosquito cells closely resembles that after infection with alphaviruses. Persistent infection is soon established and followed by viral mutagenesis. Maintenance of the persistent state can be due to curing and reinfection, or to viral transmission during cell division. Establishment of the persistent state may be due to defective interfering RNA or to a virus-specific antiviral factor leading to an inhibition of v RNA synthesis.[517-519] In any case, the ability to establish a persistent viral infection is, in part, under host control, since certain clones of *A. albopictus* cells allow for a lytic infection (see above).[130,138,239]

g. *Mosquito Cells As a Tool*

For sequencing, several flaviviruses (JE, DEN) were grown in C6/36 cells. The RNA was extracted from harvested virus or from infected cells and used for polymerase chain amplification or RNA fingerprinting.[520] Mass production of C6/36 cells adapted to grow in Eagle's medium is suitable for this.[429] C6/36 cells can also be used to identify a flavivirus directly by enzyme immunoassays. The same cells have been used to grow DEN candidate vaccine viruses,[521] or to assess the potential of flaviviruses with unclear arthropod-borne status for replication in mosquitoes,[522] or for rapid diagnosis of SLE from field-collected mosquitoes or of DEN from clinical specimens.[449] In this respect, the incubation temperatures have to be considered. Kuno and Oliver examined the upper temperature limits for maintenance of 11 mosquito cell lines.[523] Most lines could be grown at 32 to 34°C; *Anopheles stephensi* at 37°C. The yield of cells after a week was best at 28°C. Several flaviviruses grew faster in C6/36 cells adapted to 34.5°C. In uninfected *A. albopictus* cells, heat-shock proteins are induced by a brief exposure to 37 to 40°C; however, host protein synthesis is shut off at temperatures over 35°C. Finally, unknown viral strains isolated from wild mosquitoes can be assayed for growth in mammalian and mosquito cells in order to find new insect viruses.[524]

C. BUNYAVIRIDAE

Relevant papers and books are References 1,2,7,11,14,20,24,25,31,33,65,77,109,125, 127,145,147,539–556.

1. Introduction

The family Bunyaviridae is defined as a family of RNA viruses with a negative or ambisense polarity (ss) RNA genome. The genome is tripartite: the enveloped particles (80 to

100 nm diameter) contain three molecules of RNA, the large (L), the medium (M), and the small (S) RNA. The particles usually possess four proteins, the two external glycoproteins G1 and G2, the nucleocapsid protein N, and the large (L) protein, representing the transcriptase which must be associated with the virion. The following genera are listed (Table 6):

1. *Bunyamwera supergroup* (Bunyavirus). Bunyamwera virus as a prototype. L RNA 7 kb, M RNA 4.5 kb, S RNA 0.9 kb. G1 protein 108 to 120 $\times 10^3$, G2 29 to 41 $\times 10^3$, N 19 to 25 $\times 10^3$, L 200 $\times 10^3$. G1 and G2 and a 15 to 18 $\times 10^3$ NS_M nonstructural protein coded for by M RNA; N and NS_S nonstructural protein by S RNA; L protein by L RNA. There are at least 162 viruses constituting 18 antigenic groups, without serologic relation to the members of the other genera. The bunyaviruses have a host range encompassing vertebrates, mosquitoes, and culicoides. The Tete group (Oropouche, Brazil) is tick-transmitted. Transovarian infection in arthropods is known.
2. *Phlebovirus* (Sandfly fever, Uukuniemi group). Sandfly fever and Sicilian virus as prototypes. L RNA 7.5 kb, M RNA 3.8 kb, S RNA 1.8 kb. G1 protein 55 to 75 $\times 10^3$, G2 50 to 70 $\times 10^3$, N 25 $\times 10^3$, L 145 to 200 $\times 10^3$. Glycoproteins encoded by M; N and NS_S (30 $\times 10^3$) by S. The Sandfly fever group also exhibits a nonstructural protein NS_M (preglycoprotein), which is absent in the Uukuniemi group. There are 8 antigenic complexes (at least 23 viruses) within the Sandfly fever group, whereas the Uukuniemi group forms 1 serogroup (at least 12 viruses). There is no serological cross-reactivity with the other genera. The most important complexes are the following: Sandfly fever Naples and Pappataci fever (4), Punta Toro (PT) (2), Rift Valley fever (RVFV) (3), and Uukuniemi group (12). Sandfly fever group viruses are transmitted by phlebotomines or mosquitoes; Uukuviruses are carried by ticks.
3. *Nairovirus* (Nairobi sheep disease). Crimean-Congo hemorrhagic fever virus as prototype. L RNA 13 kb, M RNA 5.4 kb, S RNA 1.9 kb. G1 protein 72 to 84 $\times 10^3$, G2 30 to 40 $\times 10^3$, N 50 $\times 10^3$, L 145 to 200 $\times 10^3$. There are 7 antigenic complexes (at least 34 viruses) without serological relations to the members of the other genera. The host range encompasses various vertebrates, ticks, but also mosquitoes and culicoides. Most viruses are tick-transmitted. Important groups are Crimean-Congo hemorrhagic fever (3) and Nairobi sheep disease (2, Dugbe, Nairobi sheep disease).
4. *Hantavirus* (Hantaan and related hemorrhagic fevers with renal syndrome). Hantaan virus as a prototype. There is only one serological group, not related to the other genera, with six members (Hantaan, Seoul, Prospect Hill, Puumala, and others). The genus is most interesting; it seems to be an exception within the Bunyaviridae, because no vectors are known. The host range comprises vertebrates, primarily rodents. There is no evidence for nonstructural proteins. Hantaviruses will not be discussed in this review.
5. *Tospovirus* (Tomato spotted wilt group). Tomato spotted wilt virus as a prototype. The virus is a plant virus with an extremely large host range. It is noteworthy that several species of thrips transmit the virus. It is one of the few plant viruses with an envelope. The S RNA seems to code for the nucleocapsid and for a nonstructural protein. Tospovirus will not be treated.

To summarize, bunyaviruses constitute a huge family with an extremely large host range, infecting V, I, and P. It is one of those very important families with the VIP-host range.

The coding strategies are as follows: S genomic segment: In the case of the hantaviruses, the strategy is simple. They transcribe the segment into a mRNA to translate the N protein. There is no nonstructural protein. Bunyaviruses use overlapping reading frames in a nearly full length mRNA encoding N and the nonstructural protein NS_S. Phleboviruses have an even more complicated strategy for expressing S, the ambisense strategy. This term describes a genome (segment) containing positive- and negative- sense genes. In the phlebovirus genomic S RNA, the N gene is in negative sense and an NS_S gene is in positive sense, in nonoverlapping fashion. Therefore, two subgenomic messengers are formed: the N messenger is directly transcribed from

TABLE 6
Bunyaviridae: Vectors, Cells, Sequences

Genus and "species"	Vector	Invertebrate cells	Sequence information
Bunyamwera supergroup			
Bunyamwera (Bun)	*Culicine* mosquito	*A. albopictus*	588, 589
La Crosse (LAC) (California group)	*A. triseriatus*	*A. triseriatus* (MAT)[585]	580
Snowshoe hare (SSH) (California group)	*Culiseta inornata, Aedes* species	*A. triseriatus*[586]	
		Toxorhynchites amboinensis (TA 9)[587]	
		C6/36	
Germiston (Bunyamwera group)	*Culicine* mosquito	C6/36	
Tahyna (California group)	*Culiseta annulata, A. vexans*		
Marituba (Bwamba group)	*Culicine mosquito*	C6/36	
Akabane (Simbu group)	*Culex, Culicoides brevitarsis*		
Tete (Tete group)	*Ticks, Culicoides paraensis*		
Phlebovirus			
Sandfly fever	*Phlebotomus* species (old world)	*A. pseudoscutellaris* (AP-61)	
Naples, Sicilian (Papatacci)	*Phlebotomus papatasi, Lutzomyia* (new world)		590
Toscana	*Phlebotomines*	AP-61	
Rift valley (RVFV)	*Cx. pipiens, Cx. Theileri, Aedes* species	*Cx. Theileri* [526]	591
	Sandflies	C6/36	
Punta Toro (PT)	*Phlebotomines*		592, 593
Uukuniemi (Uuk)	*Ixodid ticks (ricinus, uriae, putus)*	*R. appendiculatus*[120]	594, 595
Nairovirus			
Crimean-Congo	*Ixodid ticks* (28 species), (*Hyalomma* species), *Argasidae* (2 species)?		596
Nairobi sheep disease (NSDV)	*Ixodid ticks, Culicoides*	*R. appendiculatus* (TTC-243)	
Dugbe	*Ixodid ticks, Amblyomma variegatum,* (transstadial persistence)	*R. appendiculatus, Boophilus microplus,*	596, 597
	Culicoides species	*D. parumapertus*	
Hazara			596

the genome (cRNA, antivirion sense), whereas the NS_S messenger is transcribed from the complementary genome (virion sense). The function(s) of the NS proteins is unknown. It is noteworthy that an ambisense strategy is also found in the S segment of the bipartite arenaviruses.

M segment: This segment encodes the G proteins and, in the case of bunya- and phleboviruses, a nonstructural protein NS_M. The open reading frame of the cRNA is continuous. The designations of the G proteins refer to electrophoresis and are probably without biological significance. In hantaviruses, the gene orders in cRNA are (Hantaan) 5′-G1-?-G2-3′ as in Uukuviruses, with a short intergenic region, which may partially represent the signal sequence of G2; in bunyaviruses (SSH) G2-NS_M-G1; in phleboviruses (RVFV) 5′-NS_M-G2-G1-3′ and in PT 5′-NS_M-G1-G2-3′. Cleavage seems to occur cotranslationally between G1 and G2 and post-translationally between NS_M and G.

The receptors on the cell surfaces are unknown. Virions may be taken up by endocytosis, thereafter fusing at acidic pH with the endosomal membrane. After uncoating, transcription into messengers is assumed to occur with the virion-associated L protein in the cytoplasm. Later on, primary transcription is switched to replication, so that full-length cRNA templates and virion RNA can be formed. The nature of the switch is unknown. Genome replication needs ongoing protein synthesis. Only the G proteins are translated on membrane-bound ribosomes; these proteins seem to have individual N-terminal signals for translocation into the membranes of the ER and they have C-terminal anchor regions. They possess N-linked glycans; 5 to 8 sites were identified.

The viral particles bud intracellularly at smooth-surface vesicles of the Golgi complex, and the envelope proteins must possess Golgi retention signals irrespective of the presence of NS_M. Interaction between the transmembrane G proteins and the nucleocapsids which accumulate on the cytoplasmic side of vesicular membranes is direct, without a bridging membrane (matrix, M) protein. After budding into the Golgi cisternae, the virions are believed to be transported by exocytotic vesicles. However, RVFV can bud from the plasma membrane as well as into Golgi cisternae of the same primary rat hepatocytes. Generally, members of the Bunyaviridae are cytolytic for vertebrate host cells. Host cell protein synthesis is reduced and finally abolished (LAC/BHK cells; RVFV/Vero cells); however, Uukuniemi or Dugbe (nairovirus) viruses do not shut off host protein synthesis; and hantaviruses provoke persistent, noncytolytic infections in mammalian cells.

2. Vectors and Cell Cultures

The growth of bunyaviruses in tick cells was treated by several authors. See References 113,116,118,120, 124,125,471–473,557.

In Table 6, a compilation of vectors and cell culture types can be found. The contents of the table are not exhaustive. Many studies have been performed with live vectors. The results have been discussed in depth by Beaty and Calisher.[31] Viral tropisms may provide the molecular basis for the restriction of virus replication to a limited number of vector species. LAC efficiently infected the *A. triseriatus* mosquito when ingested; however, a variant grown in presence of a monoclonal antibody was strongly restricted. A revertant regained the antigenic phenotype and the fusion capability of the original virus. This points to the (G1) envelope glycoprotein as attachment protein in the midgut.[558] Oral transmission of LAC-SSH reassortant viruses in *A. triseriatus* mosquitoes, followed by infection of mice, demonstrated the same: viruses with the M segment of LAC were efficiently transmitted; in contrast, viruses with the M segment of SSH were inefficiently transmitted.[559] Similarly, viruses containing the LAC M RNA segment were well transmitted by transovarian infection of *A. triseriatus* mosquitoes.[560] There are many other fields of interest in work with live vectors such as the development of ts mutants, reassortment, complementation, deletion mutants, interference, and persistent infection (reviewed in Reference 65) or the fate of viruses in ticks. For example, transovarian transmission in infected female ticks (Crimean-Congo fever[561]) or the venereal and transovarial transmission in mosquitoes (LAC, SSH[560]) are of major importance. Immunohistochemical analysis and *in situ* hybridizations were successfully applied to detect Dugbe virus products in ticks infected *in vivo* and to demonstrate the trans-stadial persistence of the virus in hemocytes.[562] The different barriers against virus propagation within the vector — both for sequential infection and escape — in the midgut, in the hematocele, and in the salivary gland are another important reason to study the interaction of the virus with the live vector.[128]

In contrast, arthropod cell cultures have rarely been used for biochemical work; rather they were applied to answer the basic question as to whether a given virus can grow in a cell line. The literature has been indicated in Section IV.B.2 and in the Introduction. Tick cell culture must be considered a forgotten art.

3. Selected Topics

a. Uptake

FFWO at low pH and a conformational change of the G1 protein of LAC have been demonstrated. On the other hand, digestion of the G1 protein by protease enhanced the binding of LAC to mosquito midgut cells.[27] Thus, there is circumstantial evidence for an endosomal uptake, but no direct studies have been performed with arthropod cells in culture.

LAC was treated with monoclonal antibodies specific for the G1 protein and then tested for neutralization on BHK- and *A. albopictus* cells. There were antibodies exerting host-dependent neutralization. One antibody neutralized LAC when assayed on BHK cells, but not on mosquito cells; another antibody binding to a different G1 epitope had the opposite effect. The latter epitope was present on all California serogroup viruses.[563] Thus, these viruses could have a common epitope in the G1 protein for interaction with the arthropods; G1 would be the binding, G2 the fusion protein. If true, this would be exciting, because with respect to the gene order, NS_M-G1 might correspond functionally to the prM protein of flaviviruses and the p62 protein of alphaviruses.

b. RNA and Protein Biosynthesis

Genome accumulation started at about 6 hpi in C6/36 cells infected with LAC (BHK cells: about 2 hpi). The L, M, and S genomic RNAs still increased in concentration at 24 hpi. There was no equimolarity (M > L > S), and there were more M genomes in C6/36 cells than in BHK cells, but only about half as many S genomes. The relative mRNA concentrations were nonequimolar (4 hpi onwards). Five S mRNAs per genome, 0.5 M mRNAs per genome, and 0.05 L mRNAs per genome were measured (BHK cells). Even lower ratios were found in C6/36 cells, and L mRNA could not be detected over background at any time. S genome levels increased over 24 hpi and then started to decrease. This is the time of transition to the persistent infection.[564]

Genome synthesis seems to be the result of a cooperative action of the v polymerase and unassembled N protein. N protein, coded by the S segment, is thought to bind to nascent chains; at suboptimal concentrations of the N protein, genome synthesis did not proceed properly.[565] Therefore, protein synthesis was also analyzed in C6/36 cells.[564] Only the N protein was detectable over the host background, which was not reduced by the infection; its synthesis was maximal at 10 to 12 hpi and then declined sharply to 5 to 10% by day 2. This decline correlated with the shutoff of genome biosynthesis. Defective genomes were not detected; however, a relative excess of S genome segments was present in the virions, as was also the case in C6/36 cells after 72 days of persistent infection.

Within 24 to 48 h of LAC infection of C6/36 cells, much of the S *mRNA* was encapsidated. This correlated with the repression of the N protein synthesis. The encapsidated S mRNA could not be translated *in vitro*.[566] The authors suggest that genome replication (exhaustion of a host factor or modification of the replicase?), but not transcription, is downregulated with time in mosquito cells. The pool of the N protein would then expand to the point at which it interacts with its own mRNA, which also contains the regular assembly signal.[566,567,580] This interaction would lead to a self-limiting infection through the unavailability of the N protein for genome synthesis and proper encapsidation. (S) mRNA encapsidation is highly cell-type dependent.

The synthesis of full-length S mRNA can be observed *in vitro* under the condition that a translation system is included. Otherwise, only short transcripts, stopped at nt 175 (may be due to base pairing of the nascent mRNA and the template), can be detected. Thus, concurrent translation is required to prevent the polymerase from premature termination. This translational requirement for LAC mRNA synthesis was cell-type dependent and was present in BHK, HEL, and Vero cells, but not in C6/36 cells. Using a cell-free approach by mixing virions and uninfected cell extracts, it could be shown in BHK reactions that the polymerase terminated prematurely in the absence of concurrent translation; in the C6/36 reactions, however, the polymerase read through independent of translation. The requirement was due to a factor present in BHK but not in C6/36 cells.[568] The host component may favor the annealing between the strands. This coupling of transcription to translation is unique amongst minus-strand RNA viruses.

The replication of Marituba virus was studied in C6/36 cells. After an initial acute phase of infection with large amounts of virus, a persistent phase with low virus yields followed. There was no CPE; no changes in host cell DNA, RNA, and protein synthesis could be observed. In contrast, the infection led to shutoff of the host macromolecular synthesis and to a CPE in mouse fibroblasts (L-A9). Two virus-specific proteins, G1 and N, accumulated in the C6/36 cells.[569] As these viruses mature in the Golgi complex of vertebrate and arthropod cells[570] and probably are released by virus-carrying vesicles fusing with the PM, no direct changes at the PM are expected. However, the electrophoretic mobility of L-A9 cells decreased progressively from about 6 hpi to 42 hpi, when the virus production was maximal. Such an effect could not be observed in case of C6/36 cells.[571] Correspondingly, infection of L-A9 cells led to an increase of the intracellular Na^+ concentration and a decrease of the intracellular K^+ concentration. All these parameters remained unaltered in C6/36 cells.[572]

I am not aware of studies concerning intracellular transport of the glycoproteins in arthropod cells; in fact, the virus assembly at intracellular membranes is a fascinating aspect, because the glycoproteins seem to determine the site of the intracellular maturation process at the membranes of the cis/medial Golgi.[554]

c. *Electron Microscopy*

Munz et al.[557] described the propagation of NSDV in tick cells (TTC-243). Three to four virus harvests could be made from one culture flask; then the virus titer dropped and no further subcultures were possible. A typical CPE could not be detected. No signs of attenuation after 45 viral passages were observed. Viral antigens were easily stained with the immunoperoxidase technique. Striking morphological differences were found between vertebrate and tick cells. The tick cells presented many electron-dense lipid structures and intracytoplasmic vacuoles. ER-like structures were discrete. Reovirus-like particles were present in noninfected and NSDV infected tick cells. Only a few particles of NSDV could be detected in infected TTC-243 cells. To summarize, except for special investigations, tick cells do not seem to provide advantages for virological work.

LAC assembled exclusively at internal membranes of *A. albopictus* cells, mainly of the Golgi complex. Later on, Golgi smooth membranes proliferated and the cisternae and vesicles were dilated. Virions budded into these structures. In a low proportion of infected mosquito cells, dense granulofibrillar masses were visible in the cytoplasm. These cells did not exhibit gross CPE; they became persistently infected. For some cells, the infection was lethal.[573]

d. *Persistent Infection*

Several groups have reported the establishment of persistence in mosquito cells. Bunyamwera virus led to a marked CPE and cell death of infected vertebrate cells; however, *A. albopictus* cells became persistently infected after a phase of acute infection. Titers reached more than 10^7 PFU/ml after 60 h of infection. Maximum rates of v RNA and v protein synthesis were low and corresponded to less than 2% of the synthetic capacities of uninfected cells. Viral protein synthesis was maximal at 12 hpi and was barely detectable at 24 hpi. The percentage of cells releasing virions fell from 100% to 1.5% after 54 days. Titers were 10^3 to 10^5 PFU/ml in persistently infected cultures after 10 days onwards. There was homologous, but not heterologous interference. A CPE could not be observed at any stage of infection. sp variants appeared after 7 days and increased to 95% of the total infectious virus at 66 dpi; ts mutants appeared between the 25th and the 50th dpi. Antiviral activities shed into the medium were never detected. In contrast, in BHK cells RNA and protein synthesis reached 20% of the synthetic capacity of uninfected control cells for up to 30 hpi. Host protein synthesis was completely shut down by 14 hpi.[574] Persistently infected cell lines from the C6/36 clone (Bunyamwera virus) were resistant to superinfection with viruses of the genus, but not Dugbe virus (Nairovirus) nor vesicular stomatitis virus (VSV, Rhabdoviridae). With increasing passage level, more and more ts virions appeared. In this study, the viruses from the persistently infected cultures interfered strongly with the replication of the standard virus in normal C6/36 cells but not with other bunyaviruses or VSV. Northern blot hybridization showed the S segment in the released autointerfering particles to be prominent whereas the L and M RNA species were barely detectable.[575]

Toscana virus also established a persistent infection of AP-61 mosquito cells. A chronic titer of 10^3 to 10^5 PFU/ml, as assayed at 37°C, was maintained. The percentage of virus-producing cells was very low (0.003 to 0.02%), the percentage of cells containing viral antigens was 10 to 50%. CPE was never observed. sp and ts mutants were apparently very rare.[576]

In situ hybridization with cDNA and RNA probes was performed using fixed Vero and C6/36 cells infected with Germiston virus. This virus also provokes an acute and then a persistent phase of infection in mosquito cells without CPE. The negative S segment and the messenger and antigenomic S RNAs were always found in the cytoplasmic compartment of acutely and persistently infected C6/36 cells. The persistently infected cells contained less genome copies than the acutely infected cells, and less than 10% of the persistently infected cells contained the segment.[577,578]

Several mechanisms may lead to the persistent infection. A general instability of the host mRNAs is induced in mammalian cells but not in mosquito cells.[579]

Viral macromolecules accumulate more slowly in mosquito cells than in mammalian cells.[564] At 24 hpi, viral protein synthesis and genome replication become restricted in the mosquito cells, though mRNA levels and synthesis rates remain unchanged. The encapsidation of the S mRNA with the N protein was described above.[566] N controls its own synthesis in the absence of sufficient genome production.

e. Mutant Bunyaviruses

As is to be expected in the case of a virus with a partite genome, reassortment has been proved to occur *in vivo*.[62] It is quite frequent but confined within viruses of one serogroup. LAC and SSH viruses reassorted in the ovaries of *A. triseriatus* mosquitoes, and the reassortants were transmitted transovarially.[63] Another example concerns RVFV; different strains reassorted in *Cx. pipiens* after sequential ingestion of parental viruses or after feeding on a single, double-infected host. The reassortants replicated in mosquitoes and were transmitted by bite to hamsters.[64] The Maguari virus (Bunyamwera serogroup) reassorted in a homologous and heterologous manner in both vertebrate (BHK-21) and *A. albopictus* cells after dual infection and represented the third reassortment group. One phenotype was growth restricted in BS-C-1 cells.[61] Possible reassortants of a bunyavirus were examined for neurovirulence by intracerebral injection in adult mice, and the results indicated that the large RNA segment is the major determinant of neuroattenuation. The segment encodes the L protein, presumed to be the viral polymerase.[581] This data seems to demonstrate that virulence is not only determined by the binding of the viral surface proteins. The occurrence of sp and ts mutants after infection of *A. albopictus* cells with Bun has been mentioned above.[574]

Defective RNAs have also been described. BHK and C6/36 cells were infected with Bun. Persistently infected C6/36 cells were passaged for over a year; northern blot analysis of viral RNAs extracted from the cells revealed the presence of subgenomic RNAs derived from the L segment. These defective RNAs were not packaged into nucleocapsids. The role of these defective RNAs is unclear, as their presence did not correlate with homologous interference.[582] However, Bun-infected BHK cells produced virions with the sp phenotype in BHK and mouse L cells. The defective L RNAs were packaged into viral particles which caused interference in mammalian cells. Novel DI-specific polypeptides were synthesized in DI-infected cells.[583]

Thus, in infected mosquito cells there is ample opportunity for viral changes (shift and drift) due to the persistence, which may create new host ranges, new cell tropisms, or altered virulence. The relative genetic stability of the bunyaviruses in nature must be explained by selective mechanisms as in the case of other arbovirus families. Specialization of a given virus to a specific vector-vertebrate pairing within defined geographical areas may represent such a selection mechanism leading to natural isolation, which precludes the full use of the potential of the overall gene pool.[24,25]

Viruses grown in mosquitoes for prolonged times might drift away from the vector-vertebrate pairing. For example, Bunyamwera and SSH viruses were passaged up to six times through mosquito cells in culture and the resultant viruses were compared to the mammalian cell-passed viruses with monoclonal antibodies raised against the mammalian types. The mosquito cell-adapted virus population consisted of mutants with improved replication in mosquito cells and with different susceptibilities to neutralization.[584] Similar findings were reported with respect to LAC and YFV grown either in SW13 or C6/36 cells.[493,585] In these cases, the presentation of the epitopes on the virion surface, eventually connected to the different glycan structures, may play a role.

D. REOVIRIDAE (ORBI- AND COLTIVIRUSES)

Relevant papers and books are in References 1,2,7,8,11,15,20,26,29,34,56,109,113,116,120, 125,145,147,161,471, 541,598–607.

1. Introduction

Reoviruses are nonenveloped, icosahedral particles with a diameter of 60 to 80 nm. They possess one or two (double capsid shell) outer protein coats and an inner core with an associated transcriptase activity. The nucleic acid consists of 10 to 12 segments of linear, ds RNA with a total M_r of 12 to 20×10^6. Each segment corresponds to one open reading frame for a corresponding protein requiring no processing. Six to ten protein types are collected in a virion, including the transcriptase and the mRNA capping enzyme. The Reovirus group consists of the Orthoreovirus; Orbivirus; Coltivirus; Rotavirus; and Aquareovirus. In addition there exist the cytoplasmic polyhedrosis virus group (Cypovirus); plant Reovirus subgroup 1 (Phytoreovirus); plant Reovirus subgroup 2 (Fijivirus); and plant Reovirus subgroup 3.

The orthoreoviruses with 10 segments of RNA and 9 proteins infect a broad range of vertebrates (man, monkeys, birds, cattle, bats). They will not be discussed.

The orbiviruses possess 10 segments of linear ds RNA (0.5 to 2.8×10^6, total M_r 15×10^6).

Seven polypeptides (M_r 35 to 150×10^3) compose the virions. The coding assignments of the RNA segments (BTV) are as follows: L1: VP1; L2: VP2; L3: VP3; M4: VP4; M5: VP5; M6: NS1; S7: VP7; S8: NS2; S9: VP6; and S10: NS3. The structural proteins are designated as VP, the nonstructural proteins as NS proteins. The inner core consists of the RNA segments and VP1, 4, and 6. The inner shell is composed of VP3 (60 copies in 12 pentameric complexes) and VP7 with 260 trimers in an icosahedral structure. The inner core and the inner shell form the core or the single-shell particle. VP2 and VP5 make up the outer shell. VP1 represents the polymerase, VP4 belongs to the transcription-replication complex and may be the capping and methylation enzyme; VP6 has as yet an unclear function (polymerase complex, helicase?). VP3 and VP7 lay on the core surface, and are major proteins with group-specific epitopes; VP2 is the hemagglutinin and the neutralization antigen. The functions of the NS proteins are not yet fully understood; NS1 is the major nonstructural protein and is involved in the tubule formation in the cytoplasm, NS2 is a phosphorylated ssRNA binding protein, and NS3 in fact forms two related proteins (NS3a and NS3b) of unknown function (for a review see Reference 34).

The host range is wide (man, horses, monkeys, rabbits, cattle, deer, suckling mice, arthropods). Orbiviruses are transmitted by arthropod vectors; they exhibit 12 serological groups. Some are listed in Table 7.

The coltiviruses with the prototype member Colorado Tick Fever virus contain 12 segments of RNA (0.24 to 2.5×10^6, total M_r 18×10^6). There are two shells. Two serotypes are known (North America and the European, Eyach). The host range involves man, deer, cattle, pigs. Vectors are ticks *(D. andersoni)* and possibly mosquitoes.

The rotaviruses with 2 shells, 11 RNA segments (0.4 to 2.1×10^6), and 6 structural proteins infect mammals and birds. There are no vectors and therefore the rotaviruses will not be discussed.

The aquareoviruses are pests of fish and shellfish; vectors have not been identified. They contain 11 segments of RNA and at least seven structural proteins.

The cypoviruses are confined to insects (Lepidoptera, Diptera, Hymenoptera) and Crustacea. They contain 10 segments of ds RNA and more than 3 to 5 structural proteins. Prototype

TABLE 7
Reoviridae: Vectors and Cells

Genera	Vector	Invertebrate cells
1. Orbivirus		
African horse sickness (AHSV)	*Culicoides*, ticks?	*A. albopictus*
Bluetongue virus (BTV)	*Culicoides: Culicoides variipennis, brevitarsis*	*Culicoides variipennis* (CuVa), C6/36, tick cells
Changuinola	*Lutzomyia mosquitoes, phlebotomines*	*A. albopictus* (no CPE)
Corriparta	Mosquitoes	
Epizootic hemorrhagic disease (EHDV)	*Culicoides*	C6/36, *Cx. quinquefasciatus*
Kemerovo virus (KEM)	Argasid and ixodid ticks, *Ixodes persulcatus, ricinus, uriae; Hyalomma anatolicum*	*A. albopictus*
Orungo	*Anopheles funestus*; *Aedes* and *culicine mosquitoes*	*A. aegypti* (intrathoracal) *A. pseudoscutellaris*
Palyam	*Culicoides*, mosquitoes, and ticks	*A. albopictus*
Warrego	*Culicoides*, mosquitoes	
2. Coltivirus		
Colorado tick fever virus (CTFV)	*Boophilus microplus, D. andersoni, variabilis, Ornithodoros, Rhipicephalus*; mosquitoes	*A. aegypti, albopictus; D. andersoni*
Eyach	*Ixodes ricinus, ventalloi*	

is the *Bombyx mori* cytoplasmic polyhedrosis virus. These viruses must be considered as arthropod viruses proper and are not treated further.

Obviously, the plant reoviruses are also outside the scope of this paper. However, it may be mentioned that phytoreoviruses (12 RNA segments, 7 structural proteins, 2 outer layers; prototype: wound tumor virus) are transmitted by cicadellid leafhoppers in a propagative manner throughout the whole of their life. Transovarian infection of offspring is possible.

Fijiviruses (10 segments, 6 structural proteins; prototype: Fiji Disease virus) are transmitted only by delphacid plant hoppers in a propagative manner for the whole of their life. Plant reoviruses of the subgroup 3 finally (10 segments; prototype: Rice ragged stunt virus) are transmitted to flowering plants and gramineae by a plant hopper, again in a propagative, life-long way.

In summary, the Reoviridae represent a very large, true VIP family, with a diversified host range within the genera. However, it is obvious that arthropods play a major role in reovirus epidemiology.

The information available about adsorption of reoviruses is controversial and results cannot be generalized. Up to 5×10^4 reovirus particles were adsorbed per L cell, and receptor density has been estimated to amount to 4×10^3 to 1×10^5 according to cell type. The erythrocyte surface receptor may be glycophorin A, and in other cell types the β-adrenergic receptor. The viral binding protein seems to be the outer capsid protein, which exhibits hemagglutinating activity. Regular uptake appears to be driven by receptor-mediated endocytosis; however, partially digested particles might penetrate L cells directly. Uncoating is temperature dependent; it starts 20 to 30 min after penetration and requires an acidic pH. However, typical fusogenic sequences have not been identified. (See also Mackow et al.,[608] with respect to rotavirus.) The virion-associated transcriptase becomes activated after digestion of the outer capsid proteins. The parental ds RNA remains within the uncoated core. The minus-strands only are then transcribed in a conservative fashion. Copies are full-length. 5′ termini of the messengers become capped. These capped "early transcripts" serve for protein synthesis and for minus-strand biosynthesis. They reach their maximum at 6 to 8 h after infection in L cells. Replication takes place within nascent progeny subviral particles, whereby the newly formed minus-strands are associated with the corresponding plus-strands. Free minus-strands and free double-strands cannot be detected in infected cells. Thus, assortment is an early event in the life of a progeny particle. "Late mRNAs" are products of immature progeny particles; they reach their maximum at about 12 hpi. Their 5′ ends are uncapped.

Newly synthesized v proteins start to appear at 2 hpi; they increase concomitantly with the gradual decrease of host protein synthesis. The reason for the shutoff is unclear; it may have to do with the fact that late v mRNAs are uncapped. As the rate of transcription of the dsRNA segments and the translation frequency of the v messengers varies, the amount of the v proteins synthesized also varies. Particles mature in the cytoplasm in sequential steps. The earliest particles seem to contain the mRNA set and to synthesize ds RNA. Then they start producing new mRNA. The structural polypeptides can self-assemble. Cell lysis by unknown mechanisms finally results in release of mature virions. However, virions budding at the PM can sometimes be observed. This may explain why several viruses of the KEM serogroup are sensitive to ether. Cells infected with reovirus contain cytoplasmic inclusions ("viral factories") containing microtubules and intermediate filaments (vimentin). Inhibition of DNA synthesis can be detected at 8 to 10 hpi in L cells. The hemagglutinin seems to be responsible. Persistent infection has been described in embryonic fibroblasts, CHO cells, and mouse L cells.

In terms of the topic being reviewed, the orbiviruses are the most important.

The genome of BTV type 10 has been completely sequenced, and full-size morphological structures representing cores and virions were produced with baculovirus expression vectors.[34] BTV segments are monocistronic. In contrast to other Reoviridae, the temperature optimum of orbivirus transcriptase is 28°C.[609] Correlations between BTV of sheep, EHDV of deer, and AHSV of horse are evident.[724] However, significant evolutionary divergence must have taken place between these viruses transmitted by *Culicoides* species and the tick-borne Kemerovo viruses.[727]

Tubules and granular inclusion bodies are typical for the cytoplasm of cells infected with orbiviruses. These structures are linked to vimentin and may play a role in virus assembly. The tubules consist of NS1.[610,611]

NS2 is the only viral phosphoprotein which is phosphorylated at serine residues in mammalian and insect cells. It binds ss RNA and is part of the inclusion bodies in *Spodoptera frugiperda* cells.[612] Coexpression of two or more proteins with the baculovirus system allowed formation of empty cores (VP3 and VP7) and empty double-shell virus-like particles (VP2, VP3, VP5, and VP7) in insect cells.[613] In such systems, VP1 can be incorporated into virus-like particles, showing that VP1 also has a structural role.[34]

2. Vectors and Cell Cultures

Table 7 represents a compilation of vectors and cell culture types.

3. Selected Topics

Much work has been carried out on the replication of BTV in live *Culicoides* vectors.[607] Apparently, *C. variipennis* midges can be infected for prolonged times; however, there seems to be a series of barriers against the viral distribution in the organism (mesenteron escape barrier, transovarian barrier?). Treatment of BTV with proteolytic enzymes (cleavage of VP2) gives rise to subviral particles which are highly infectious for *Culicoides* and for *A. albopictus* cells in culture (cited according to Reference 607). Core particles (outer shell removed) are also orally infectious for the vector but uninfectious for mammalian cells in culture.

After ingestion, the midgut cells are infected and the virus replicates. Viruses then pass through the basement membrane into the hemocele, and the salivary gland becomes infected. The time required from infection to transmission competence is 10 to 15 days. Variable proportions of refractory animals exist within one vector species; this phenotype is under genetic control.[607] Unfortunately, this important phenomenon has not been studied in cell culture.

Much work has also been done with respect to expression of v proteins by recombinant baculovirus expression systems. The relevant papers can be found.[34]

a. Entry

VP2 is the hemagglutinin and seems to be the binding protein in case of vertebrate cells.

b. Virus Biosynthesis

To the best of my knowledge, the biochemical events associated with orbivirus biosynthesis have not been investigated in arthropod cells, apart from the baculovirus expression systems mentioned above. Some knowledge has been gained by feeding *C. variipennis* with BTV and by then following the course of infection with a direct immunofluorescence technique.[614] After 6 days of incubation, the virus antigen was detected in large parts of the midgut and after 10 days in head squash preparations. Other organs such as the ovaries rarely contained detectable concentrations of antigens. This is consistent with the lack of transovarian transmission of BTV in *C. variipennis.*

c. Cytopathic Effects

In vertebrate cells, CPE by orbiviruses result in cell rounding, detachment of the cells, inclusion bodies, pyknosis of nuclei, and karyorrhexis. Orbiviruses replicating in *A. albopictus* (Singh) cells did not provoke a CPE.[231,615] Viruses of the Kemerovo group and BTV produced plaques in these cells.[142,244,616,617]

An orbivirus (JKT-7400) isolated from *Culex* mosquitoes replicated to high titers and induced CPE in C6/36 cells. The virus produced lethal sensitivity to CO_2 in *Culex* and *Aedes* mosquitoes and *D. melanogaster.* The sensitivity arises after the injection of the virus into the animals: after chilling the insects to 1°C and then exposing them to pure CO_2 for 15 min, they will remain paralyzed after transfer to room temperature. When injected intravenously, the virus was lethal for rabbits. The mechanism of the phenomenon which can also be elicited by rhabdo- and bunyaviruses is unknown.[618] The etiologic agent of hereditary CO_2 sensitivity in *D. melanogaster* is the rhabdovirus-like sigma. Recently, the putative etiologic agent for the hereditary sensitivity in *Cx. quinquefasciatus* was adapted to C6/36 cells. This virus — Matsu — exhibits a most unusual structure.[726]

d. Persistent Infection and Viral Mutants

A. albopictus cells can develop persistent infection carrying viruses of the Palyam subgroup and AHSV.[615,619] In the live vector *C. variipennis,* a high frequency reassortment of BTV serotypes 10 and 17 has been described. The proportion of reassortant progeny varied from 7 to 78% depending on the individual fly after oral infection. Reassortment of most segments was random.[57] Reassortment of genome segments has also been demonstrated in vertebrate cells.[58] If the second virus followed the first by more than 4 h, the superinfecting virus failed to contribute genome segments. Reassortments showed that certain combinations of segments 4 and 6 of Kemerovo viruses, originally isolated from *Ixodes uriae,* resulted in nonpathogenic intra-subgroup reassortants.[59] Data on reassortants, ts and sp mutants obtained in arthropod cell cultures, are eagerly awaited.

e. Arthropod Systems As a Tool

Insect and tick cell cultures were mostly used to assay for susceptibility to orbi- and coltiviruses, to compare such susceptible cell lines as a tool with vertebrate cell lines and to find out whether they would develop a CPE in a short period of time. Yunker and Cory reported the growth of CTFV in cell cultures of *D. andersoni* and of *A. albopictus* (Singh).[620,621] In the latter cells, the titers reached their maximum between 10^6 and 10^7 PFU/ml after 3 to 4 weeks. No CPE was visible. Similar studies investigated the growth characteristics of BTV in various continuous cell lines to find sensitive and reliable systems for virus isolation. All serotypes of BTV tested multiplied in Singh's *A. albopictus* without any CPE;[622] however, best yields were attained from BHK cells. Again, a large series of cell lines was tested to investigate the susceptibilities to BTV. Virus did grow in C6/36 cells and formed discrete plaques; it also grew in lines of *C. variipennis* without plaque formation, but sometimes monolayer destruction, to reasonably high titers in the range of 25 to 32°C. The bovine endothelial cell line (CPAE) produced a higher titer than BHK-21 cells and perished within 2 to 5 dpi.[616,617] Continuous cell lines from various ticks, C6/36, *Cx. quinquefasciatus*-, and *Xenopus laevis* cells in culture allowed the replication of BTV, whereas EHDV failed to replicate in tick cells. C6/36 and *Xenopus* cells were as sensitive to infection with these viruses as Vero and BHK cells. At 27°C, persistent infection of *Xenopus laevis* cells occurred and low yields of BTV and EHDV were produced. Tick cells survived without CPE, whereas the mosquito cells detached from the monolayer within 3 to 6 days.[142]

Finally, it may be mentioned that a panel of arthropod cells may be useful to detect new orbiviruses or to give evidence for hidden contaminating viruses.[623-625] The potential of invertebrate cell lines to grow plant reoviruses has also been demonstrated. For example, inoculation of the leafhopper cell line AC-20 with wound tumor virus resulted in a productive noncytopathic infection with no detectable alteration of cellular protein synthesis.[626]

E. RHABDOVIRIDAE

Relevant papers and books are References 1,2,3a,7,10,11,21,23,109,113,116,118,120, 123–125,145,147,231,627–649.

1. Introduction

The vertebrate and invertebrate viruses are usually bullet-shaped, the plant viruses rather bacilliform. The inner part, the nucleocapsid, has a helical symmetry and consists of one molecule of negative polarity (noninfectious) ss monopartite linear RNA complexed with the N protein. In the inner part, the L and NS proteins are also present. The virions are enveloped; the envelope contains the M protein and protruding spikes built up by the transmembrane G protein. The transcriptase activity is associated with the nucleocapsid. The G protein defines the serotype and exhibits the neutralization epitopes. The Rhabdoviridae consist of the Vesiculovirus, the Lyssavirus, and the plant Rhabdovirus subgroups A and B.

The Vesiculovirus (VSV as a prototype) has a size of 170 × 70 nm. The gene order is 3′-N-NS-M-G-L-5′ without overlaps (11 161 nt in the prototype). The L and NS (better P) proteins occur in small amounts in the virion, together forming the transcriptase. The NS and the M proteins are phosphoproteins. The host range of the genus is wide. Of the approximately 30 members, about one half has been isolated from invertebrates and/or vertebrates. The prototype grows well in mosquito cells. Some members have been isolated from fish. However, there is no clear evidence of transmission of VSV by invertebrates.

The Lyssavirus (Rabies virus as a prototype) is similar to the vesiculoviruses, but antigenically distinct. The genome organization is the same as in VSV. Four serotypes have been distinguished. Rabies virus multiplies in neurons and myotubes of vertebrates; it also grows in insects. Members have been isolated from vertebrates and invertebrates, and the host range of the genus is very wide.

The plant Rhabdovirus subgroups A (prototype Lettuce necrotic yellows virus) and B (prototype Potato yellow dwarf virus) will not be discussed here. It suffices to say that most members are transmitted by leafhoppers, planthoppers, or aphids. The viruses replicate in both plant and insect cells. Nonenveloped rhabdovirus-like particles seem to occur (e.g., Citrus leprosis virus).

In summary, rhabdoviruses again represent a huge VIP family with many members. In the epidemiology, insects play a major role in the transmission of the plant viruses. The rhabdoviruses of plants are the second example of plant viruses with an envelope. However, the only relative significance of the envelope as a taxonomic criterium is demonstrated by the fact that transitions to nonenveloped rhabdovirus-like particles seem to be possible.

The receptor-recognition protein of VSV is the G spike already binding at 4°C. VSV G (Indiana serotype) contains 511 amino acids and 2 N-glycans, which, however, do not contribute to adsorption. The cellular receptors have not been identified; saturable and nonsaturable sites may exist in Vero cells. Phosphatidylserine completely inhibited the high affinity (saturable) binding. VSV penetrates by endocytosis and is fusogenic at acidic pH. The N-terminal amino acids of the G protein may represent the fusogenic peptide stretch. After ejection of the nucleocapsid into the cytoplasm, primary transcription takes place in a manner not disturbed by actinomycin D and inhibitors of protein biosynthesis. Five polyadenylated and capped mRNAs are produced. The L protein seems to possess protein kinase activity, phosphorylating the NS(P) protein. It may also be responsible for capping, methylation, and polyadenylation. The polymerase enters the template at the extreme 3′ end and starts by synthesizing a leader of 48 nt, then transcribes sequentially and allows for polyadenylation of the individual messengers at intercistronic regions. The M protein, being the link between the nucleocapsid core and the virion membrane, also has a transcription-inhibitory function. The messengers are translated immediately; translation is continuous throughout the infection. There is a gradient of quantities, N yielding the highest amounts. The G protein is made on ribosomes of the RER and is translocated by means of a signal. The other four proteins are synthesized in the cytoplasm.

Replication requires biosynthesis of v proteins and can be inhibited with cycloheximide. Again, as for transcription, the RNP template together with L and NS(P) forms the essential component. Full-length plus- and minus-strands are formed. Thus, the enzyme(s) must switch from the transcription mode to a replication mode for readthrough. RNA/N/L/NS cores are continuously formed and coated with M. The G proteins move by vesicular transport into the PM and contact the M protein molecules which leads to budding. VSV causes CPE mainly in vertebrate cells, and kills them. This may result from inhibition of cellular RNA, DNA, and protein synthesis. Sequences have been reported.[650-652]

2. Sources and Cell Cultures

The sources of some viruses and the cell cultures in which they can grow are listed in Table 8.

3. Selected Topics

Mainly for epidemiological reasons, it is still a matter of debate as to whether rhabdoviruses can be considered to be arthropod-transmitted viruses. A glance at tables indicating the organisms in nature from which the viruses were isolated shows a perplexing picture.[1,2] Many viruses were isolated from invertebrates. Considering the quasi-species nature, eventually changing the host range or the sero-specificity by discrete substitutions/mutations, one should have extensive sequence comparisons in order to arrive at firm conclusions about relationships. In the meantime, infection of live vector candidates remains a promising approach. The sandfly *Lutzomyia shannoni* was successfully infected with VSV NJ by oral and by intrathoracic inoculation and a persistent infection occurred. Virus was transmitted by bite, and transovarially to progeny.[653] After a blood meal, the virus was first detected in the midgut after 34 h; after 5 to 6 days, it could be found in salivary glands. The virions matured by budding from the PM. In the midgut, budding took place at the basolateral side, in the salivary glands at the apical side.[654] In these studies, no CPE was observed. Similarly, Chandipura virus multiplied in *Cx. tritaeniorhynchus, bitaeniorhynchus, quinquefasciatus,* and *A. aegypti* after parenteral injection.[655]

a. Entry and Uncoating

As in all cases of viral families with extremely large host ranges, it is tempting to assume an ubiquitous cell surface component as receptor. On the basis of enzymatic digestions,

TABLE 8
Rhabdoviridae: Sources and Cells[1,21,120,653,688]

Genus serocomplex, member	Sources of virus	Invertebrate cells
1. Vesiculovirus		
Vesicular stomatitis virus (VSV)	Bovines, equines, swine; black flies, culicoid midges, eye gnats, mosquitoes, mites, muscoid flies, *phlebotomines*[654,681]	*A. aegypti* *A. albopictus* (Singh), C6/36, C7-10 77 OM 3[663] *Antheraea eucalypti moth*[687]
New Jersey (NJ)		
Indiana		
Alagoas		
Cocal		
San Juan		
Orsay		
Chandipura	Mammals; *Culex* species, sandflies, *A. aegypti*	*A. aegypti, A. albopictus, A. pseudoscutellaris, Anopheles stephensi, R. appendiculatus*[113]
Isfahan	Sandflies, ticks	
2. Lyssavirus		
Lyssa (rabies) virus	Bats, cats, cattle, dogs, humans, wild carnivores; insects	*A. albopictus* (Singh), C6/36 AP 61
Bovine ephemeral fever	Insects, vertebrates	
Eel virus	Vertebrates	
Kimberley virus	Insects, vertebrates	
Kotonkan virus	*Culicoides*, mosquitoes	*A. albopictus*
Obodhiang virus	*Mansonia*, mosquitoes	*A. albopictus*
3. Probable members		
Bahia grande serogroup	Mosquitoes	
Hart Park serogroup	Vertebrates, invertebrates	
Flanders virus	Birds, mosquitoes	
Infectious hematopoietic necrosis virus (IHNV)	Vertebrates, invertebrates (salmonids and shrimp)	
Sigma	Invertebrates, *Drosophila*	*D. melanogaster*[123,678]

surface proteins and sialic acids were not considered to represent receptor candidates for VSV. Detergent extracts from the cells inhibited the binding. The inhibitor was destroyed by phospholipase C and identified as phosphatidylserine. Removal of fatty acids from HeLa, CER, EPC, and *A. albopictus* cells followed by extraction of the lysoderivatives caused a strong restriction of infectibility with VSV. Surprisingly, removal of galactose also reduced VSV attachment to *A. albopictus* cells to 50%. Obviously, neuraminidase treatment had no effect on binding in the case of the mosquito cells, a finding confirming earlier data. Thus, a common component for VSV attachment may be a glycolipid.[656,657] In the same cell lines, lysosomotropic agents blocked the early stages of infection. Virions were bound to the cell surface at 0°C. After a shift to 26°C, virions could be seen within 1 min by electron microscopy in coated pits and within 5 min in coated vesicles of mosquito cells. It took then about 30 min until virions appeared in larger smooth vesicles. These findings are arguments in favor of receptor-mediated endocytosis.[658] Other authors rather took the view that receptor-mediated endocytosis is not essential for uptake of VSV.[207]

Rabies virus concentrates at neuromuscular junctions. The nicotinic acetylcholine receptor was postulated to represent the receptor, but cells lacking this receptor were also susceptible to rabies virus infection and further search implied a lipidic compound (gangliosides). On the other hand, sialic acid was shown to be necessary for the attachment of rabies virus to myotubes in culture.[659] The number of virions bound to IMR 32, HeLa, CER, EPC, and *A. albopictus* cells was similar. The lipidic fractions (total lipids, glycolipids) from the PM of all these cell lines prevented the infection of CER cells. Removal of sialic acid from PM caused a dramatic reduction of the susceptibility of the vertebrate cells, but not of mosquito cells. The vertebrate cells were widely susceptible to infection, whereas a lower level of specific viral antigens was detectable in *A. albopictus* cells. At 37°C, only 30% of the *A. albopictus* cells showed a positive immunofluorescence after 48 and 72 h; virus was not detected in the culture fluid. It is important that the differences in permissiveness did not correlate with the insignificant differences in number and types of cell surface binding structures.[660]

A. pseudoscutellaris cells were analyzed. Acid glycosphingolipids from mosquito cells and (neutral) galacto- and glucocerebrosides from bovine brain and human spleen, respectively, prevented viral antigen synthesis to a large extent; gangliosides from bovine brain completely inhibited the synthesis. Only β-galactosidase proved to be moderately effective among a series of glycosidases used to treat the cell surface. Ricinus communis lectin (binding GlcNAc and Gal residues of glycoproteins and glycolipids) had a noticeable inhibitory action on binding.[661] All this data is corroborated by the fact that rabies virus infects many cell types in culture. The presence of the nicotinic acetylcholine receptor on the PM is not an obligatory factor for rabies susceptibility (L 8 myoblasts, L 8 Cl 3 U, BHK-21, NA, CER, C6/36 cells).[662] The entry of rabies virus seems to occur by receptor-mediated endocytosis, the penetration through the membrane being acid-dependent.

b. RNA and Protein Biosynthesis

Usually, cultured insect cells produce low yields of infectious VSV;[656] however, *Drosophila* cells in culture, and the *A. albopictus* clones LT-C-7 and C6/36 (own unpublished data) produce high titers.[663,664] In LT-C-7 cells, yields between 6×10^8 and 3×10^9 PFU/ml were obtained by 18 to 24 hpi. At 34°C, the rate of production, but not the final yield, was greater than at 28°C. At 12 hpi, a massive CPE started to appear. The virions, as in case of Sin or SFV, were biologically equivalent to those produced in vertebrate cells. However, the G protein of VSV grown in the mosquito cells was devoid of sialic acid. The virus contained less G and L molecules on a molar basis as compared to virions grown in BHK cells. In infected mosquito cells, mainly the G, the N, and the M proteins were visible; the L and NS proteins were faint at 28°C. With cell fractionation, the G protein was found exclusively in the microsomal fraction, the NS protein predominantly in the soluble fraction, as in vertebrate cells.

Six RNAs corresponding to the 42S genomic RNA and to the five mRNAs could be extracted from LT-C-7 cells and resolved in denaturing gels, migrating with the same relative mobilities as those extracted from BHK cells. However, the proportion of G mRNA was increased in the mosquito cells; in view of the relative paucity of G proteins in these cells and their virions, the increased proportion of G protein mRNA suggests a translational control.[664] In contrast to alphaviruses, actinomycin D did not inhibit v protein synthesis.[141,664]

In *Drosophila* cells, only the VSV G, N, NS, and M mRNAs could be detected by a hybridization assay; they appeared as early as 3 hpi. The rate of positive-stranded RNA synthesis increased in vertebrate cells up to 3 hpi and then remained constant up to cell lysis. In *Drosophila* cells, it increased, and started to drop after 3 hpi to a minimum already reached 7 hpi. It is unlikely that the mRNAs were rapidly degraded as v protein synthesis was going on for many hours; G protein synthesis came to a stop at 15 hpi, when N, M, and NS protein synthesis still continued. VSV leader RNA could be detected in *Drosophila* cells and it seemed to migrate into the nucleus, albeit in very low amounts as compared to vertebrate cells. In vertebrate cells, the leader RNA is involved in shut off of transcription.[123]

Rabies virus was used to infect *A. pseudoscutellaris* cells, incubated at both 26 and 33°C. At 33°C, infected and non-infected cells were heavily damaged after 12 days. At 26°C, the production of v material was slow; after 60 days of incubation 70% of the cells were positive for the N protein as assayed by immunofluorescence. The envelope protein could not be detected before 30 dpi. Virions were not found in the supernatant. CPE was rare.[661] Thus, a persistent infection seems to be established very early after infection.

As in the case of alphaviruses grown in mosquito cells, there is an absolute need for methionine.[407] LT-C-7 cells infected with VSV produced 1000 times less infectious virus than cultures kept in a complete medium. The reduction in virus yield was directly correlated with the reduction of v RNA and v protein synthesis. The v mRNA produced in methionine-starved cells was not polysome-associated in contrast to that of cells maintained in complete medium. This effect was reversible within 15 min after addition of methionine. These and other data suggest that methionine starvation of *A. albopictus* cells prevents the methylation of the caps, thereby reducing the efficiency of translation. Cycloleucine reduces the methylation of RNA; it inhibits the biosynthesis of adomet and also inhibits the replication of Sin and VSV in *A. albopictus* cells. Note that the enzymatic functions have to be coded for by the virus in case of all those RNA viruses which replicate in the cytoplasm.[665] The authors think that the VSV RNA methyltransferase may have a higher Michaelis constant for adomet than the corresponding (nuclear) host enzyme(s). In contrast to results with cellular mRNAs, cap methylation of VSV mRNA *in vitro* is as follows: at low concentrations of adomet, it is Gppp Nmp; at higher concentrations it becomes m^7Gppp Nmp. Only the latter structure allows translation *in vitro.* The question remains why there is such a drastic difference for the virus grown in mosquito cells as compared to vertebrate cells (BHK, CEF) when the medium lacks methionine.

In this respect, host range mutants of VSV which have an mRNA methylation defect are interesting. HEp-2 cells are nonpermissive *in vivo:* the synthesis of mRNAs is reduced and the synthesis of genome-length RNA is abolished. The mRNAs are not translated either *in vivo* or *in vitro.* There is an unmethylated cap, GpppA, which can be methylated *in vitro* when high levels of adomet are added. The permissive BHK cells have a tenfold higher level of endogenous adomet than the nonpermissive HEp-2 cells.[666]

Thus the wt VSV (Indiana serotype) reacting so drastically to methionine deprivation in mosquito cells closely mimicks a host-range mutant.

c. *Cytopathic Effects*

To produce a prominent CPE in VSV infected C-7 cells, serum had to be added to the medium and the temperature had to be increased to 34°C. However, virus yields were similar at 28°C without serum to those at 34°C with serum. Under CPE-provoking conditions, host RNA and protein synthesis were markedly inhibited and polysomes decayed into 80S monosomes. On the other hand, the rate of v macromolecule synthesis was relatively stimulated. All these findings suggest that at 28°C, competition between v and host mRNAs exists for ribosomes and that at elevated temperatures there is a block at the level of initiation for (host) mRNAs, so that the competitive restriction of v mRNAs is released.[667] The phosphorylation of polysome-associated proteins was reduced under conditions of overall inhibition of protein biosynthesis.[23,667] Cell-free systems for protein translation prepared from infected or noninfected cells kept under the conditions mentioned mimicked the cell culture system. The activity of an extract from infected cells maintained at 34°C in the presence of serum could be partially restored by addition of a 100,000-g supernatant from uninfected cells.[668,669] Incubation at 34°C seems to predispose to CPE linked to an inhibition of host protein synthesis by increasing the amounts of v gene products.

Replication of VSV was studied in several *Drosophila* cell lines, but also in 18 clones from the *Drosophila* line 77 OM 3; however, the responses of all the clones were identical and no CPE was visible. The production was quite low at the maximum (about 8 hpi, 17 PFU per cell per 2 h).[123]

d. *Post-Translational Modifications, Transport, and Maturation*

Unfortunately, to my knowledge, mosquito cells have not been used for studies of post-translational modifications (apart from the glycan analysis) of transport of the G protein from the RER through the Golgi complex to the PM, or of maturation. In contrast, there exists a large body of literature concerning these topics in vertebrate cells. The NJ serotype of VSV grown in

BHK cells contains two identical oligosaccharides with three conventional outer chains at the trimannosyl core.[670] The G protein contains one palmitate linked to the cysteine residue of the C-terminal cytoplasmic domain.[671] The rabies virus glycoprotein contains three potential glycosylation sites, but only two appear to be glycosylated. Extensive work with G protein mutants (complementation group V) of VSV and rabies depict a complicated situation.[672,673]

Three conclusions may be drawn:

1. The best guess as to the role of glycans in viral glycoproteins is that they increase the solubility and promote the correct conformation; however, the consequences of the lack at a given glycosylation site are unpredictable, because the amino acid sequence context governing folding seems to play a role. Similar conclusions as to the role of fatty acids may be drawn.
2. In the case of VSV, possessing only one type of spike protein, particles without the G protein may be formed. One can hypothesize that the M protein is sufficient to drive the budding process. In fact, even under normal conditions, VSV particles (Indiana serotype, Vero cells) are heterogeneous; early virions contain only one third the amount of G protein relative to M and N. This also explains the incorporation of VSV cores into virion glycoproteins of unrelated viruses (pseudotypes).[674]
3. This situation is different from that of alphaviruses where the heterooligomeric spike must interact directly with the nucleocapsid without the help of a bridging matrix protein.

A promising approach is the comparative investigation of invertebrate rhabdoviruses such as Sigma (a probable member of the genus lyssavirus) and vertebrate rhabdoviruses such as VSV or Piry (a member of the genus vesiculovirus) which all grow for example in *D. melanogaster* cells.[123,663,675,676] This has been analyzed in 77 OM 3 cells and has been reviewed elsewhere.[123] It is interesting to note that the G protein of VSV (Indiana) was synthesized in only small amounts throughout the cycle (see above for LT-C-7 cells[664]) and that the M protein was hyperphosphorylated with quantitative and qualitative changes as compared to that generated in CEF cells. The G synthesis rate relative to that of the N protein was fivefold lower than in CEF cells, and the G protein deficiency was also observed in the released virions. The synthesis of each v protein was increased upon actinomycin D treatment, but virions still contained as little G protein as in the absence of the drug.[663]

e. Persistent Infection and Host Factors

The highly cytopathogenic VSV usually leads to cell death in vertebrate cells. Persistent infections are mostly the result of unusual conditions (nonpermissive cells, infection with ts mutants or with wt virus in presence of a high proportion of DI particles, or after interferon treatment). Some cells survive and the viral population then evolves rapidly: phenotypes with reduced cytopathogenicity will become dominant.[677] In contrast, in insect cells, there is usually no CPE and after an acute phase, the virus production declines and a persistent infection is established.[663] Sublines and clones of insect cells as described above, reacting strongly to the infection, are rather exceptions.[664] In cloned *Drosophila* cells, a CPE was never observed after infection with VSV.[123] In these cells, the rate of synthesis of v proteins increased for 7 hpi and then declined to the persistence level; the production of virions followed this curve with some delay.[663] Transcription in the 77 OM 3 line was already strongly inhibited at 7 hpi, whereas genomes and antigenomes accumulated for 15 h to decline afterwards. Short plus-strand molecules (5′ end of antigenomes; leader and variable parts of N mRNA) were encapsidated. Assembly and release of virions were not restricted, but the infectivity was extremely reduced (0.05% as compared to 10% in CER cells).[676] The positive-stranded leader RNA seems to be involved in the inhibition of host synthesis in vertebrate cells; however, in *Drosophila* cells it was found in low amounts only in the cytoplasm and it did not migrate into the cell nucleus.[678]

Host factors, such as tubulin and/or microtubule-associated proteins, seem to play an important role in VSV multiplication. Several refractory genes of *D. melanogaster* interfere with the multiplication of Sigma virus and Piry virus;[679,680] five have been localized.[123] However, the exact relations between the functions encoded and the phenotypic symptoms as described above are unclear.

The living *Drosophila* flies also develop a persistent infection and shed Sigma virus throughout their life. Hereditary transmission via germ cells seems to be the only known transmission pathway. VSV was transmitted to the progeny in the case of sandflies.[681]

In summary, the infected arthropod seems to have the ability to neutralize the killing factor, e.g., by reducing the biosynthesis or the concentration of the leader RNA and/or by excluding this RNA from the nucleus. The genetic drift of the viral population is an epiphenomenon automatically following the establishment of a persistent infection.

f. Mutant Rhabdoviruses

In vertebrate cells, VSV and related viruses grow rapidly. Because of the relatively small genome encoding only five proteins, rhabdoviruses are ideal for the isolation of mutants. In fact, due to the genetic lability, ts mutants and revertants occur spontaneously and many have been described for all sorts of purposes. The high substitution frequency may also explain why fewer than 1 in 10 morphologically indistinguishable progeny virions in a cloned VSV population are infectious.[55,648]

All VSV (Indiana) ts mutants (nonpermissive temperature: 37 to 40°C, permissive temperature: 30°C; vertebrate cells) (for a review see Reference 639) fall into five complementation groups according to the five cistrons. Mostly the basis is a single (nonconservative) base substitution in a gene, leading to temperature-sensitive conformational changes, which in turn leads to an abolished function of the respective protein at the nonpermissive temperature.

DI particles with large deletions of parts of the genome are extremely frequent. Due to the envelope proteins provided by coinfecting intact virions, such deleted RNAs can be packaged and released and may be infectious. All these types, generated mainly during undiluted transfer of virus suspensions, interfere with and dampen the replication of standard infectious virus. For generation of DI-RNA, a copy-choice mechanism of replication as in RNA-RNA recombination is most probable.[682] Infection with high proportions of DI particles can lead to persistent infections in vertebrate cells.

More exotic changes are known; for example, some ts mutants of VSV in the M protein gene (complementation group III) form noninfectious particles which contain the M and G proteins but no RNA. This demonstrates again the importance of the M protein-nucleocapsid interaction and the key role of M in bud formation.

Pseudotype formation has also been described. The G protein is incorporated into virions of an unrelated virus, or the genome of VSV is decorated by the envelope proteins of certain unrelated viruses under conditions of dual infection of a cell.[683]

Temperature-sensitive mutants of the three RNA complementation groups of VSV (I, II, IV) can interfere with wt VSV in vertebrate cells. Simultaneous infection with the mutants and the wt led to a 10- to 10,000-fold reduction of wt progeny yields, depending on the mutant. After a long-term persistent infection of L cells, one of these mutants became dominant; this mutant, when assayed on uninfected cells, was very inefficient in shutoff of host protein synthesis and even interfered with the wt in double infections with respect to shutoff.[684]

Other variants of VSV were found to generate and amplify DI particles at a rate greatly exceeding the rates normally observed for wt VSV or other variants. One single undiluted passage of the first clonal pool of such a variant produced masses of DI particles.[685]

Thus, the interplay between mutant and deleted genotypes/phenotypes with the wt virions is extremely complicated and variable depending on the constellation. In addition, the host cell can influence the generation and replication of DI particles. An example of a host-cell effect on the replication of DI particles of Sin has been given in the corresponding section.[377]

Another example has been reported.[686] In this case, serial undiluted passages of a VSV clone in LT-C-7 cells resulted in a rapidly declining titer of infectious virus. Virus after two passages (VSV_{AP2}) contained truncated, interfering, noninfectious particles. VSV produced by three undiluted passages in BHK cells (VSV_{BP3}) interfered with the replication of VSV equally well in mosquito and BHK cells; however, VSV_{AP2} interfered efficiently only in the former. When VSV_{AP2} or VSV_{BP3} was passaged once in the heterologous cell type the interference pattern was reversed; thus, the specificity of interference with VSV_{STD} could be modified. Interference led to a decrease in synthesis of standard genome RNA (42S RNA). The major DI particles in VSV_{AP2} contained covalently linked sequences of positive and negative polarity which were highly self-complementary ("snapback" RNA). Such RNAs were not detectable in VSV_{BP3} DI particles. This is an interesting example of a host influence, since the replication of DI particles was associated with an interference with the production of standard virus (mosquito cells infected with VSV_{AP} DI particles) in one case only. The rapid modification of the interfering specificity is difficult to explain. It is reasonable to assume that incoming inner proteins (L, N, NS) play a role in interfering at the level of replication (where interference occurs), because they could be modified differently according to the cell type.

F. FAMILIES BEYOND THE ARBOVIRUSES[1,2,689-691]

These families will not be treated in depth in this paper. However, they deserve to be mentioned, because they demonstrate again the relativity of host ranges. In these families, the important genera with members infecting man and animals do not infect arthropods and are not transmitted by arthropods. However, they possess genera which might be described as insect (arthropod) viruses, some with arbovirus-like members.

1. The Positive-Sense Small RNA Viruses of Insects[689-694]

a. Nodaviridae

Nodaviruses are considered to represent true insect viruses and thus do not seem to fit into the topic of this review. However, two outstanding characteristics justify a brief mention of these viruses. (1) The host range of some species is amazing. (2) Nodaviridae are the (only) positive-sense, ss, bipartite animal RNA viruses. They seem to have a position between the Poty-like (Picorna-like) and the Flavi-like supergroups[75] and may therefore represent a link between the animal and the plant plus-sense RNA viruses. As stated in Section II, partite genomic arrangements are frequent in plant viruses. The particles are small (30 nm) and nonenveloped. They contain two proteins, β and γ, in a T = 3 icosahedral surface lattice. The two capped, nonpolyadenylated RNAs 1 (about 3100 nt) and 2 (about 1400 nt) are copackaged within the same particle; they are only infectious in combination. Replication occurs in the cytoplasm, where three RNAs, 1, 2, and 3, can be found. The nonpackaged RNA 3 is a subgenomic mRNA product derived from the 3′ end of RNA 1. RNA 1 has two open reading frames: one encodes protein A, and the other, in a different reading frame, encodes the small protein B. A is probably the catalytic part of the replicase; it contains the GDD amino acid sequence motif. RNA 3 is the messenger for the B protein which has an unknown function.[695] RNA 2 encodes the coat protein precursor α. It is cleaved autocatalytically into the coat proteins β and γ after assembly of the precursor into provirions.[695] Cleavage-deficient provirions are noninfectious.[696] After transfection, RNA 1 can direct its own replication in absence of RNA 2. RNA 1 is therefore a replication module and with its encoded enzyme activities forms an autonomous RNA replicon. In the course of multiple passages of the v RNAs in BHK cells by transfection at limiting dilutions, RNA 1 was capable of prolonged autonomous replication accompanied by synthesis of RNA 3.[697,698] All species — Nodamura, Black beetle virus (BBV),

Flock House virus (FHV), Gypsy moth virus, and others — were isolated from Diptera, Coleoptera (BBV, FHV), or Lepidoptera (Boolarra, Nodamura virus). All form plaques on *Drosophila* cell monolayers (mostly Schneider 1 cell lines) except Nodamura virus. This virus was originally isolated from *Cx. tritaeniorhynchus* mosquitoes. It is transmissible by *A. aegypti* to suckling mice; in infant mice, it produces an ascending paralysis and death. But it also kills adult *honey bees* and *wax moth* larvae when injected. Nodamura virus replicated in BHK cells, in *A. aegypti,* and *A. albopictus* cell cultures with a persistent infection and in AP-61 cultures.[699,700] It multiplied and caused death when injected into the thoraces of adult *A. albopictus* and *Toxorhynchites amboinensis* mosquitoes, but not in *Cx. quinquefasciatus.*[700] The virus is unique among the members of the Nodaviridae in its ability to infect lethally both insects and higher animals. BBV did not replicate in BHK cells or kill suckling mice. It is really awesome that FHV (transfected RNA) produced infectious cycles in plants and massively in mammalian cells in culture (BHK cells).[697,701] FHV RNA infected whole plants or protoplasts of barley, cowpea, and tobacco after conventional inoculation and produced virions which were able to move. In addition, reassortment in *Drosophila* cells seems to be possible.[702]

b. *Tetraviridae*[1,48,49,693]

These icosahedral (T = 4) virions contain 1 (to 2) ss positive-sense molecules of an unpolyadenylated RNA and (1 to) 2 protein types. The viruses replicate primarily in the cytoplasm of gut cells of Lepidoptera. The seven members propagate exclusively in insect hosts. The type member Nudaurelia capensis β virus was isolated from the pine emperor moth *Nudaurelia cytharea capensis*; it has a diameter of 35 to 40 nm and encapsidates a ss RNA with an M_r of 1.8 to 1.9×10^6 Da in a dual polypeptide type shell. Often the larvae are coinfected with Nudaurelia capensis ω virus, also integrated in the group. This virus contains 2 species of ss RNA in a dual polypeptide type shell. RNA 2 has recently been sequenced (despite the fact that there is no suitable cell culture system), and the organization is similar to that of the RNA 2 of the T = 3 Nodaviridae.[703] At this time, the ω virus is unique among the Tetraviridae because of the bipartite genome. Thus, this fascinating family might represent a transition from the monopartite to the rare partite ss plus-sense animal viral RNA genomes within the Poty-like (Picorna-like) superfamily.

c. *Picornaviridae*[689,690,693]

The virions are nonenveloped, icosahedral particles with a diameter of about 30 nm. They contain one molecule of monopartite, ss, positive-sense, linear RNA (M_r 2.4 to 2.7×10^6) with a Poly(A) tract at the 3′ terminal and a small protein, VPg, covalently linked to the 5′ end. Four proteins, derived from a single polyprotein, are present in the capsid. The following genera have been described: Enterovirus (e.g., human poliovirus), Hepatovirus (e.g., human hepatitis A virus), Cardiovirus (e.g., encephalomyocarditis virus), Rhinovirus (e.g., human rhinovirus), and Aphthovirus (e.g., foot and mouth disease virus). In these genera, most species are host-specific with a small host range in mammals. No vectors are known. However, the interesting picornaviruses have not been assigned to genera; they are insect viruses such as the *cricket* paralysis virus and the *Drosophila* C virus.[704] They may be related to unclassified small RNA viruses (bee acute paralysis, bee virus X, *Drosophila* virus A and P, aphid lethal paralysis virus). These viruses are mentioned here because again a viral superfamily (the Potyvirus-like or Picornavirus-like) possesses a host range from plants and insects to mammals, even if the individual members cannot be considered as arboviruses.

2. Double-Stranded RNA Viruses: the Birnaviridae[1,705,706]

The Birnaviridae represent the bisegmented ds RNA virus group. It is most significant that this exotic animal virus family also has representatives infecting arthropods. Prototype(s) is infectious pancreatic necrosis virus (IPNV) infecting mostly salmonids, or infectious bursal disease virus (IBDV) infecting the bursa *Fabricius* of chickens, ducks, and turkeys. The host range of the family encompasses fish, mollusks, and birds; Drosophila X virus infects *Drosophila*. Transmission occurs horizontally and vertically and no vectors are known.

The particles are icosahedral, 60 nm in diameter. There is no envelope. Two segments of linear ds RNA, each containing an RNA-linked protein (VP1), make up the genome. Segment A encodes proteins in the order 5′-pre VP2-NS-VP3-3′; segment B encodes the putative RNA polymerase VP1. VP4 is a truncated version of VP3 in case of IPNV. VP2 is glycosylated and is the major capsid protein containing the neutralization epitopes. Transcription produces 2 full-length messengers without a 3′-poly(A) tail. Translation of segment A yields a formal polyprotein which seems to be cotranslationally cleaved into pre VP2, NS, and VP3 (VP4); pre VP2 is cleaved later during virus maturation. Cells lyse. Reassortment has been described.

3. The Minus-Sense RNA Viruses: the Orthomyxoviridae

The orthomyxoviruses (Influenza viruses A, B, C, and D) are enveloped viruses with helical nucleocapsids containing a segmented single-stranded linear genome of negative polarity. Virions are pleomorphic to filamentous. In influenza A and B, the M_1 protein is believed to form a layer inside the bilayer membrane. Homooligomeric hemagglutinin and neuraminidase spikes in a ratio of about 4 to 5:1 protrude from the membrane. The nucleocapsids contain a core protein and the P proteins. Influenza A and B viruses possess 8 RNA segments (900 to 2350 nt, total M_r 4.5×10^6).

The P proteins PB1, PB2, PA, and the nucleoprotein NP are found in the transcriptase complex. The phosphorylated NP is a species-specific antigen. Influenza A viruses infect man, other mammals, and birds, but mutation and reassortment make the host range quite unpredictable. Influenza B seems to infect only man.

In influenza C, 7 RNA molecules are present. Segments 1 to 3 code for 3 polymerase proteins (PB1, PB2, PB3); segments 4, 5, and 6 code for the hemagglutinin-esterase (HEF), the nucleoprotein (NP) and the membrane protein M, respectively. Segment 7 codes for 2 nonstructural proteins (NS1 and NS2). Correspondingly, virions are composed of six proteins.

In terms of the topic reviewed here, Influenza D is the most interesting. It comprises tick-borne viruses (Dhori and Thogoto) which occasionally infect man. The virions contain 6 (Thogoto) or 7 (Dhori) RNA segments. Thogoto virus is of medical and veterinary significance and is transmitted by *R. appendiculatus* to cattle. In the laboratory, the virus grows in hamsters, in the mouse, and in Vero and BHK cells in culture. It is interesting to note that all three stages of *R. appendiculatus* can acquire the virus while cofeeding with infected ticks on an apparently nonviremic host (guinea pig).[707]

The sequences of RNA segments 5 (NP), 4 (single envelope protein, gp 65), 2 (PB1), and 6 (M) of Dhori (India) have been elucidated. gp 65 does not share amino acid sequence homologies with other orthomyxovirus envelope proteins, but it shares homologies with the baculovirus envelope protein gp 64 (cited according to Reference 708). A similar situation pertains in the case of Thogoto virus.[728]

4. DNA Viruses

a. The Single-Stranded Parvoviridae[1]

The parvoviruses are nonenveloped icosahedral particles with a diameter of 18 to 22 nm. They contain 1 molecule of monopartite ss DNA (M_r 1.5 to 2.0×10^6) and 3 to 4 proteins. The family consists of the Parvovirus group, the Adeno-associated group (Dependovirus), and the insect Parvovirus group (Densovirus). The parvoviruses exhibit a relatively large host range within vertebrates. The helper-dependent dependoviruses also infect vertebrates. The host range of the densoviruses encompasses Diptera, Lepidoptera, and Orthoptera. Infection of and multiplication in crabs and shrimps is possible.

b. The Double-Stranded Poxviridae[1,709,710]

This very large family has virions of a diameter of 220 to 450 × 140 to 260 nm, containing 1 monopartite double-stranded DNA genome (130 to 375 kbp) and more than 100 polypeptides. The subfamilies Chordopoxvirinae (poxviruses of vertebrates) and Entomopoxvirinae (poxviruses of insects) are distinguished. The host ranges are generally narrow. Poxviruses of Coleoptera (genus A), of Lepidoptera and Orthoptera (genus B), and of Diptera (genus C) are known. Cotia and Embu viruses are transmitted to rodents by mosquitoes; they are not yet integrated into genera.

V. CONCLUSIONS

Unquestionably, virology has arrived at a crossroad. The gain of new knowledge concerning taxonomy, interviral relationships, types of cell cultures which can be used for growth of viruses, and analysis of viral components in cell culture will soon come to an end. At the turn of the millennium, we will know the sequences of the genomes of most or all of the important viruses. Alphaviruses may represent the first viruses providing us with a clear picture as to how structure and function are connected and how proteins and nucleic acids interact to form a neat geometrical structure. New virological problems will arise. One of these problems will be the prediction of the three-dimensional configuration of a protein on basis of its amino acid sequence. A related problem will consider the interactions of viral proteins and of viral and host proteins, respectively. Investigations on the reactions of host cells to viral infections will be important, and this review mentions a series of corresponding white spots. What is the nature of invertebrate antiviral factors? What is the nature of the factors cooperating in virus formation? Are there host proteins neighboring viral proteins in a regular fashion, for example, in the ER or in the PM? What is the reason of the expansion and remodeling of the ER, occurring in invertebrate cells in several cases of viral infections? What is the role of stress proteins induced by viruses and how is transcription triggered? How can it be explained that virions of the same strain mature sometimes at the PM and sometimes at intracellular locations in different invertebrate cell lines? A puzzling aspect should be mentioned here. On the one hand, many viral families infect the cells through the endocytotic pathway, making use of the acidic endosomal pH to fuse the virion membrane with the endosomal membrane. On the other hand, the spike proteins of these viruses have to travel after their biosynthesis through acidic compartments of the exocytic pathway. This poses a dilemma. It seems reasonable to assume that transient protection mechanisms must exist which shield the spikes during the exocytic transport against the fusogenic or inactivating influence of an acidic pH. A systematic analysis of the families of enveloped RNA viruses presents the following picture.

1. The group of families/genera not transmitted by arthropod vectors (Myxoviruses A, B, C; paramyxoviruses; coronaviruses; filoviruses; arenaviruses) synthesizes monocistronic or quasimonocistronic mRNAs. Accordingly, it forms one or two types of homooligomeric spikes. The envelope protein containing the fusogenic sequence such as HA_0 of Influenza may undergo a post-translational cleavage (e.g., $HA_1 + HA_2$) which is not mandatory for morphopoiesis but rather for activation of infectivity and fusogenicity. Thus, the fusogenic sequence is embedded as long as possible in the interior of the fusogenic homooligomer. In this group, matrix proteins are mostly present. There is convincing evidence that the M_2 protein of some influenza A strains can act as a proton channel, increasing the pH in the exocytotic pathway and protecting thereby the fusogenic spike.

2. The group of families/genera transmitted by arthropods makes use of polygenic messengers (togaviruses, flaviviruses, bunyaviruses). Accordingly, one type of heterooligomeric spike is formed. Post-translational cleavage events (see above) concern the nonfusogenic partner in the spike; after this (late) cleavage one of the products may be lost. This piece seems to represent the protective shield for the fusogenic protein in the spike. The fusogenic sequences seem to be buried in the depth of the sequence of the fusogenic proteins. A matrix protein is not present. The group of genera which may be derivatives of arthropod-borne families, which, however, are not transmitted by arthropods (rubi-, pesti-, hepatitis C-, and hantaviruses), seems to have lost the information for synthesizing the cleavable propiece of the nonfusogenic precursor.

It is unknown whether these organizations of messengers and spikes represent true needs with respect to the host or whether they are fortuitous. In addition, the pH of the exocytic compartments of invertebrates is unknown. However, the interesting question deserves some attention. In any case, arthropod cell lines will remain most worthful tools for basic and applied virus research.

ACKNOWLEDGMENTS

The Swiss National Science Foundation supported my group in the work with vector-borne alphaviruses for many years. I am grateful to my collaborators Dr. C. Schoch for the literature search and Dr. R. Julen and C. Schärer for help with the tables, and all of them for discussions. I thank Mrs C. Lam and Mrs U. Brönnimann for excellent secretarial work and colleagues for preprints. This review is dedicated to my wife Catherine and to Miss U. Kohler, my technician-in-chief, for their endless patience.

The literature search was extended until 1992. Data bases were MEDLINE and BIOSIS under the headings: "viral family/genus/cell line/cell culture/insect vectors/tissue culture/virus replication"; and "arthropod cell lines/viruses", respectively.

NOTE ADDED IN PROOF

Additional information concerning the inter-viral relationships of Alpha-virus (IV.A.1.) can be found in Weaver et al., *Virology,* 197, 375, 1993. Rubella virus is placed in a group with hepatitis E virus. Cholesterol (IV.A.3.c.) was also needed for the *exit* pathway (C6/36) of viruses (SFV) that require cholesterol for uptake (Marquardt et al., *J. Cell Biol.,* 123, 57, 1993; Justman et al., *J. Virol.,* 67, 7597, 1993). The old question concerning the triangulation number of the Alphavirus - nucleocapsid (IV.A.3.k.) has been decided; it is T = 4.[336a] The binding domain of the Sin C protein for the RNA packaging signal (IV.A.3.k.) lies between aa 76 and 116 (Geigenmüller-Gnirke et al., *J. Virol.,* 67, 1620, 1993). Luo and Brown (*Virology,* 194, 44, 1993) have now purified the nativiral protein (IV.A.3.m.); it is a hydrophobic peptide of 3200 Da. The 3′ terminal 250 nt of minus-strands (Sindbis, SFV, RRV) contain four domains binding the same 50- and 52 kDa host factors in extracts of C6/36 cells. Corresponding probes from Rubella virus bound these proteins only poorly (IV.A.3.n.,o.) (Pardigon et al., *J. Virol.,* 67, 5003, 1993). Sindbis infecton of various mammalian cell types led to apoptosis; however, the lytic infection was converted to a persistent infection by the cellular oncogene *bcl*-2 (Levine et al., *Nature,* 361, 739, 1993). Tick-borne flaviviruses (IV.B.) share a common hexapeptide in their envelope glycoprotein (Gritsun et al., *J. Gen. Virol.,* 74, 287, 1993). A glycoprotein corresponding to C-prM could be detected in lysats of *A. albopictus* infected with DEN-2 (IV.B.3.d.); such a protein was not found in infected Vero cells. The pro-fragment pr was cleaved at the time of virion release and was shed into the medium (Murray et al., *J. Gen. Virol.,* 74, 175, 1993).

REFERENCES

1. **Francki, R. I. B., Fauquet, C. M., Knudson, D. L., and Brown, F., Eds.,** Classification and nomenclature of viruses. *Arch. Virol. Suppl.,* 2, 1991.
2. **Fields, B. N. and Knipe, D. M., Eds.,** *Virology,* Vol. 1 and 2. 2nd ed., Raven Press, New York, 1990.
3. **Hurlbut, H. S. and Thomas, J. I.,** The experimental host range of the arthropod-borne animal viruses in arthropods, *Virology,* 12, 391, 1960.

3a. **Tesh, R. B. and Johnson, K. M.,** Vesicular stomatitis, in *Diseases Transmitted from Animals to Man,* 6th ed., Hubbert, W. T., McCullock, W. F., and Schnurrenberger, R. P., Eds., C. C. Thomas, Springfield, IL, 1975, 879.

4. **Chamberlain, R. W.,** Epidemiology of arthropod-borne Togaviruses: the role of arthropods as hosts and vectors and of vertebrate hosts in natural transmission cycles, in *The Togaviruses,* Schlesinger, R. W., Ed., Academic Press, New York, 1980, 175.
5. **Schlesinger, R. W., Ed.,** *The Togaviruses,* Academic Press, New York, 1980.
6. **Schlesinger, R. W.,** Virus-host interactions in natural and experimental infections with Alphaviruses and Flaviviruses, in *The Togaviruses,* Schlesinger, R. W., Ed., Academic Press, New York, 1980, 83.
7. **Steele, J. H., Ed.,** *Viral Zoonoses,* Vol. 1 and 2, CRC Handbook Series in Zoonoses, CRC Press, Boca Raton, FL, 1981.
8. **Joklik, W. K., Ed.,** *The Reoviridae,* Plenum Press, New York, 1983.
9. **Schlesinger, S. and Schlesinger, M. J., Eds.,** *The Togaviridae and Flaviviridae,* Plenum Press, New York, 1986.
10. **Wagner, R. R.,** Rhabdovirus biology and infection: an overview, in *The Rhabdoviruses,* Wagner, R. R., Ed., Plenum Press, New York, 1987, 9.
11. **Monath, T. P., Ed.,** *The Arboviruses: Epidemiology and Ecology,* CRC Press, Boca Raton, FL, 1988.
12. **Hurlbut, H. S.,** Arthropod transmission of animal viruses, *Adv. Virus Res.,* 11, 277, 1965.
13. **Schlesinger, R. W.,** Some speculations on the possible role of arthropods in the evolution of arboviruses, *Curr. Top. Microbiol. Immunol.,* 55, 241, 1971.
14. **Bishop, D. H. L. and Shope, R. E.,** Bunyaviridae, in *Comprehensive Virology,* Vol. 14, Fraenkel-Conrat, H. and Wagner, R. R., Eds., Plenum Press, New York, 1979, 1.
15. **Verwoerd, D. W., Huismans, H., and Erasmus, B. J.,** Orbiviruses, in *Comprehensive Virology,* Vol. 14, Fraenkel-Conrat, H. and Wagner, R. R., Eds., Plenum Press, New York-London, 1979, 285.
16. **Rosen, L.,** Arthropods as hosts and vectors of Alphaviruses and Flaviviruses — experimental infections, in *The Togaviruses,* Schlesinger, R. W., Ed., Academic Press, New York, 1980, 229.
17. **Grimstad, P. R.,** Mosquitoes and the incidence of encephalitis, *Adv. Virus Res.,* 28, 357, 1983.
18. **Leake, C. J.,** Transovarial transmission of arboviruses by mosquitoes, in *Vectors in Virus Biology,* Mayo, M. A. and Harrap, K. A., Eds., Academic Press, London, 1984, 63.
19. **Nuttall, P. A.,** Transmission of viruses to wild life by ticks, in *Vectors in Virus Biology,* Mayo, M. A. and Harrap, K. A., Eds., Academic Press, London, 1984, 135.
20. **Karabatsos, N.,** International catalogue of arboviruses including certain other viruses of vertebrates, *Am. Soc. Trop. Med. Hyg.,* San Antonio, TX, 1985.
21. **Shope, R. E. and Tesh, R. B.,** The ecology of Rhabdoviruses that infect vertebrates, in *The Rhabdoviruses,* Wagner, R. R., Ed., Plenum Press, New York, 1987, 509.
22. **Stollar, V.,** Approaches to the study of vector specificity for arboviruses-model systems using cultured mosquito cells, *Adv. Virus Res.,* 33, 327, 1987.
23. **Stollar, V.,** Arbovirus-vector cell interactions and host range viral mutants, in *Arboviruses in Arthropod Cells in Vitro,* Vol. 2, Yunker, C. E., Ed., CRC Press, Boca Raton, FL, 1987, 91.
24. **Beaty, B. J. and Bishop, D. H. L.,** Bunyavirus-vector interactions, *Virus Res.,* 10, 289, 1988.
25. **Calisher, C. H.,** Evolutionary significance of the taxonomic data regarding bunyaviruses of the family Bunyaviridae, *Intervirology,* 29, 268, 1988.
26. **Emmons, R. W.,** Ecology of Colorado tick fever, *Am. Rev. Microbiol.,* 42, 49, 1988.
27. **Ludwig, G. V., Christensen, B. M., Yuill, T. M., and Schultz, K. T.,** Enzyme processing of La Crosse virus glycoprotein G1: a bunyavirus vector infection model, *Virology,* 171, 108, 1989.
28. **Eldridge, B. F.,** Evolutionary relationships among California serogroup viruses (Bunyaviridae) and *Aedes* mosquitoes (Diptera: Culicidae), *J. Med. Entomol.,* 27, 738, 1990.
29. **Roy, P. and Gorman, B. M., Eds.,** Bluetongue viruses, in *Current Topics Microbiology Immunology,* Vol. 162, Springer Verlag, Heidelberg, 1990.
30. **Scott, T. W. and Weaver, S. C.,** Eastern equine encephalomyelitis virus: epidemiology and evolution of mosquito transmission, *Adv. Virus Res.,* 37, 277, 1990.
31. **Beaty, B. J. and Calisher, C. H.,** Bunyaviridae-natural history, *Curr. Top. Microbiol. Immunol.,* 169, 27, 1991.
32. **Labuda, M.,** Arthropod vectors in the evolution of bunyaviruses, *Acta virol.,* 35, 98, 1991.

33. **Nuttall, P. A., Jones, L. D., and Davies, C. R.,** The role of arthropod vectors in arbovirus evolution, *Adv. Dis. Vector Res.*, 8, 15, 1991.
34. **Roy, P.,** *Towards the Control of Emerging Bluetongue Disease*, Oxford Virology Publishing, 1991, 71pp.
35. **Halstead, S. B.,** Pathogenesis of Dengue: challenges to molecular biology, *Science*, 239, 476, 1988.
36. **Service, M. W., Ed.,** *Demography and Vector-Borne Diseases*, CRC Press, Boca Raton, FL, 1989.
37. **Culliton, B. J.,** Emerging viruses, emerging threat, *Science*, 247, 279, 1990.
38. **Kilbourne, E. D.,** New viral diseases. A real and potential problem without boundaries, *J. Am. Med. Assoc.*, 264, 68, 1990.
39. **Kilbourne, E. D.,** New viruses and new diseases: mutation, evolution and ecology, *Curr. Opinion Immunol.*, 3, 518, 1991.
40. **Morse, S. S. and Schluederberg, A.,** Emerging viruses: the evolution of viruses and viral diseases, *J. Infect. Dis.*, 162, 1, 1990.
41. **Livdahl, T. P. and Willey, M. S.,** Prospects for an invasion: competition between *Aedes albopictus* and native *Aedes triseriatus*, *Science*, 253, 189, 1991.
42. **Morse, S. S.,** Emerging viruses: defining the rules for viral traffic, *Persp. Biol. Med.*, 34, 387, 1991.
43. **Raymond, M., Callaghan, A., Fort, P., and Pasteur, N.,** Worldwide migration of amplified insecticide resistance genes in mosquitoes, *Nature*, 350, 151, 1991.
44. **Koblet, H.,** New ecological niches for viruses in a future europe. I. New diseases, changing viruses, *Med. Microbiol. Lett.*, 1, 135, 1992.
45. **Koblet, H.,** New ecological niches for viruses in a future europe. II. Changing ecology and viral traffic, *Med. Microbiol. Lett.*, 1, 185, 1992.
46. **Koblet, H.,** Invertebrate tissue culture as a tool to study insect-transmitted viruses: a brief introduction, *In Vitro Cell. Dev. Biol.*, in press.
47. **Koblet, H.,** Viral evolution and insects as a possible virological turning table, *In Vitro Cell. Dev. Biol.*, in press.
48. **Adams, J. R. and Bonami, J. R., Eds.,** *Atlas of Invertebrate Viruses*, CRC Press, Boston, London, 1991.
49. **Kurstak, E., Ed.,** *Viruses of Invertebrates*, Marcel Dekker, New York, 1991.
50. **Holland, J. J., Spindler, K., Horodyski, F., Grabau, E., Nichol, S., and Vande Pol, S.,** Rapid evolution of RNA genomes, *Science*, 215, 1577, 1982.
51. **Reanney, D. C.,** The evolution of RNA viruses, *Annu. Rev. Microbiol.*, 36, 47, 1982.
52. **Gibbs, A.,** Molecular evolution of viruses, 'trees', 'clocks' and 'modules', *J. Cell Sci.*, Suppl. 7, 319, 1987.
53. **Steinhauer, D. A. and Holland, J. J.,** Rapid evolution of RNA viruses, *Annu. Rev. Microbiol.*, 41, 409, 1987.
54. **Holland, J. J., Ed.,** Genetic diversity of RNA viruses, *Current Topics Microbiological Immunology*, Springer Verlag, Heidelberg, 226pp., 1992.
55. **Steinhauer, D. A. and Holland, J. J.,** Direct method for the quantitation of extreme polymerase error frequency at selected single base sites in viral RNA, *J. Virol.*, 57, 219, 1986.
56. **Gorman, B. M.,** *Molecular Structure of Bluetongue and Related Orbiviruses*, Alan R. Liss, New York, 1985, 329.
57. **Samal, S. K., El-Hussein, A., Holbrook, F. R., Beaty, B. J., and Ramig, R. F.,** Mixed infection of *Culicoides variipennis* with Bluetongue virus serotypes 10 and 17: evidence for high frequency reassortment in the vector, *J. Gen. Virol.*, 68, 2319, 1987.
58. **Ramig, R. F., Garrison, C., Chen, D., and Bell-Robinson, D.,** Analysis of reassortment and superinfection during mixed infection of Vero cells with Bluetongue virus serotypes 10 and 17, *J. Gen. Virol.*, 70, 2595, 1989.
59. **Nuttall, P. A., Moss, S. R., Carey, D., Jones, L. D., and Jacobs, S. C.,** Genetic determinants modulating the pathogenic phenotype of tick-borne Orbiviruses, *Virology*, 174, 430, 1990.
60. **Ramig, R. F. and Ward, R. L.,** Genomic segment reassortment in Rotaviruses and other Reoviridae, *Adv. Virus Res.*, 39, 163, 1991.
61. **Pringle, C. R. and Iroegbu, C. U.,** Mutant identifying a third recombination group in a Bunyavirus, *J. Virol.*, 42, 873, 1982.
62. **Beaty, B. J., Sundin, D. R., Chandler, L. J., and Bishop, D. H. L.,** Evolution of Bunyaviruses by genome reassortment in dually infected mosquitoes *(Aedes triseriatus)*, *Science*, 230, 548, 1985.
63. **Chandler, L. J., Beaty, B. J., Baldridge, G. D., Bishop, D. H. L., and Hewlett, M. J.,** Heterologous reassortment of Bunyaviruses in *Aedes triseriatus* mosquitoes and transovarial and oral transmission of newly evolved genotypes, *J. Gen. Virol.*, 71, 1045, 1990.
64. **Turell, M. J., Saluzzo, J.-F., Tammariello, R. F., and Smith, J. F.,** Generation and transmission of Rift Valley fever viral reassortants by the mosquito *Culex pipiens*, *J. Gen. Virol.*, 71, 2307, 1990.
65. **Pringle, C. R.,** The Bunyaviridae and their genetics —an overview, *Curr. Top. Microbiol. Immunol.*, 169, 2, 1991.
66. **Perry, K. L. and Francki, R. I. B.,** Insect-mediated transmission of mixed and reassorted Cucumovirus genomic RNAs, *J. Gen. Virol.*, 73, 2105, 1992.

67. **Jarvis, T. C. and Kirkegaard, K.,** The polymerase in its labyrinth: mechanisms and implications of RNA recombination, *Trends Genet.*, 7, 186, 1991.
68. **Lai, M. M. C.,** Genetic recombination in RNA viruses, *Curr. Top. Microbiol. Immunol.*, 176, 21, 1992.
69. **Lai, M. M. C.,** RNA recombination in animal and plant viruses, *Microbiol. Rev.*, 56, 61, 1992.
70. **Dominguez, G., Wang, Ch.-Y., and Frey, T. K.,** Sequence of the genome RNA of Rubella virus: evidence for genetic rearrangement during Togavirus evolution, *Virology*, 177, 225, 1990.
71. **Weaver, S. C., Rico-Hesse, R., and Scott, T. W.,** Genetic diversity and slow rates of evolution in new world Alphaviruses, *Curr. Top. Microbiol. Immunol.*, 176, 99, 1992.
72. **Geigenmüller-Gnirke, U., Weiss, B., Wright, R., and Schlesinger, S.,** Complementation between Sindbis viral RNAs produces infectious particles with a bipartite genome, *Proc. Natl. Acad. Sci. U.S.A.*, 88, 3253, 1991.
73. **King, A. M. Q., Lomonossoff, G. P., and Ryan, M. D.,** Picornaviruses and their relatives in the plant kingdom, *Semin. Virol.*, 2, 11, 1991.
74. **Goldbach, R. W., Le Gall, O., and Wellink, J.,** Alpha-like viruses in plants, *Semin. Virol.*, 2, 19, 1991.
75. **Bruenn, J. A.,** Relationships among the positive strand and double-strand RNA viruses as viewed through their RNA-dependent RNA polymerases, *Nucleic Acids Res.*, 19, 217, 1991.
76. **Pereira, H. G.,** Double-stranded RNA viruses, *Semin. Virol.*, 2, 39, 1991.
77. **Peters, D.,** Divergent evolution of Rhabdoviridae and Bunyaviridae in plants and animals, *Semin. Virol.*, 2, 27, 1991.
78. **Doolittle, R. F. and Feng, D. F.,** Tracing the origin of retroviruses, *Curr. Top. Microbiol. Immunol.*, 176, 195, 1992.
79. **Calisher, C. H.,** Making sense of viral gene sequences, *New Biol.*, 4, 299, 1992.
80. **Koonin, E. V., Ed.,** Evolution of viral genomes, *Semin. Virol.*, 3, 311, 1992.
81. **Hubbard, S. C. and Ivatt, R. J.,** Synthesis and processing of asparagine-linked oligosaccharides, *Annu. Rev. Biochem.*, 50, 555, 1981.
82. **Kornfeld, R. and Kornfeld, S.,** Assembly of asparagine-linked oligosaccharides, *Annu. Rev. Biochem.*, 54, 631, 1985.
83. **Abeijon, C. and Hirschberg, C. B.,** Topography of glycosylation reactions in the endoplasmic reticulum, *Trends Biochem. Sci.*, 17, 32, 1992.
84. **Grand, R. J. A.,** Acylation of viral and eukaryotic proteins, *Biochem. J.*, 258, 625, 1989.
85. **Schmidt, M. F. G.,** Fatty acylation of proteins, *Biochim. Biophys. Acta*, 988, 411, 1989.
86. **Magee, A. I.,** Lipid modification of proteins and its relevance to protein targeting, *J. Cell Sci.*, 97, 581, 1990.
87. **Warren, L.,** The distribution of sialic acids in nature, *Comp. Biochem. Physiol.*, 10, 153, 1963.
88. **Wiegandt, H.,** Insect glycolipids, *Biochim. Biophys. Acta*, 1123, 117, 1992.
89. **Thompson, S. N.,** A review and comparative characterisation of the fatty acid compositions of seven insect orders, *Comp. Biochem. Physiol.*, 45 B, 467, 1973.
90. **Mitsuhashi, J., Nakasone, S., and Horie, Y.,** Total fatty acids of some insect cell lines, *Appl. Entomol. Zool.*, 20, 8, 1985.
91. **Stanley-Samuelson, D. W., Keddie, B. A., and Volkman, L. E.,** Fatty acid composition of whole bodies, specific tissues and cell lines of two lepidopteran insects, *Comp. Biochem. Physiol.*, 85 B, 369, 1986.
92. **Stanley-Samuelson, D. W., Jurenka, R. A., Cripps, C., Blomquist, G. J., and de Renobales, M.,** Fatty acids in insects: composition, metabolism, and biological significance, *Arch. Insect Biochem. Physiol.*, 9, 1, 1988.
93. **Beenakkers, M. T., Van der Horst, D. J., and Van Marrewijk, W. J. A.,** Insect lipids and lipoproteins, and their role in physiological processes, *Prog. Lipid Res.*, 24, 19, 1985.
94. **Bloch, K. E.,** Sterol structure and membrane function, *CRC Crit. Rev. Biochem.*, 14, 47, 1983.
95. **Mitsuhashi, J., Nakasone, S., and Horie, Y.,** Sterol-free eukaryotic cells from continuous cell lines of insects, *Cell Biol. Intern. Rep.*, 7, 1057, 1983.
96. **Olafsen, J. A.,** Invertebrate lectins: biochemical heterogeneity as a possible key to their biological function, in *Immunity in Invertebrates*, Brehélin, M., Ed., Springer Verlag, Berlin, 1986, 94.
97. **Sagami, H. and Lennarz, W. J.,** Glycoprotein synthesis in *Drosophila Kc* cells. Biosynthesis of dolichol-linked saccharides, *J. Biol. Chem.*, 262, 15610, 1987.
98. **Fyrberg, E. A. and Goldstein, L. S. B.,** The *Drosophila* cytoskeleton, *Annu. Rev. Cell Biol.*, 6, 559, 1990.
99. **Clements, A. A.,** *The Biology of Mosquitoes*, Vol. 1, Chapman and Hall, London, 1992.
100. **James, A. D.,** Mosquito molecular genetics: the hands that feed bite back, *Science*, 257, 37, 1992.
101. **Fallon, A. M. and Stollar, V.,** The biochemistry and genetics of mosquito cells in culture, *Adv. Cell Cult.*, 5, 97, 1987.
102. **Stollar, V., Stollar, B. D., Koo, R., Harrap, K. A., and Schlesinger, R. W.,** Sialic acid contents of Sindbis virus from vertebrate and mosquito cells, *Virology*, 69, 104, 1976.
103. **Butters, T. D., Hughes, R. C., and Vischer, P.,** Steps in the biosynthesis of mosquito cell membrane glycoproteins and the effects of tunicamycin, *Biochim. Biophys. Acta*, 640, 672, 1981.
104. **Hsieh, P. and Robbins, P. W.,** Regulation of asparagine-linked oligosaccharide processing, *J. Biol. Chem.*, 259, 2375, 1984.

105. **Dunphy, W. G., Brands, R., and Rothman, J. E.,** Attachment of terminal N-acetylglucosamine to asparagine-linked oligosaccharides occurs in central cisternae of the Golgi stack, *Cell,* 40, 463, 1985.
106. **Elbein, A. D.,** Inhibitors of the biosynthesis and processing of N-linked oligosaccharide chains, *Annu. Rev. Biochem.,* 56, 497, 1987.
107. **Elbein, A. D.,** Glycosidase inhibitors: inhibitors of N-linked oligosaccharide processing, *FASEB J.,* 5, 3055, 1991.
108. **Hink, W. F.,** The 1979 compilation of invertebrate cell lines and culture media, in *Invertebrate Systems In Vitro,* Kurstak, E., Maramorosch, K., and Dübendorfer, A., Eds., Elsevier, Amsterdam, 1980, 553.
109. **Maramorosch, K. and Mitsuhashi, J., Eds.,** *Invertebrate Cell Culture Applications,* Academic Press, New York, 1982.
110. **Hink, W. F., Ralph, D. A., and Joplin, K. H.,** Metabolism and characterization of insect cell cultures, *Compr. Insect Physiol. Biochem. Pharmacol.,* 10, 547, 1985.
111. **Hink, W. F. and Hall, R. L.,** Recently established invertebrate cell lines, in *Invertebrate Cell System Applications,* Vol. 2, Mitsuhashi, J., Ed., CRC Press, Boca Raton, FL, 1989, 269.
112. **Kurtti, T. J. and Munderloh, U. G.,** Mosquito cell culture, *Adv. Cell Cult.,* 3, 259, 1984.
113. **Pudney, M.,** Tick cell lines for the isolation and assay of arboviruses, in *Arboviruses in Arthropod Cells In Vitro,* Vol. 1, Yunker, C. E., Ed., CRC Press, Boca Raton, FL, 1987, 87.
114. **Weiss, E., Ed.,** Arthropod cell cultures and their application to the study of viruses, *Curr. Top. Microbiol. Immunol.,* 55, 1971.
115. **Yunker, C. E.,** Arthropod tissue culture in the study of arboviruses and rickettsiae: a review, *Curr. Top. Microbiol. Immunol.,* 55, 113, 1971.
116. **Buckley, S. M., Hayes, C. G., Maloney, J. M., Lipman, M., Aitcken, T. H. G., and Casals, J.,** Arbovirus studies in invertebrate cell lines, in *Invertebrate Tissue Culture,* Kurstak, E. and Maramorosch, K., Eds., Academic Press, New York, 1976, 3.
117. **Stollar, V., Shenk, T. E., Koo, R., Igarashi, A., and Schlesinger, R. W.,** Established mosquito cell lines and the study of Togaviruses, in *Invertebrate Tissue Culture,* Kurstak, E. and Maramorosch, K., Eds., Academic Press, New York, 1976, 49.
118. **Leake, C. J., Pudney, M., and Varma, M. G. R.,** Studies on arboviruses in established tick cell lines, in *Invertebrate Systems In Vitro,* Kurstak, E., Maramorosch, K., and Dübendorfer, A., Eds., Elsevier, Amsterdam, 1980, 327.
119. **Stollar, V.,** Togaviruses in cultured arthropod cells, in *The Togaviruses,* Schlesinger, R. W., Ed., Academic Press, New York, 1980, 583.
120. **Pudney, M., Leake, C. J., and Buckley, S. M.,** Replication of arboviruses in arthropod in vitro systems: an overview, in *Invertebrate Cell Culture Applications,* Maramorosch, K. and Mitsuhashi, J., Eds., Academic Press, New York, 1982, 159.
121. **Brown, D. T.,** Alphavirus growth in cultured vertebrate and invertebrate cells, in *Vectors in Virus Biology,* Mayo, M. A. and Harrap, K. A., Eds., Academic Press, London, 1984, 113.
122. **Igarashi, A.,** Mosquito cell cultures and the study of arthropod-borne Togaviruses, *Adv. Virus Res.,* 30, 21, 1985.
123. **Dezélée, S., Wyers, F., Blondel, D., Petitjean, A. M., Contamine, D., Bras, F., and Teninges, D.,** Rhabdoviruses in *Drosophila* cells, in *Arboviruses in Arthropod Cells In Vitro,* Vol. 2, Yunker, C. E., Ed., CRC Press, Boca Raton, FL, 1987, 111.
124. **Leake, C. J.,** Comparative growth of arboviruses in cell lines derived from *Aedes* and *Anopheles mosquitoes* and from the tick *Boophilus microplus,* in *Arboviruses in Arthropod Cells In Vitro,* Vol. 2, Yunker, C. E., Ed., CRC Press, Boca Raton, FL, 1987, 25.
125. **Yunker, C. E., Ed.,** *Arboviruses in Arthropod Cells In Vitro,* Vol. 1 and 2, CRC Press, Boca Raton, FL, 1987.
126. **Kuroda, Y., Kurstak, E., and Maramorosch, K., Eds.,** *Invertebrate and Fish Tissue Culture,* Japan Scientific Societies Press, Tokyo, and Springer Verlag, Berlin, 1988.
127. **Mitsuhashi, J., Ed.,** *Invertebrate Cell System Applications,* Vol. 1 and 2, CRC Press, Boca Raton, FL, 1989.
128. **Paulson, S. L. and Grimstad, P. R.,** Replication and dissemination of La Crosse virus in the competent vector *Aedes triseriatus* and the incompetent vector *Aedes hendersoni* and evidence for transovarial transmission by *Aedes hendersoni* (Diptera: Culicidae), *J. Med. Entomol.,* 26, 602, 1989.
129. **Miller, M. L. and Brown, D. T.,** Alphavirus infection in cultured tissue cells, *Adv. Dis. Vector Res.,* 8, 107, 1991.
130. **Igarashi, A.,** Isolation of a Singh's *Aedes albopictus* cell clone sensitive to Dengue and Chikungunya viruses, *J. Gen. Virol.,* 40, 531, 1978.
131. **White, L. A.,** Susceptibility of *Aedes albopictus* C6/36 cells to viral infection, *J. Clin. Microbiol.,* 25, 1221, 1987.
132. **Hirumi, H.,** Viral, microbial, and extrinsic cell contamination of insect cell cultures, in *Invertebrate Tissue Culture. Research Applications,* Maramorosch, K., Ed., Academic Press, New York, 1976, 233.

133. **Hirumi, H., Hirumi, K., Speyer, G., Yunker, C. E., Thomas, L. A., Cory, J., and Sweet, B. H.,** Viral contamination of a mosquito cell line, *Aedes albopictus*, associated with syncytium formation, *In Vitro Cell. Dev. Biol.*, 12, 83, 1976.
134. **Grace, T. D.,** Establishment of four strains of cells from insect tissues grown in vitro, *Nature (London)*, 195, 788, 1962.
135. **Grace, T. D.,** Establishment of a line of mosquito *(Aedes aegypti L.)* cells grown in vitro, *Nature (London)*, 211, 366, 1966.
136. **Mitsuhashi, J. and Maramorosch, K.,** Leafhopper tissue culture: embryonic, nymphal and imaginal tissues from aseptic insects, *Contrib. Boyce Thompson Inst.*, 22, 435, 1964.
137. **Varma, M. G. R., Pudney, M., and Leake, C. J.,** Cell lines from larvae of *Aedes (Stegomyia) malayensis* Colless and *Aedes(S) pseudoscutellaris* (Theobald) and their infection with some arboviruses, *Trans. R. Soc. Trop. Med. Hyg.*, 68, 374, 1974.
138. **Sarver, N. and Stollar, V.,** Sindbis virus-induced cytopathic effect in clones of *Aedes albopictus* (Singh) cells, *Virology*, 80, 390, 1977.
139. **Kurtti, T. J., Munderloh, U. G., and Samish, M.,** Effect of medium supplements on tick cells in culture, *J. Parasitol.*, 68, 930, 1982.
140. **Kurtti, T. J. and Munderloh, U. G.,** Tick cell culture: characteristics, growth requirements, and applications to parasitology, in *Invertebrate Cell Culture Applications*, Maramorosch, K. and Mitsuhashi, J., Eds., Academic Press, New York, 1982, 195.
141. **Condreay, L. D., Adams, R. H., Edwards, J., and Brown, D. T.,** Effect of actinomycin D and cyclohex-imide on replication of Sindbis virus in *Aedes albopictus* (mosquito) cells, *J. Virol.*, 62, 2629, 1988.
142. **Homan, E. J. and Yunker, C. E.,** Growth of Bluetongue and epizootic hemorrhagic disease of deer viruses in poikilothermic cell systems, *Vet. Microbiol.*, 16, 15, 1988.
143. **Mitsuhashi, J.,** Media for insect cell cultures, *Adv. Cell Cult.*, 2, 133, 1982.
144. **Brown, D. T., Smith, J. F., Gliedman, J. B., Riedel, B., Filtzer, D., and Renz, D.,** Morphogenesis of Sindbis virus in cultured mosquito cells, in *Invertebrate Tissue Culture*, Kurstak, E. and Maramorosch, K., Eds., Academic Press, New York, 1976, 35.
145. **Kurstak, E. and Maramorosch, K., Eds.,** *Invertebrate Tissue Culture*, Academic Press, New York, 1976.
146. **Stollar, V.,** Defective interfering particles of Togaviruses, *Curr. Top. Microbiol. Immunol.*, 86, 35, 1979.
147. **Kurstak, E., Maramorosch, K., and Dübendorfer, A.,** Eds., *Invertebrate Systems In Vitro*, Elsevier, Amsterdam, 1980.
148. **Späth, P. J. and Koblet, H.,** Alphavirus induced syncytium formation in *Aedes albopictus* cell cultures, in *Invertebrate Systems In Vitro*, Kurstak, E., Maramorosch, K., and Dübendorfer, Eds., Elsevier, Amsterdam, 1980, 375.
149. **Stollar, V.,** Defective interfering Alphaviruses, in *The Togaviruses*, Schlesinger, R. W., Ed., Academic Press, New York, 1980, 427.
150. **Strauss, E. G. and Strauss, J. H.,** Mutants of Alphaviruses: genetics and physiology, in *The Togaviruses*, Schlesinger, R. W., Ed., Academic Press, New York, 1980, 393.
151. **Garoff, H., Kondor-Koch, C., and Riedel, H.,** Structure and assembly of Alphaviruses, *Curr. Top. Microbiol. Immunol.*, 99, 1, 1982.
152. **Brown, D. T. and Condreay, L. D.,** Replication of Alphaviruses in mosquito cells, in *The Togaviridae and Flaviviridae*, Schlesinger, S. and Schlesinger, M. J., Eds., Plenum Press, New York, 1986, 171.
153. **Enzmann, P. J.,** A comparative view of the replication of Togaviruses in vertebrate and mosquito cells, in *Arboviruses in Arthropod Cells In Vitro*, Vol. 2, Yunker, C. E., Ed., CRC Press, Boca Raton, FL, 1987, 53.
154. **Enzmann, P. J.,** Persistent infections of mosquito cells by Togaviruses, in *Arboviruses in Arthropod Cells In Vitro*, Vol. 2, Yunker, C. E., Ed., CRC Press, Boca Raton, FL, 1987, 67.
155. **Kääriäinen, L., Takkinen, K., Keränen, S., and Söderlund, H.,** Replication of the genome of Alphaviruses, *J. Cell Sci. Suppl.*, 7, 231, 1987.
156. **Koblet, H., Omar, A., and Kempf, C.,** Fusion of Alphavirus-infected mosquito cells, in *Arboviruses in Arthropod Cells In Vitro*, Vol. 2, Yunker, C. E., Ed., CRC Press, Boca Raton, FL, 1987, 77.
157. **Kräusslich, H. G. and Wimmer, E.,** Viral proteinases, *Annu. Rev. Biochem.*, 57, 701, 1988.
158. **Wellink, J. and van Kammen, A.,** Proteases involved in the processing of viral polyproteins, *Arch. Virol.*, 98, 1, 1988.
159. **Koblet, H.,** The "merry-go-round": Alphaviruses between vertebrate and invertebrate cells, *Adv. Virus Res.*, 38, 343, 1990.
160. **Peters, C. J. and Dalrymple, J. M.,** Alphaviruses, in *Virology*, Fields, B. N. and Knipe, D. M., Eds., 2nd ed., Vol. 1, Raven Press, New York, 1990, 713.
161. **Van Regenmortel, M. H. V. and Neurath, A. R., Eds.,** *Immunochemistry of Viruses II*, Elsevier, Amsterdam, 1990.
162. **Schlesinger, S. and Schlesinger, M. J.,** Replication of Togaviridae and Flaviviridae, in *Virology*, Fields, B. N. and Knipe, D. M., Eds., 2nd ed., Vol. 1, Raven Press, New York, 1990, 697.

163. **Strauss, J. H. and Strauss, E. G.,** Alphavirus proteinases, *Semin. Virol.,* 1, 347, 1990.
164. **Brown, D. T. and Edwards, J.,** Structural changes in Alphaviruses accompanying the process of membrane penetration, *Semin. Virol.,* 3, 519, 1992.
165. **Singh, I. and Helenius, A.,** Nucleocapsid uncoating during entry of enveloped animal RNA viruses into cells, *Semin. Virol.,* 3, 511, 1992.
166. **Stollar, V.,** Insect transmitted vertebrate viruses: Alphatogaviruses, *In Vitro Cell. Dev. Biol.,* 29A, 289, 1993.
167. **Froshauer, S., Kartenbeck, J., and Helenius, A.,** Alphavirus RNA replicase is located on the cytoplasmic surface of endosomes and lysosomes, *J. Cell. Biol.,* 107, 2075, 1988.
168. **Singh, I. and Helenius, A.,** Role of ribosomes in Semliki Forest virus nucleocapsid uncoating, *J. Virol.,* 66, 7049, 1992.
169. **Singh, K. R. P.,** Cell cultures derived from larvae of *Aedes albopictus* (Skuse) and *Aedes aegypti*(L), *Curr. Sci.,* 36, 506, 1967.
170. **Renkonen, O., Kääriäinen, L., Simons, K., and Gahmberg, C. G.,** The lipid class composition of Semliki Forest virus and of plasma membranes of the host cells, *Virology,* 46, 318, 1971.
171. **Luukkonen, A., Kääriäinen, L., and Renkonen, O.,** Phospholipids of Semliki Forest virus grown in cultured mosquito cells, *Biochim. Biophys. Acta,* 450, 109, 1976.
172. **Moore, N. F., Barenholz, Y., and Wagner, R. R.,** Microviscosity of Togavirus membranes studied by fluorescence depolarization: influence of envelope proteins and the host cell, *J. Virol.,* 19, 126, 1976.
173. **Burge, B. W. and Strauss, J. H.,** Glycopeptides of the membrane glycoprotein of Sindbis virus, *J. Mol. Biol.,* 47, 449, 1970.
174. **Strauss, J. H., Burge, B. W., and Darnell, J. E.,** Carbohydrate content of the membrane protein of Sindbis virus, *J. Mol. Biol.,* 47, 437, 1970.
175. **Keegstra, K., Sefton, B., and Burke, D.,** Sindbis virus glycoproteins: effect of the host cell on the oligosaccharides, *J. Virol.,* 16, 613, 1975.
176. **Sefton, B. M.,** Virus-dependent glycosylation, *J. Virol.,* 17, 85, 1976.
177. **Pesonen, M. and Renkonen, O.,** Serum glycoprotein-type sequence of monosaccharides in membrane glycoproteins of Semliki Forest virus, *Biochim. Biophys. Acta,* 455, 510, 1976.
178. **Keegstra, K. and Burke, D.,** Comparison of the carbohydrate of Sindbis virus glycoproteins with the carbohydrate of host glycoproteins, *J. Supramol. Struct.,* 7, 371, 1977.
179. **Luukkonen, A., von Bonsdorff, C.-H., and Renkonen, O.,** Characterization of Semliki Forest virus grown in mosquito cells. Comparison with the virus from hamster cells, *Virology,* 78, 331, 1977.
180. **Naim, H. Y. and Koblet, H.,** Investigation of the role of glycans for the biological activity of Semliki Forest virus grown in *Aedes albopictus* cells using inhibitors of asparagine-linked oligosaccharides trimming, *Arch. Virol.,* 102, 73, 1988.
181. **Wang, K.-S. and Strauss, J. H.,** Use of a λgt 11 expression library to localize a neutralizing antibody-binding site in glycoprotein E2 of Sindbis virus, *J. Virol.,* 65, 7037, 1991.
182. **Meyer, W. J., Gidwitz, S., Ayers, V. K., Schoepp, R. J., and Johnston, R. E.,** Conformational alteration of Sindbis virion glycoproteins induced by heat, reducing agents, or low pH, *J. Virol.,* 66, 3504, 1992.
183. **Hase, T., Summers, P. L., and Cohen, W. H.,** A comparative study of entry modes into C6/36 cells by Semliki Forest and Japanese encephalitis viruses, *Arch. Virol.,* 108, 101, 1989.
184. **Julen, R., Wiesmann, U. N., and Koblet, H.,** Chemically induced plasma membrane vesicles as an useful tool for the investigation of virus binding to susceptible cells, *J. Virol. Methods,* 42, 147, 1993.
185. **Lustig, S., Jackson, A., Hahn, C. S., Griffin, D. E., Strauss, E. G., and Strauss, J. H.,** Molecular basis of Sindbis virus neurovirulence in mice, *J. Virol.,* 62, 2329, 1988.
186. **Tucker, P. C. and Griffin, D. E.,** Mechanism of altered Sindbis virus neurovirulence associated with a single-amino acid change in the E2 glycoprotein, *J. Virol.,* 65, 1551, 1991.
187. **Ubol, S. and Griffin, D. E.,** Identification of a putative Alphavirus receptor on mouse neural cells, *J. Virol.,* 65, 6913, 1991.
188. **Wickham, T. J., Granados, R. R., Wood, H. A., Hammer, D. A., and Shuler, M. L.,** General analysis of receptor-mediated viral attachment to cell surfaces, *Biophys. J.,* 58, 1501, 1990.
189. **Birdwell, C. R. and Strauss, J. H.,** Distribution of the receptor sites for Sindbis virus on the surface of chicken and BHK cells, *J. Virol.,* 14, 672, 1974.
190. **Houk, E. J., Arcus, Y. M., Hardy, J. L., and Kramer, L. D.,** Binding of western equine encephalomyelitis virus to brush border fragments isolated from mesenteronal epithelial cells of mosquitoes, *Virus Res.,* 17, 105, 1990.
191. **Pierce, J. S., Strauss, E. G., and Strauss, J. H.,** Effect of ionic strength on the binding of Sindbis virus to chick cells, *J. Virol.,* 13, 1030, 1974.
192. **Smith, A. L. and Tignor, G. H.,** Host cell receptors for two strains of Sindbis virus, *Arch. Virol.,* 66, 11, 1980.
193. **Mooney, J. J., Dalrymple, J. M., Alving, C. R., and Russell, P. K.,** Interaction of Sindbis virus with liposomal model membranes, *J. Virol.,* 15, 225, 1975.

194. **Mastromarino, P., Conti, C., Rieti, S., and Orsi, N.,** Identification of lipid components of human serum lipoproteins involved in the inhibition of Sindbis virus infectivity, hemagglutination, and hemolysis, *Arch. Virol.,* 103, 243, 1988.
195. **Mastromarino, P., Conti, C., Rieti, S., and Orsi, N.,** Involvement of lipids in the interaction of Sindbis virus with goose erythrocytes, *Microbiologica,* 12, 113, 1989.
196. **Mastromarino, P., Conti, C., Petruzziello, R., Lapadula, R., and Orsi, N.,** Effect of polyions on the early events of Sindbis virus infection of Vero cells, *Arch. Virol.,* 121, 19, 1991.
197. **Sanders, B. G., Wan, K. M., Kline, K., Garry, R. F., and Bose, H. R.,** Chicken fetal antigens: role as cell surface receptors for Sindbis virus hemagglutinin, *Virology,* 106, 183, 1980.
198. **Heizmann, C. W. and Hunziker, W.,** Intracellular calcium-binding proteins: more sites than insights, *Trends Biochem. Sci.,* 16, 98, 1991.
199. **Helenius, A., Morein, B., Fries, E., Simons, K., Robinson, P., Schirrmacher, V., Terhorst, C., and Strominger, J. L.,** Human (HLA-A and HLA-B) and murine (H-2K and H-2D) histocompatibility antigens are cell surface receptors for Semliki Forest virus, *Proc. Natl. Acad. Sci. U.S.A.,* 75, 3846, 1978.
200. **Oldstone, M. B. A., Tishon, A., Dutko, F. J., Kennedy, S. I. T., Holland, J. J., and Lampert, P. W.,** Does the major histocompatibility complex serve as a specific receptor for Semliki Forest virus?, *J. Virol.,* 34, 256, 1980.
201. **Maassen, J. A. and Terhorst, C.,** Identification of a cell-surface protein involved in the binding site of Sindbis virus on human lymphoblastoic cell lines using a heterobifunctional cross-linker, *Eur. J. Biochem.,* 115, 153, 1981.
202. **Wang, K.-S., Schmaljohn, A. L., Kuhn, R. J., and Strauss, J. H.,** Antiidiotypic antibodies as probes for the Sindbis virus receptor, *Virology,* 181, 694, 1991.
203. **Wang, K.-S., Kuhn, R. J., Strauss, E. G., Ou, S., and Strauss, J. H.,** High-affinity laminin receptor is a receptor for Sindbis virus in mammalian cells, *J. Virol.,* 66, 4992, 1992.
204. **Kielian, M. C. and Jungerwirth, S.,** Mechanisms of enveloped virus entry into cells, *Mol. Biol. Med.,* 7, 17, 1990.
205. **Marsh, M. and Helenius, A.,** Adsorptive endocytosis of Semliki Forest virus, *J. Mol. Biol.,* 142, 439, 1980.
206. **Helenius, A., Marsh, M., and White, J.,** Inhibition of Semliki Forest virus penetration by lysosomotropic weak bases, *J. Gen. Virol.,* 58, 47, 1982.
207. **Coombs, K., Mann, E., Edwards, J., and Brown, D. T.,** Effects of chloroquine and cytochalasin B on the infection of cells by Sindbis virus and vesicular stomatitis virus, *J. Virol.,* 37, 1060, 1981.
208. **Omar, A. and Koblet, H.,** Application of mosquito cell culture and Toga virus for studying the mechanism of membrane fusion, in *Invertebrate Cell System Applications,* Mitsuhashi, J., Ed., Vol. 2, CRC Press, Boca Raton, FL, 1989, 151.
209. **Nes, M. R. and McKean, M. L.,** *Biochemistry of Steroids and Other Isopentenoids,* University Park Press, Baltimore, MD, 1977, 412.
210. **Silberkang, M., Havel, C. M., Friend, D. S., McCarthy, B. J., and Watson, J. A.,** Isoprene synthesis in isolated embryonic *Drosophila* cells. I. Sterol-deficient eukaryotic cells, *J. Biol. Chem.,* 258, 8503, 1983.
211. **White, J. and Helenius, A.,** pH-dependent fusion between the Semliki Forest virus membrane and liposomes, *Proc. Natl. Acad. Sci. U.S.A.,* 77, 3273, 1980.
212. **Kielian, M. C. and Helenius, A.,** The role of cholesterol in the fusion of Semliki Forest virus with membranes, *J. Virol.,* 52, 281, 1984.
213. **Kielian, M. C. and Helenius, A.,** pH-induced alterations in the fusogenic spike protein of Semliki Forest virus, *J. Cell Biol.,* 101, 2284, 1985.
214. **Phalen, T. and Kielian, M. C.,** Cholesterol is required for infection by Semliki Forest virus, *J. Cell Biol.,* 112, 615, 1991.
215. **Brown, D. T.,** The assembly of Alphaviruses, in *The Togaviruses,* Schlesinger, R. W., Ed., Academic Press, New York, 1980, 473.
216. **Murphy, F. A.,** Togavirus morphology and morphogenesis, in *The Togaviruses,* Schlesinger, R. W., Ed., Academic Press, New York, 1980, 241.
217. **Raghow, R. S., Davey, M. W., and Dalgarno, L.,** The growth of Semliki Forest virus in cultured mosquito cells: ultrastructural observations, *Arch. Gesamte Virusforsch.,* 43, 165, 1973.
218. **Grimley, P. M., Berezesky, I. M., and Friedman, R. M.,** Cytoplasmic structures associated with an arbovirus infection: Loci of viral ribonucleic acid synthesis, *J. Virol.,* 2, 1326, 1968.
219. **Peränen, J. and Kääriäinen, L.,** Biogenesis of type I cytopathic vacuoles in Semliki Forest virus-infected BHK cells, *J. Virol.,* 65, 1623, 1991.
220. **Raghow, R. S., Grace, T. D. C., Filshie, B. K., Bartley, W., and Dalgarno, L.,** Ross River virus replication in cultured mosquito and mammalian cells: virus growth and correlated ultrastructural changes, *J. Gen. Virol.,* 21, 109, 1973.
221. **Gliedman, J. B., Smith, J. F., and Brown, D. T.,** Morphogenesis of Sindbis virus in cultured *Aedes albopictus* cells, *J. Virol.,* 16, 913, 1975.

222. **Miller, M. L. and Brown, D. T.**, Morphogenesis of Sindbis virus in three subclones of *Aedes albopictus* (mosquito) cells, *J. Virol.*, 66, 4180, 1992.
223. **Mezencio, J. M. S., de Souza, W., Fonseca, M. E. F., and Rebello, M. A.**, Replication of Mayaro virus in *Aedes albopictus* cells: an electron microscopic study, *Arch. Virol.*, 104, 299, 1989.
224. **Lehane, M. J. and Leake, C. J.**, A kinetic and ultrastructural comparison of Alphavirus infection of cultured mosquito and vertebrate cells, *J. Trop. Med. Hyg.*, 85, 229, 1982.
225. **Birdwell, C. R. and Strauss, J. H.**, Replication of Sindbis virus. IV. Electron microscope study of the insertion of viral glycoproteins into the surface of infected chick cells, *J. Virol.*, 14, 366, 1974.
226. **Wengler, G. and Wengler, G.**, Protein synthesis in BHK-21 cells infected with Semliki Forest virus, *J. Virol.*, 17, 10, 1976.
227. **Singh, K. R. P. and Paul, S. D.**, Multiplication of arboviruses in cell lines from *Aedes albopictus* and *Aedes aegypti*, *Curr. Sci.*, 37, 65, 1968.
228. **Peleg, J.**, Growth of arboviruses in monolayers from subcultured mosquito embryo cells, *Virology*, 35, 617, 1968.
229. **Řeháček, J.**, The growth of arboviruses in mosquito cells in vitro, *Acta Virol.*, 12, 241, 1968.
230. **Suitor, E. C.**, Plaque formation by an arbovirus in a mosquito cell line, *J. Gen. Virol.*, 5, 545, 1969.
231. **Buckley, S. M.**, Susceptibility of *Aedes albopictus* and *Aedes aegypti* cell lines to infection with arboviruses, *Proc. Soc. Exp. Biol. Med.*, 131, 625, 1969.
232. **Stevens, T. M.**, Arbovirus replication in mosquito cell lines (Singh) grown in monolayer or suspension culture, *Proc. Soc. Exp. Biol. Med.*, 134, 356, 1970.
233. **Sweet, B. H. and Unthank, H. D.**, A comparative study of the viral susceptibility of monolayer and suspended mosquito cell lines, *Curr. Top. Microbiol. Immunol.*, 55, 150, 1971.
234. **Peleg, J.**, Inapparent persistent virus infection in continuously grown *Aedes aegypti* mosquito cells, *J. Gen. Virol.*, 5, 463, 1969.
235. **Peleg, J.**, Studies on the behavior of arboviruses in an *Aedes aegypti* mosquito cell line (Peleg), *Arch. Gesamte Virusforsch.*, 37, 54, 1972.
236. **Davey, M. W., Dennett, D. P., and Dalgarno, L.**, The growth of two Togaviruses in cultured mosquito and vertebrate cells, *J. Gen. Virol.*, 20, 225, 1973.
237. **Davey, M. W. and Dalgarno, L.**, Semliki Forest virus replication in cultured *Aedes albopictus cells*: studies on the establishment of persistence, *J. Gen. Virol.*, 24, 453, 1974.
238. **Richardson, M. A., Boulton, R. W., Raghow, R. S., and Dalgarno, L.**, Polypeptide synthesis in Alphavirus-infected *Aedes albopictus* cells during the establishment of persistent infection, *Arch. Virol.*, 63, 263, 1980.
239. **Tooker, P. and Kennedy, S. I. T.**, Semliki Forest virus multiplication in clones of *Aedes albopictus* cells, *J. Virol.*, 37, 589, 1981.
240. **Tatem, J. and Stollar, V.**, Dominance of the CPE(+) phenotype in hybrid *Aedes albopictus* cells infected with Sindbis virus, *Virus Res.*, 5, 121, 1986.
241. **Koblet, H., Kempf, C., Kohler, U., and Omar, A.**, Conformational changes at pH 6 on the cell surface of Semliki Forest virus infected *Aedes albopictus* cells, *Virology*, 143, 334, 1985.
242. **Simizu, B. and Maeda, S.**, Growth patterns of temperature-sensitive mutants of Western equine encephalitis virus in cultured *Aedes albopictus* (mosquito) cells, *J. Gen. Virol.*, 56, 349, 1981.
243. **Omar, A., Flaviano, A., Kohler, U., and Koblet, H.**, Fusion of Semliki Forest virus infected *Aedes albopictus* cells at low pH is a fusion from within (FFWI), *Arch. Virol.*, 89, 145, 1986.
244. **Yunker, C. E. and Cory, J.**, Plaque production by arboviruses in Singh's *Aedes albopictus* cells, *Appl. Microbiol.*, 29, 81, 1975.
245. **Lachmi, B.-E. and Kääriäinen, L.**, Control of protein synthesis in Semliki Forest virus infected cells, *J. Virol.*, 22, 142, 1977.
246. **Garry, R. F., Bishop, J. M., Parker, S., Westbrook, K., Lewis, G., and Waite, M. R. F.**, Na^+ and K^+ concentrations and the regulation of protein synthesis in Sindbis virus-infected chick cells, *Virology*, 96, 108, 1979.
247. **Ulug, E. T., Garry, R. F., and Bose, H. R.**, The role of monovalent cation transport in Sindbis virus maturation and release, *Virology*, 172, 42, 1989.
248. **Clewley, J. P. and Kennedy, S. I. T.**, Purification and polypeptide composition of Semliki Forest virus RNA polymerase, *J. Gen. Virol.*, 32, 395, 1976.
249. **Stalder, J., Reigel, F., Flaviano, A., and Koblet, H.**, Infection of the *Aedes albopictus* cell clone C6/36 with Semliki Forest virus, *Experientia*, 37, 1229, 1981.
250. **Ahmed, R., Canning, W. M., Kauffman, R. S., Sharpe, A. H., Hallum, J. V., and Fields, B. N.**, Role of the host cell in persistent viral infection: coevolution of L cells and reovirus during persistent infection, *Cell*, 25, 325, 1981.
251. **Omar, A. and Koblet, H.**, Semliki Forest virus particles containing only the E_1 envelope glycoprotein are infectious and can induce cell-cell fusion, *Virology*, 166, 17, 1988.

252. **Koblet, H., Omar, A., Kohler, U., and Kempf, C.,** Investigation of cell-cell fusion in Semliki Forest virus (SFV) infected C6/36 (mosquito) cells, in *Invertebrate and Fish Tissue Culture,* Kuroda, Y., Kurstak, E., and Maramorosch, K., Eds., Japan Scientific Societies Press, Tokyo, and Springer Verlag, Berlin, 1988, 140.
253. **Omar, A., Flaviano, A., Reigel, F., Kohler, U., and Koblet, H.,** Syncytium formation and inhibition in Semliki Forest virus-infected *Aedes albopictus* cells at low pH, in *Invertebrate Cell System Applications,* Vol. 2, Mitsuhashi, J., Ed., CRC Press, Boca Raton, FL, 1989, 147.
254. **Kielian, M. C.,** Membrane fusion activity of Alphaviruses, in *Viral Fusion Mechanisms,* Bentz, J., Ed., CRC Press, Boca Raton, FL, 1993, 385.
255. **Stollar, V., Shenk, T. E., and Stollar, B. D.,** Double-stranded RNA in hamster, chick and mosquito cells infected with Sindbis virus, *Virology,* 47, 122, 1972.
256. **Stollar, V., Stevens, T. M., and Shenk, T., III.** RNA of uninfected and Sindbis virus-infected *Aedes albopictus* cells, *Curr. Top. Microbiol. Immunol.,* 55, 164, 1971.
257. **Shine, J. and Dalgarno, L.,** Occurrence of heat-dissociable ribosomal RNA in insects: the presence of three polynucleotide chains in 26 S RNA from cultured *Aedes aegypti* cells, *J. Mol. Biol.,* 75, 67, 1973.
258. **Kennedy, S. I. T.,** Isolation and identification of the virus-specific RNA species found on membrane bound polyribosomes of chick embryo cells infected with Semliki Forest virus, *Biochem. Biophys. Res. Commun.,* 48, 1254, 1972.
259. **Moshowitz, D.,** Identification of polysomal RNA in BHK cells infected with Sindbis virus, *J. Virol.,* 11, 535, 1973.
260. **Rosemond, H. and Sreevalsan, T.,** Viral RNAs associated with ribosomes in Sindbis virus-infected HeLa cells, *J. Virol.,* 11, 399, 1973.
261. **Eaton, B. T. and Regnery, R. L.,** Polysomal RNA in Semliki Forest virus infected *Aedes albopictus* cells, *J. Gen. Virol.,* 29, 35, 1975.
262. **Bruton, C. J., Porter, A., and Kennedy, S. I. T.,** Defective-interfering particles of Semliki Forest virus: intracellular events during interference, *J. Gen. Virol.,* 31, 397, 1976.
263. **Strauss, J. H. and Strauss, E. G.,** Togaviruses, in *The Molecular Biology of Animal Viruses,* Vol. 1, Nayak, D. P., Ed., Marcel Dekker, New York, 1977, 111.
264. **Stark, C. and Kennedy, S. I. T.,** The generation and propagation of defective-interfering particles of Semliki Forest virus in different cell types, *Virology,* 89, 285, 1978.
265. **Frey, T. K. and Strauss, J. H.,** Replication of Sindbis virus. VI. Poly (A) and Poly (U) in virus-specific RNA species, *Virology,* 86, 494, 1978.
266. **Wengler, G., Wengler, G., and Gross, H. J.,** Terminal sequences of Sindbis virus-specific nucleic acids: identity in molecules synthesized in vertebrate and insect cells and characteristic properties of the replicative form RNA, *Virology,* 123, 273, 1982.
267. **Wengler, G.,** Effects of Alphavirus on host cell macromolecular synthesis, in *The Togaviruses: Biology, Structure, Replication,* Schlesinger, R. W., Ed., Academic Press, New York, 1980, 459.
268. **Saborio, J. L., Pong, S. S., and Koch, G.,** Selective and reversible inhibition of initiation of protein synthesis in mammalian cells, *J. Mol. Biol.,* 85, 195, 1974.
269. **Ng, M. L. and Westaway, E. G.,** Proteins specified by Togaviruses in infected *Aedes albopictus* (Singh) mosquito cells, *J. Gen. Virol.,* 43, 91, 1979.
270. **Jalanko, A.,** Expression of Semliki Forest virus capsid protein from SV 40 recombinant virus, *FEBS Lett.,* 186, 59, 1985.
271. **Elgizoli, M., Dai, Y., Kempf, Ch., Koblet, H., and Michel, M. R.,** Semliki Forest virus capsid protein acts as a pleiotropic regulator of host cellular protein synthesis, *J. Virol.,* 63, 2921, 1989.
272. **Michel, M. R., Elgizoli, M., Dai, Y., Jakob, R., Koblet, H., and Arrigo, A.-P.,** Karyophilic properties of Semliki Forest virus nucleocapsid protein, *J. Virol.,* 64, 5123, 1990.
272a. **Faure, D., Studer, E., Nishimura, T., Weitz, M., and Michel, M. R.,** Semliki Forest virus capsid protein expressed by a baculovirus recombinant, *Arch. Virol.,* 132, 307, 1993.
273. **Eaton, B. T.,** Transient enhanced synthesis of a cellular protein following infection of *Aedes albopictus* cells with Sindbis virus, *Virology,* 117, 307, 1982.
274. **Garry, R. F. and Bostick, D. A.,** Induction of the stress response: alterations in membrane-associated transport systems and protein modification in heat-shocked or Sindbis virus-infected cells, *Virus Res.,* 8, 245, 1987.
275. **Reigel, F. and Koblet, H.,** Selective disappearance of two secreted host proteins in the course of Semliki Forest virus infection of *Aedes albopictus* cells, *J. Virol.,* 39, 321, 1981.
276. **Reigel, F. and Koblet, H.,** Viral proteins from SFV infected *Aedes* cells with unusual mol. wt., *Experientia,* 37, 1220, 1981.
277. **Mayne, J. T., Bell, J. R., Strauss, E. G., and Strauss, J. H.,** Pattern of glycosylation of Sindbis virus envelope proteins synthesized in hamster and chicken cells, *Virology,* 142, 121, 1985.
278. **Hubbard, C. S.,** Regulation of glycosylation, *J. Biol. Chem.,* 263, 19303, 1988.
279. **Naim, H. Y. and Koblet, H.,** Asparagine-linked oligosaccharides of Semliki Forest virus grown in mosquito cells, *Arch. Virol.,* 122, 45, 1992.

280. **Schlesinger, S., Gottlieb, C., Feil, P., Gelb, N., and Kornfeld, S.,** Growth of enveloped RNA viruses in a line of *Chinese hamster* ovary cells with deficient N-acetyl-glucosaminyltransferase activity, *J. Virol.*, 17, 239, 1976.
281. **Schlesinger, S., Koyama, A. H., Malfer, C., Gee, S. L., and Schlesinger, M. J.,** The effects of inhibitors of glucosidase I on the formation of Sindbis virus, *Virus Res.*, 2, 139, 1985.
282. **Mc Dowell, W., Romero, P. A., Datema, R., and Schwarz, R. T.,** Glucose trimming and mannose trimming affect different phases of the maturation of Sindbis virus in infected BHK cells, *Virology*, 161, 37, 1987.
283. **Kaluza, G., Repges, S., and Mc Dowell, W.,** The significance of carbohydrate trimming for the antigenicity of the Semliki Forest glycoprotein E_2, *Virology*, 176, 369, 1990.
284. **Hsieh, P., Rosner, M. R., and Robbins, P. W.,** Host-dependent variation of asparagine-linked oligosaccharides at individual glycosylation sites of Sindbis virus glycoproteins, *J. Biol. Chem.*, 258, 2548, 1983.
285. **Hsieh, P., Rosner, M. R., and Robbins, P. W.,** Selective cleavage by endo-β-*N*-acetylglucosaminidase H at individual glycosylation sites of Sindbis virion envelope glycoproteins, *J. Biol. Chem.*, 258, 2555, 1983.
286. **Butters, T. D. and Hughes, R. C.,** Isolation and characterization of mosquito cell membrane glycoproteins, *Biochim. Biophys. Acta*, 640, 655, 1981.
287. **Weber, A., März, L., and Altmann, F.,** Characteristics of the asparagine-linked oligosaccharide from honey-bee venom phospholipase A_2, *Comp. Biochem. Physiol.*, 83B, 321, 1986.
288. **Staudacher, E., Kubelka, V., and März, L.,** Distinct N-glycan fucosylation potentials of three lepidopteran cell lines, *Eur. J. Biochem.*, 207, 987, 1992.
289. **Prenner, C., Mach, L., Glössl, J., and März, L.,** The antigenicity of the carbohydrate moiety of an insect glycoprotein, honey-bee *(Apis mellifera)* venom phospholipase A_2, *Biochem. J.*, 284, 377, 1992.
290. **Kuroda, K., Geyer, H., Geyer, R., Doerfler, W., and Klenk, H.-D.,** The oligosaccharides of Influenza virus hemagglutinin expressed in insect cells by a Baculovirus vector, *Virology*, 174, 418, 1990.
291. **Wathen, M. W., Aeed, P. A., and Elhammer, A. P.,** Characterization of oligosaccharide structures on a chimeric respiratory syncytial virus protein expressed in insect cell line Sf9, *Biochemistry*, 30, 2863, 1991.
292. **Davidson, D. J., Fraser, M. J., and Castellino, F. J.,** Oligosaccharide processing in the expression of human plasminogen cDNA by lepidopteran insect *(Spodoptera frugiperda)* cells, *Biochemistry*, 29, 5584, 1990.
293. **Davidson, D. J. and Castellino, F. J.,** Asparagine-linked oligosaccharide processing in lepidopteran insect cells, *Biochemistry*, 30, 6167, 1991.
294. **Davidson, D. J., Bretthauer, R. K., and Castellino, F. J.,** α-mannosidase-catalyzed trimming of high-mannose glycans in noninfected and Baculovirus-infected *Spodoptera frugiperda* cells (IPLB-SF-21 AE), *Biochemistry*, 30, 9811, 1991.
295. **Davidson, D. J. and Castellino, F. J.,** Structures of the asparagine-289-linked oligosaccharides assembled on recombinant human plasminogen expressed in a *Mamestra brassicae* cell line (IZD-MBO 503), *Biochemistry*, 30, 6689, 1991.
296. **Pelham, H. R. B.,** Control of protein exit from the endoplasmic reticulum, *Annu. Rev. Cell Biol.*, 5, 1, 1989.
297. **Hurtley, S. M. and Helenius, A.,** Protein oligomerization in the endoplasmic reticulum, *Annu. Rev. Cell Biol.*, 5, 277, 1989.
298. **Ellis, R. J. and van der Vies, S. M.,** Molecular chaperones, *Annu. Rev. Biochem.*, 60, 321, 1991.
299. **Marquardt, T. and Helenius, A.,** Misfolding and aggregation of newly synthesized proteins in the endoplasmic reticulum, *J. Cell Biol.*, 117, 505, 1992.
300. **Garoff, H.,** Cross-linking of the spike glycoproteins in Semliki Forest virus with dimethylsuberimidate, *Virology*, 62, 385, 1974.
301. **Garoff, H. and Simons, K.,** Location of the spike glycoproteins in the Semliki Forest virus membrane, *Proc. Natl. Acad. Sci. U.S.A.*, 71, 3988, 1974.
302. **von Bonsdorff, C.-H. and Harrison, S. C.,** Sindbis virus glycoproteins form a regular icosahedral surface lattice, *J. Virol.*, 16, 141, 1973.
303. **Ziemiecki, A. and Garoff, H.,** Subunit composition of the membrane glycoprotein complex of Semliki Forest virus, *J. Mol. Biol.*, 122, 259, 1978.
304. **Enzmann, P.J. and Weiland, F.,** Studies on the morphology of Alphaviruses, *Virology*, 95, 501, 1979.
305. **Rice, C. M. and Strauss, J. H.,** Association of Sindbis virion glycoproteins and their precursors, *J. Mol. Biol.*, 154, 325, 1982.
306. **Vogel, R. H., Provencher, S. W., von Bonsdorff, C.-H., Adrian, M., and Dubochet, J.,** Envelope structure of Semliki Forest virus reconstructed from cryo-electron micrographs, *Nature*, 320, 533, 1986.
307. **Fuller, S. D. and Argos, P.,** Is Sindbis a simple picornavirus with an envelope?, *EMBO J.*, 6, 1099, 1987.
308. **Naim, H. Y. and Koblet, H.,** The cleavage of p62, the precursor of E_2 and E_3, is an early and continuous event in Semliki Forest virus infected *Aedes albopictus* cells, *Arch. Virol.*, 110, 221, 1990.
309. **Syväoja, P., Peränen, J., Suomalainen, M., Keränen, S., and Kääriäinen, L.,** A single amino acid change in E_3 of ts 1 mutant inhibits the intracellular transport of SFV envelope protein complex, *Virology*, 179, 658, 1990.
310. **Lobigs, M., Zhao, H., and Garoff, H.,** Function of Semliki Forest virus E_3 peptide in virus assembly: replacement of E_3 with an artificial signal peptide abolishes spike heterodimerization and surface expression of E_1, *J. Virol.*, 64, 4346, 1990.

311. **Green, J., Griffiths, G., Louvard, D., Quinn, P., and Warren, G.,** Passage of viral membrane proteins through the Golgi complex, *J. Mol. Biol.*, 152, 663, 1981.
312. **Schmidt, M. F. G. and Schlesinger, M. J.,** Relation of fatty acid attachment to the translation and maturation of vesicular stomatitis and Sindbis virus membrane glycoproteins, *J. Biol. Chem.*, 255, 3334, 1980.
313. **Schlesinger, M. J. and Malfer, C.,** Cerulenin blocks fatty acid acylation of glycoproteins and inhibits vesicular stomatitis and Sindbis virus particle formation, *J. Biol. Chem.*, 257, 9887, 1982.
314. **Schmidt, M. F. G.,** Acylation of viral spike glycoproteins: a feature of enveloped RNA viruses, *Virology*, 116, 327, 1982.
315. **Schmidt, M. F. G.,** The transfer of myristic and other fatty acids on lipid and viral protein acceptors in cultured cells infected with Semliki Forest and Influenza virus, *EMBO J.*, 3, 2295, 1984.
316. **Berger, M. and Schmidt, M. F. G.,** Identification of acyl donors and acceptor proteins for fatty acid acylation in BHK cells infected with Semliki Forest virus, *EMBO J.*, 3, 713, 1984.
317. **Koblet, H., Schärer, C. G., Vallan, C., and Naim, H. Y.,** Exocytotic transport of Semliki Forest virus (SFV) envelope proteins in mosquito (C6/36) cells, in *Proc. the 8th Int. Conf. Invertebrate and Fish Tissue Culture,* Fraser, M. J., Jr., Ed., Tissue Culture Association, Columbia, MD, 1991, 195.
317a. **Schärer, C. G., Naim, H. Y., and Koblket, H.,** Palmitoylation of Semliki Forest virus glycoproteins in insect cells (C6/36) occurs in an early compartment and is coupled to the cleavage of the precursor p62, *Arch. Virol.*, 132, 237, 1993.
318. **Schmidt, M., Schmidt, M. F. G., and Rott, R.,** Chemical identification of cysteine as palmitoylation site in a transmembrane protein (Semliki Forest virus E_1), *J. Biol. Chem.*, 263, 18635, 1988.
319. **Gaedigk-Nitschko, K. and Schlesinger, M. J.,** The Sindbis virus 6K protein can be detected in virions and is acylated with fatty acids, *Virology*, 175, 274, 1990.
320. **Gaedigk-Nitschko, K., Ding, M., Levy, M. A., and Schlesinger, M. J.,** Site-directed mutations in the Sindbis virus 6K protein reveal sites for fatty acylation and the underacylated protein affects virus release and virion structure, *Virology*, 175, 282, 1990.
321. **Liljeström, P., Lusa, S., Huylebroeck, D., and Garoff, H.,** In vitro mutagenesis of a full-length cDNA clone of Semliki Forest virus: the small 6000 molecular weight membrane protein modulates virus release, *J. Virol.*, 65, 4107, 1991.
322. **Lusa, S., Garoff, H., and Liljeström, P.,** Fate of the 6K membrane protein of Semliki Forest virus during virus assembly, *Virology*, 185, 843, 1991.
323. **Presley, J. F., Polo, J. M., Johnston, R. E., and Brown, D. T.,** Proteolytic processing of the Sindbis virus membrane protein precursor PE2 is nonessential for growth in vertebrate cells but is required for efficient growth in invertebrate cells, *J. Virol.*, 65, 1905, 1991.
324. **De Curtis, I. and Simons, K.,** Dissection of Semliki Forest virus glycoprotein delivery from the trans-Golgi network to the cell surface in permeabilized BHK cells, *Proc. Natl. Acad. Sci. U.S.A.*, 85, 8052, 1988.
325. **Garoff, H., Frischauf, A. M., Simons, K., Lehrach, H., and Delius, H.,** Nucleotide sequence of cDNA coding for Semliki Forest virus membrane glycoproteins, *Nature*, 288, 236, 1980.
326. **Lobigs, M. and Garoff, H.,** Fusion function of the Semliki Forest virus spike is activated by proteolytic cleavage of the envelope glycoprotein precursor p62, *J. Virol.*, 64, 1233, 1990.
327. **Jain, S. K., De Candido, S., and Kielian, M.,** Processing of the p62 envelope precursor protein of Semliki Forest virus, *J. Biol. Chem.*, 266, 5756, 1991.
328. **Knipfer, M. E. and Brown, D. T.,** Intracellular transport and processing of Sindbis virus glycoproteins, *Virology*, 170, 117, 1989.
329. **Russell, D. L., Dalrymple, J. M., and Johnston, R. E.,** Sindbis virus mutations which coordinately affect glycoprotein processing, penetration, and virulence in mice, *J. Virol.*, 63, 1619, 1989.
330. **Presley, J. F. and Brown, D. T.,** The proteolytic cleavage of PE2 to envelope glycoprotein E2 is not strictly required for the maturation of Sindbis virus, *J. Virol.*, 63, 1975, 1989.
331. **Watson, D. G., Moehring, J. M., and Moehring, T. J.,** A mutant CHO-K1 strain with resistance to Pseudomonas exotoxin A and Alphaviruses fails to cleave Sindbis virus glycoprotein PE2, *J. Virol.*, 65, 2332, 1991.
332. **Erwin, C. and Brown, D. T.,** Intracellular distribution of Sindbis virus membrane proteins in BHK-21 cells infected with wild-type virus and maturation defective mutants, *J. Virol.*, 36, 775, 1980.
333. **Fries, E., Gustafson, L., and Peterson, P. A.,** Four secretory proteins synthesized by hepatocytes are transported from endoplasmic reticulum to Golgi complex at different rates, *EMBO J.*, 3, 147, 1984.
334. **Fuller, S. D.,** The T = 4 envelope of Sindbis virus is organized by interactions with a complementary T = 3 capsid, *Cell*, 48, 923, 1987.
335. **Vaux, D. J. T., Helenius, A., and Mellman, I.,** Spike-nucleocapsid interaction in Semliki Forest virus reconstructed using network antibodies, *Nature*, 336, 36, 1988.
336. **Paredes, A. M., Simon, M. N., and Brown, D. T.,** The mass of the Sindbis virus nucleocapsid suggests it has T = 4 icosahedral symmetry, *Virology*, 187, 329, 1992.
336a. **Paredes, A. M., Brown, D. T., Rothnagel, R., Chiu, W., Schoepp, R. J., Johnston, R. E., and Prasad, B. V. V.,** Three-dimensional structure of a membrane-containing Virus, *Proc. Natl. Acad. Sci. U.S.A.*, 90, 9095, 1993.

337. **Weiss, B., Nitschko, H., Ghattas, I., Wright, R., and Schlesinger, S.,** Evidence for specificity in the encapsidation of Sindbis virus RNAs, *J. Virol.*, 63, 5310, 1989.
338. **Strong, R. K. and Harrison, S. C.,** Proteolytic dissection of Sindbis virus core protein, *J. Virol.*, 64, 3992, 1990.
339. **Metsikkö, K. and Garoff, H.,** Oligomers of the cytoplasmic domain of the p62/E_2 membrane protein of Semliki Forest virus bind to the nucleocapsid in vitro, *J. Virol.*, 64, 4678, 1990.
340. **Kail, M., Hollinshead, M., Ansorge, W., Pepperkok, R., Frank, R., Griffiths, G., and Vaux, D.,** The cytoplasmic domain of Alphavirus E_2 glycoprotein contains a short linear recognition signal required for viral budding, *EMBO J.*, 10, 2343, 1991.
341. **Suomalainen, M. and Garoff, H.,** Alphavirus spike-nucleocapsid interaction and network antibodies, *J. Virol.*, 66, 5106, 1992.
342. **Suomalainen, M., Liljeström, P., and Garoff, H.,** Spike protein-nucleocapsid interactions drive the budding of Alphaviruses, *J. Virol.*, 66, 4737, 1992.
343. **Stollar, V., Harrap, K. A., Thomas, V., and Sarver, N.,** Observations related to cytopathic effect in *Aedes albopictus* cells infected with Sindbis virus, in *Arctic and Tropical Arboviruses*, Kurstak, E., Ed., Academic Press, New York, 1979, 277.
344. **Renz, D. and Brown, D. T.,** Characteristics of Sindbis virus temperature-sensitive mutants in cultured BHK-21 and *Aedes albopictus* (mosquito) cells, *J. Virol.*, 19, 775, 1976.
345. **Riedel, B. and Brown, D. T.,** Role of extracellular virus in the maintenance of the persistent infection induced in *Aedes albopictus* (mosquito) cells by Sindbis virus, *J. Virol.*, 23, 554, 1977.
346. **Houk, E. J., Kramer, L. D., Hardy, J. L., and Chiles, R. E.,** Western equine encephalomyelitis virus: in vivo infection and morphogenesis in mosquito mesenteronal epithelial cells, *Virus Res.*, 2, 123, 1985.
347. **Griffiths, G. and Simons, K.,** The trans-Golgi network: sorting at the exit site of the Golgi complex, *Science*, 234, 438, 1986.
348. **Kuismanen, E. and Saraste, J.,** Low temperature-induced transport blocks as tools to manipulate membrane traffic, *Methods Cell Biol.*, 32, 257, 1989.
349. **Saraste, J. and Kuismanen, E.,** Pre- and post-Golgi vacuoles operate in the transport of Semliki Forest virus membrane glycoproteins to the cell surface, *Cell*, 38, 535, 1984.
350. **Matlin, K. S. and Simons, K.,** Reduced temperature prevents transfer of a membrane glycoprotein to the cell surface but does not prevent terminal glycosylation, *Cell*, 34, 233, 1983.
351. **Kuismanen, E., Jäntti, J., Mäkiranta, V., and Sariola, M.,** Effect of caffeine on intracellular transport of Semliki Forest virus membrane glycoproteins, *J. Cell Sci.*, 102, 505, 1992.
352. **Turell, M. J.,** Effect of environmental temperature on the vector competence of *Aedes fowleri* for Rift Valley fever virus, *Res. Virol.*, 140, 147, 1989.
353. **Turell, M. J., Rossi, C. A., and Bailey, C. L.,** Effect of extrinsic incubation temperature on the ability of *Aedes taeniorhynchus* and *Culex pipiens* to transmit Rift Valley fever virus, *Am. J. Trop. Med. Hyg.*, 34, 1211, 1985.
354. **Vallan, C., Schärer, C. G., and Koblet H.,** Temperature-sensitive steps in the transport of Semliki Forest virus envelope proteins in mosquito C6/36 cells, *Arch. Virol.*, in press.
355. **Ceriotti, A. and Colman, A.,** Trimer formation determines the rate of Influenza virus hemagglutinin transport in the early stages of secretion in *Xenopus* oocytes, *J. Cell Biol.*, 111, 409, 1990.
356. **Banerjee, K. and Singh, K. R. P.,** Establishment of carrier cultures of *Aedes albopictus* cell line infected with arboviruses, *Indian J. Med. Res.*, 56, 812, 1968.
357. **Banerjee, K. and Singh, K. R. P.,** Loss of mouse virulence in Chikungunya virus from the carrier culture of *Aedes albopictus* cell line, *Indian J. Med. Res.*, 57, 1003, 1969.
358. **Stollar, V. and Shenk, T. E.,** Homologous viral interference in *Aedes albopictus* cultures chronically infected with Sindbis virus, *J. Virol.*, 11, 592, 1973.
359. **Peleg, J.,** Attenuation of Semliki Forest virus in continuously cultured *Aedes aegypti* mosquito cells (Peleg) as a step in production of vaccine, *Curr. Top. Microbiol. Immunol.*, 55, 155, 1971.
360. **Peleg, J. and Pecht, M.,** Adaptation of an *Aedes aegypti* mosquito cell line to growth at 15°C and its response to infection by Sindbis virus, *J. Gen. Virol.*, 38, 231, 1978.
361. **Eaton, B. T.,** Heterologous interference in *Aedes albopictus* cells infected with Alphaviruses, *J. Virol.*, 30, 45, 1979.
362. **Igarashi, A., Koo, R., and Stollar, V.,** Evolution and properties of *Aedes albopictus* cell cultures persistently infected with Sindbis virus, *Virology*, 82, 69, 1977.
363. **Enzmann, P.-J.,** Induction of an interferon-like substance in persistently infected *Aedes albopictus* cells, *Arch. Gesamte Virusforsch.*, 41, 382, 1973.
364. **Riedel, B. and Brown, D. T.,** Novel antiviral activity found in the media of Sindbis virus-persistently infected mosquito (*Aedes albopictus*) cell cultures, *J. Virol.*, 29, 51, 1979.
365. **Newton, S. E. and Dalgarno, L.,** Antiviral activity released from *Aedes albopictus* cells persistently infected with Semliki Forest virus, *J. Virol.*, 47, 652, 1983.
366. **Condreay, L. D. and Brown, D. T.,** Exclusion of superinfecting homologous virus by Sindbis virus-infected *Aedes albopictus* (mosquito) cells, *J. Virol.*, 58, 81, 1986.

367. **Peleg, J. and Stollar, V.**, Homologous interference in *Aedes aegypti* cell cultures infected with Sindbis virus, *Arch. Gesamte Virusforsch.*, 45, 309, 1974.
368. **Shenk, T. E., Koshelnyk, K. A., and Stollar, V.**, Temperature-sensitive virus from *Aedes albopictus* cells chronically infected with Sindbis virus, *J. Virol.*, 13, 439, 1974.
369. **Hahn, Y. S., Strauss, E. G., and Strauss, J. H.**, Mapping of RNA temperature-sensitive mutants of Sindbis virus: assignment of complementation groups A, B, and G to nonstructural proteins, *J. Virol.*, 63, 3142, 1989.
370. **Lundqvist, B. H., Di Salvo, J., Rice, C. M., Strauss, J. H., and Strauss, E. G.**, Sindbis virus mutant ts 20 of complementation group E contains a lesion in glycoprotein E_2, *Virology*, 151, 10, 1986.
371. **Stollar, V., Peleg, J., and Shenk, T. E.**, Temperature sensitivity of a Sindbis virus mutant isolated from persistently infected *Aedes aegypti* cell culture, *Intervirology*, 2, 337, 1973/74.
372. **Huang, A. S. and Baltimore, D.**, Defective viral particles and viral disease processes, *Nature*, 226, 325, 1970.
373. **Schlesinger, S., Schlesinger, M., and Burge, B. W.**, Defective virus particles from Sindbis virus, *Virology*, 48, 615, 1972.
374. **Eaton, B. T.**, Defective interfering particles of Semliki Forest virus generated in BHK cells do not interfere with viral RNA synthesis in *Aedes albopictus* cells, *Virology*, 68, 534, 1975.
375. **Steacie, A. D. and Eaton, B. T.**, Properties of defective interfering particles of Sindbis virus generated in vertebrate and mosquito cells, *J. Gen. Virol.*, 65, 333, 1984.
376. **Stollar, V., Shenk, T. E., Koo, R., Igarashi, A., and Schlesinger, R. W.**, Observations on *Aedes albopictus* cell cultures persistently infected with Sindbis virus, *Ann. N. Y. Acad. Sci.*, 266, 214, 1975.
377. **Igarashi, A. and Stollar, V.**, Failure of defective interfering particles of Sindbis virus produced in BHK or chicken cells to affect viral replication in *Aedes albopictus* cells, *J. Virol.*, 19, 398, 1976.
378. **King, C.-C., King, M. W., Garry, R. F., Wan, K. M.-M., Ulug, E. T., and Waite, M. R. F.**, Effect of incubation time on the generation of defective-interfering particles during undiluted serial passage of Sindbis virus in *Aedes albopictus* and chick cells, *Virology*, 96, 229, 1979.
379. **Stalder, J., Reigel, F., and Koblet, H.**, Defective viral RNAs in *Aedes albopictus* C6/36 cells persistently infected with Semliki Forest virus, *Virology*, 129, 247, 1983.
380. **Eaton, B. T.**, Evidence for the synthesis of defective interfering particles by *Aedes albopictus* cells persistently infected with Sindbis virus, *Virology*, 77, 843, 1977.
381. **Logan, K. B.**, Generation of defective interfering particles of Semliki Forest virus in a clone of *Aedes albopictus* (mosquito) cells, *J. Virol.*, 30, 38, 1979.
382. **Eaton, B. T.**, Viral interference and persistence in Sindbis virus infected *Aedes albopictus* cells, *Can. J. Microbiol.*, 27, 563, 1981.
383. **Stollar, V. and Hardy, J. L.**, Host-dependent mutants of Sindbis virus whose growth is restricted in cultured *Aedes albopictus* cells produce normal yields of virus in intact mosquitoes, *Virology*, 134, 177, 1984.
384. **Brown, D. T. and Gliedman, J. B.**, Morphological variants of Sindbis virus obtained from infected mosquito tissue culture cells, *J. Virol.*, 12, 1534, 1973.
385. **Hashimoto, K., Suzuki, K., and Simizu, B.**, Morphological and physical properties of a multiploid-forming mutant of Western equine encephalitis virus, *J. Virol.*, 15, 1454, 1975.
386. **Strauss, E. G.**, Mutants of Sindbis virus. III. Host polypeptides present in purified HR and ts 103 virus particles, *J. Virol.*, 28, 466, 1978.
387. **Hahn, C. S., Rice, C. M., Strauss, E. G., Lenches, E. N., and Strauss, J. H.**, Sindbis virus ts 103 has a mutation in glycoprotein E_2 that leads to defective assembly of virions, *J. Virol.*, 63, 3459, 1989.
388. **Maeda, S., Hashimoto, K., and Simizu, B.**, Complementation between temperature-sensitive mutants isolated from *Aedes albopictus* cells persistently infected with Western equine encephalitis virus, *Virology*, 92, 532, 1979.
389. **Theilmann, D., Eaton, B. T., and Downe, A. E. R.**, Properties of Sindbis virus variants from infected *Culex tarsalis* mosquitoes, *J. Gen. Virol.*, 65, 945, 1984.
390. **Durbin, R. K. and Stollar, V.**, A mutant of Sindbis virus with a host-dependent defect in maturation associated with hyperglycosylation of E_2, *Virology*, 135, 331, 1984.
391. **Durbin, R. K. and Stollar, V.**, Sequence analysis of the E_2 gene of hyperglycosylated, host-restricted mutant of Sindbis virus and estimation of mutation rate from frequency of revertants, *Virology*, 154, 135, 1986.
392. **Simizu, B., Hashimoto, K., and Ishida, I.**, A variant of Western equine encephalitis virus with nonglycosylated E_3 protein, *Virology*, 125, 99, 1983.
393. **Eaton, B. T.**, Sindbis virus variants from a persistently infected mosquito cell culture contain an altered E_2 glycoprotein, *Virology*, 122, 486, 1982.
394. **Kowal, K. J. and Stollar, V.**, Temperature-sensitive host-dependent mutants of Sindbis virus, *Virology*, 114, 140, 1981.
395. **Lemm, J. A., Durbin, R. K., Stollar, V., and Rice, C. M.**, Mutations which alter the level or structure of nsP4 can affect the efficiency of Sindbis virus replication in a host-dependent manner, *J. Virol.*, 64, 3001, 1990.
396. **Woodward, T. M., Miller, B. R., Beaty, B. J., Trent, D. W., and Roehrig, J. T.**, A single amino acid change in the E_2 glycoprotein of Venezuelan equine encephalitis virus affects replication and dissemination in *Aedes aegypti* mosquitoes, *J. Gen. Virol.*, 72, 2431, 1991.

397. **Hertz, J. M. and Huang, H. V.**, Utilization of heterologous alphavirus junction sequences as promoters by Sindbis virus, *J. Virol.*, 66, 857, 1992.
398. **Niesters, H. G. M. and Strauss, J. H.**, Defined mutations in the 5′ nontranslated sequence of Sindbis virus RNA, *J. Virol.*, 64, 4162, 1990.
399. **Kuhn, R. J., Niesters, H. G. M., Hong, Z., and Strauss, J. H.**, Infectious RNA transcripts from Ross River virus cDNA clones and the construction and characterization of defined chimeras with Sindbis virus, *Virology*, 182, 430, 1991.
400. **Kuhn, R. J., Hong, Z., and Strauss, J. H.**, Mutagenesis of the 3′ nontranslated region of Sindbis virus RNA, *J. Virol.*, 64, 1465, 1990.
401. **Nakhasi, H. L., Cao, X.-Q., Rouault, T. A., and Liu, T. Y.**, Specific binding of host cell proteins to the 3′-terminal stem-loop structure of Rubella virus negative-strand RNA, *J. Virol.*, 65, 5961, 1991.
402. **Pardigon, N. and Strauss, J. H.**, Cellular proteins bind to the 3′ end of Sindbis virus minus-strand RNA, *J. Virol.*, 66, 1007, 1992.
403. **Niesters, H. G. M. and Strauss, J. H.**, Mutagenesis of the conserved 51-nucleotide region of Sindbis virus, *J. Virol.*, 64, 1639, 1990.
404. **Kuhn, R. J., Griffin, D. E., Zhang, H., Niesters, H. G. M., and Strauss, J. H.**, Attenuation of Sindbis virus neurovirulence by using defined mutations in nontranslated regions of the genome RNA, *J. Virol.*, 66, 7121, 1992.
405. **Li, G. and Rice, C. M.**, Mutagenesis of the in-frame opal termination codon preceding nsP4 of Sindbis virus: studies of translational readthrough and its effect on virus replication, *J. Virol.*, 63, 1326, 1989.
406. **Dubin, D. T., Stollar, V., Hsuchen, C. C., Timko, K., and Guild, G. M.**, Sindbis virus messenger RNA: The 5′ termini and methylated residues of 26 S and 42 S RNA, *Virology*, 77, 457, 1977.
407. **Stollar, V.**, Inhibition of Sindbis virus replication in *Aedes albopictus* cells deprived of methionine, *Virology*, 91, 504, 1978.
408. **Malinoski, F. and Stollar, V.**, Inhibitors of IMP dehydrogenase prevent Sindbis virus replication and reduce GTP levels in *Aedes albopictus* cells, *Virology*, 110, 281, 1981.
409. **Stollar, V. and Malinoski, F.**, The effects of adenosine and guanosine on the replication of Sindbis and vesicular stomatitis viruses in *Aedes albopictus* cells, *Virology*, 115, 57, 1981.
410. **Malinoski, F. and Stollar, V.**, Inhibition of Sindbis virus replication by ribavirin: influence of cultural conditions and of the host cell phenotype, *Antiviral Res.*, 1, 287, 1981.
411. **Durbin, R. K. and Stollar, V.**, Sindbis virus mutants able to replicate in methionine-deprived *Aedes albopictus* cells, *Virology*, 144, 529, 1985.
412. **Mi, S., Durbin, R. K., Huang, H. V., Rice, C. M., and Stollar, V.**, Association of the Sindbis virus RNA methyltransferase activity with the nonstructural protein nsP1, *Virology*, 170, 385, 1989.
413. **Scheidel, L. M., Durbin, R. K., and Stollar, V.**, SV_{LM21}, a Sindbis virus mutant resistant to methionine deprivation, encodes an altered methyltransferase, *Virology*, 173, 408, 1989.
414. **Mi, S. and Stollar, V.**, Both amino acid changes in nsP1 of Sindbis virus LM_{21} contribute to and are required for efficient expression of the mutant phenotype, *Virology*, 178, 429, 1990.
415. **Mi, S. and Stollar, V.**, Expression of Sindbis virus nsP1 and methyltransferase activity in Escherichia coli, *Virology*, 184, 423, 1991.
416. **Scheidel, L. M., Durbin, R. K., and Stollar, V.**, Sindbis virus mutants resistant to mycophenolic acid and ribavirin, *Virology*, 158, 1, 1987.
417. **Scheidel, L. M. and Stollar, V.**, Mutations that confer resistance to mycophenolic acid and ribavirin on Sindbis virus map to the nonstructural protein nsP1, *Virology*, 181, 490, 1991.
418. **Ahlquist, P., Strauss, E. G., Rice, C. M., Strauss, J. H., Haseloff, J., and Zimmern, D.**, Sindbis virus proteins nsP1 and nsP2 contain homology to nonstructural proteins from several RNA plant viruses, *J. Virol.*, 53, 536, 1985.
419. **Wang, Y.-F., Sawicki, S. G., and Sawicki, D. L.**, Sindbis virus nsP1 functions in negative-strand RNA synthesis, *J. Virol.*, 65, 985, 1991.
420. **Rozanov, M. N., Koonin, E. V., and Gorbalenya, A. E.**, Conservation of the putative methyltransferase domain: a hall-mark of the 'Sindbis-like' supergroup of positive-strand RNA viruses, *J. Gen. Virol.*, 73, 2129, 1992.
421. **Erwin, C. and Brown, D. T.**, Requirement of cell nucleus for Sindbis virus replication in cultured *Aedes albopictus* cells, *J. Virol.*, 45, 792, 1983.
422. **Scheefers-Borchel, U., Scheefers, H., Edwards, J., and Brown, D. T.**, Sindbis virus maturation in cultured mosquito cells is sensitive to actinomycin D, *Virology*, 110, 292, 1981.
423. **Baric, R. S., Carlin, L. J., and Johnston, R. E.**, Requirement for host transcription in the replication of Sindbis virus, *J. Virol.*, 45, 200, 1983.
424. **Baric, R. S., Lineberger, D. W., and Johnston, R. E.**, Reduced synthesis of Sindbis virus negative strand RNA in cultures treated with host transcription inhibitors, *J. Virol.*, 47, 46, 1983.
425. **Sawicki, D. L. and Sawicki, S. G.**, Short-lived minus-strand polymerase for Semliki Forest virus, *J. Virol.*, 34, 108, 1980.

426. **Gould, E. A. and Clegg, J. C. S.**, Growth, titration and purification of Togaviruses, in *Virology: A Practical Approach*, Mahy, B. W. J., Ed., IRL Press, Oxford, 1985, 43.
427. **Igarashi, A., Sasao, F., Fukai, K., Buei, K., Ueba, N., and Yoshida, M.**, Mutants of Getah and Japanese encephalitis viruses isolated from field-caught *Culex tritaeniorhynchus* using *Aedes albopictus* clone C6/36 cells, *Ann. Virol. Inst. Pasteur*, 132 E, 235, 1981.
428. **Igarashi, A.**, A mutant of Chikungunya virus isolated from a line of Singh's *Aedes albopictus* cells by plaque formation on virus-sensitive cloned cells obtained from another Singh's *Aedes albopictus* cell line, *Virology*, 98, 385, 1979.
429. **Igarashi, A. and Srivastava, A. K.**, Suspension culture of *Aedes albopictus*, clone C6/36, cells and its application to the production of Flaviviruses, in *Invertebrate and Fish Tissue Culture*, Kuroda, Y., Kurstak, E., and Maramorosch, K., Eds., Japan Scientific Societies Press, Tokyo/Springer-Verlag, Berlin, 1988, 137.
430. **Riedel, H., Lehrach, H., and Garoff, H.**, Nucleotide sequence at the junction between the nonstructural and the structural genes of the Semliki Forest virus genome, *J. Virol.*, 42, 725, 1982.
431. **Takkinen, K.**, Complete nucleotide sequence of the nonstructural protein genes of Semliki Forest virus, *Nucleic Acids Res.*, 14, 5667, 1986.
432. **Ou, J.-H., Rice, C. M., Dalgarno, L., Strauss, E. G., and Strauss, J. H.**, Sequence studies of several Alphavirus genomic RNAs in the region containing the start of the subgenomic RNA, *Proc. Natl. Acad. Sci. U.S.A.*, 79, 5235, 1982.
433. **Dalgarno, L., Rice, C. M., and Strauss, J. H.**, Ross River virus 26 S RNA: complete nucleotide sequence and deduced sequence of the encoded structural proteins, *Virology*, 129, 170, 1983.
434. **Faragher, S. G., Hutchison, C. A., and Dalgarno, L.**, Analysis of Ross River virus genomic RNA using Hae III digests of single-stranded cDNA to infected-cell RNA and virion RNA, *Virology*, 141, 248, 1985.
435. **Strauss, E. G., Levinson, R., Rice, C. M., Dalrymple, J., and Strauss, J. H.**, Nonstructural proteins nsP3 and nsP4 of Ross River and O'Nyong-nyong viruses: sequence and comparison with those of other alphaviruses, *Virology*, 164, 265, 1988.
436. **Levinson, R. S., Strauss, J. H., and Strauss, E. G.**, Complete sequence of the genomic RNA of O'Nyong-nyong virus and its use in the construction of Alphavirus phylogenetic trees, *Virology*, 175, 110, 1990.
437. **Bell, J. R., Kinney, R. M., Trent, D. W., Strauss, E. G., and Strauss, J. H.**, An evolutionary tree relating eight Alphaviruses, based on amino-terminal sequences of their glycoproteins, *Proc. Natl. Acad. Sci. U.S.A.*, 81, 4702, 1984.
438. **Hahn, C. S., Lustig, S., Strauss, E. G., and Strauss, J. H.**, Western equine encephalitis virus is a recombinant virus, *Proc. Natl. Acad. Sci. U.S.A.*, 85, 5997, 1988.
439. **Rice, C. M. and Strauss, J. H.**, Synthesis, cleavage and sequence analysis of DNA complementary to the 26S messenger RNA of Sindbis virus, *J. Mol. Biol.*, 150, 315, 1981.
440. **Rice, C. M. and Strauss, J. H.**, Nucleotide sequence of the 26 S mRNA of Sindbis virus and deduced sequence of the encoded virus structural proteins, *Proc. Natl. Acad. Sci. U.S.A.*, 78, 2062, 1981.
441. **Strauss, E. G., Rice, C. M., and Strauss, J. H.**, Sequence coding for the Alphavirus nonstructural proteins is interrupted by an opal termination codon, *Proc. Natl. Acad. Sci. U.S.A.*, 80, 5271, 1983.
442. **Strauss, E. G., Rice, C. M., and Strauss, J. H.**, Complete nucleotide sequence of the genomic RNA of Sindbis virus, *Virology*, 133, 92, 1984.
443. **Shirako, Y., Niklasson, B., Dalrymple, J. M., Strauss, E. G., and Strauss, J. H.**, Structure of the Ockelbo virus genome and its relationship to other Sindbis viruses, *Virology*, 182, 753, 1991.
444. **Ou, J.-H., Strauss, E. G., and Strauss, J. H.**, The 5′-terminal sequences of the genomic RNAs of several Alphaviruses, *J. Mol. Biol.*, 168, 1, 1983.
445. **Chang, G.-J. and Trent, D. W.**, Nucleotide sequence of the genome region encoding the 26 S mRNA of Eastern equine encephalomyelitis virus and the deduced amino acid sequence of the viral structural proteins, *J. Gen. Virol.*, 68, 2129, 1987.
446. **Kinney, R. M., Johnson, B. J. B., Welch, J. B., Tsuchiya, K., and Trent, D. W.**, The full-length nucleotide sequences of the virulent Trinidad donkey strain of Venezuelan equine encephalitis virus and its attenuated vaccine derivative, strain TC-83, *Virology*, 170, 19, 1989.
447. **Weaver, S. C., Bellew, L. A., and Rico-Hesse, R.**, Phylogenetic analysis of alphaviruses in the Venezuelan equine encephalitis complex and identification of the source of epizootic viruses, *Virology*, 191, 282, 1992.
448. **De Mattos, I., Vier, E. J., Kobilka, B., Beaty, B. J., and Buckley, S. M.**, Comparative infectivity assays of Dengue virus in mosquitoes and mosquito cell lines, in *Invertebrate Systems In Vitro*, Kurstak, E., Maramorosch, K., and Dübendorfer, A., Eds., Elsevier, Amsterdam, 1980, 341.
449. **De Mattos, I., Jozan-Work, M., Work, T. H., Casals, J., and Buckley, S. M.**, Rapid diagnosis of St. Louis encephalitis virus in *Aedes albopictus* cells, in *Invertebrate Systems In Vitro*, Kurstak, E., Maramorosch, K., and Dübendorfer, A., Eds., Elsevier, Amsterdam, 1980, 347.
450. **Ng, M. L. and Westaway, E. G.**, Establishment of persistent infections by Flaviviruses in *Aedes albopictus* cells, in *Invertebrate Systems In Vitro*, Kurstak, E., Maramorosch, K., and Dübendorfer, A., Eds., Elsevier, Amsterdam, 1980, 389.

451. **Chambers, T. J. and Rice, C. M.,** Molecular biology of the Flaviviruses, *Microbiol. Sci.*, 4, 219, 1987.
452. **Westaway, E. G.,** Flavivirus replication strategy, *Adv. Virus Res.*, 33, 45, 1987.
453. **Brandt, W. E.,** Current approaches to the development of Dengue vaccines and related aspects of the molecular biology of Flaviviruses, *J. Inf. Dis.*, 157, 1105, 1988.
454. **Calisher, C. H., Karabatsos, N., Dalrymple, J. M., Shope, R. E., Porterfield, J. S., Westaway, E. G., and Brandt, W. E.,** Antigenic relationships between Flaviviruses as determined by cross-neutralization: tests with polyclonal antisera, *J. Gen. Virol.*, 70, 37, 1989.
455. **Hase, T., Summers, P. L., Eckels, K. H., and Putnak, J. R.,** Morphogenesis of Flaviviruses, *Subcell. Biochem.*, 15, 275, 1989.
456. **Chambers, T. J., Hahn, C. S., Galler, R., and Rice, C. M.,** Flavivirus genome organization, expression and replication, *Annu. Rev. Microbiol.*, 44, 649, 1990.
457. **Heinz, F. X. and Roehrig, J. T.,** Flaviviruses, in *Immunochemistry of Viruses*, Vol. 2, van Regenmortel, M. H. V. and Neurath, A. R., Eds., Elsevier, Amsterdam, 1990, 289.
458. **Monath, T. P.,** Flaviviruses, in *Virology*, 2nd ed., Vol. 1, Fields, B. N. and Knipe, D. M., Eds., Raven Press, New York, 1990, 763.
459. **Heinz, F. X.,** Comparative molecular biology of Flaviviruses and hepatitis C virus, *Arch. Virol. Suppl.*, 4, 163, 1992.
460. **Ludwig, G. V. and Iacono-Connors, L. C.,** Insect-transmitted vertebrate viruses: Flaviviridae, *In Vitro Cell. Dev. Biol.*, 29A, 296, 1993.
461. **Cammisa-Parks, H., Cisar, L. A., Kane, A., and Stollar, V.,** The complete nucleotide sequence of cell fusing agent (CFA): homology between the nonstructural proteins encoded by CFA and the nonstructural proteins encoded by arthropod-borne Flaviviruses, *Virology*, 189, 511, 1992.
462. **Koonin, E. V.,** The phylogeny of RNA-dependent RNA polymerases of positive-strand RNA viruses, *J. Gen. Virol.*, 72, 2197, 1991.
463. **Speight, G. and Westaway, E. G.,** Carboxyterminal analysis of nine proteins specified by the Flavivirus Kunjin: evidence that only the intracellular core protein is truncated, *J. Gen. Virol.*, 70, 2209, 1989.
464. **Nowak, T., Färber, P. M., Wengler, G., and Wengler, G.,** Analyses of the terminal sequences of West Nile virus structural proteins and of the in vitro translation of these proteins allow the proposal of a complete scheme of the proteolytic cleavages involved in their synthesis, *Virology*, 169, 365, 1989.
465. **Wengler, G., Czaya, G., Färber, P. M., and Hagemann, J. H.,** In vitro synthesis of West Nile virus proteins indicates that the amino terminal segment of the NS 3 protein contains the active centre of the protease which cleaves the viral polyprotein after multiple basic amino acids, *J. Gen. Virol.*, 72, 851, 1991.
466. **Winkler, G., Randolph, V. B., Cleaves, G. R., Ryan, T. E., and Stollar, V.,** Evidence that the mature form of the Flavivirus nonstructural protein NS 1 is a dimer, *Virology*, 162, 187, 1988.
467. **Nowak, T. and Wengler, G.,** Analysis of disulfides present in the membrane proteins of the West Nile flavivirus, *Virology*, 156, 127, 1987.
468. **Wengler, G. and Wengler, G.,** Cell-associated West Nile flavivirus is covered with E + pre-M protein heterodimers which are destroyed and reorganized by proteolytic cleavage during virus release, *J. Virol.*, 63, 2521, 1989.
469. **Wengler, G., Wengler, G., Nowak, T., and Wahn, K.,** Analysis of the influence of proteolytic cleavage on the structural organization of the surface of the West Nile flavivirus leads to the isolation of a protease-resistant E protein oligomer from the viral surface, *Virology*, 160, 210, 1987.
470. **Kamer, G. and Argos, P.,** Primary structural comparison of RNA-dependent polymerases from plant, animal and bacterial viruses, *Nucleic Acids Res.*, 12, 7269, 1984.
471. **Rehacek, J.,** Tick tissue culture and arboviruses, in *Invertebrate Tissue Culture*, Kurstak, E. and Maramorosch, K., Eds., Academic Press, New York, 1976, 21.
472. **Munz, E., Reimann, M., Munderloh, U., and Settele, U.,** The susceptibility of the tick cell line TTC 243 for Nairobi sheep disease virus and some other important species of mammalian RNA and DNA viruses, in *Invertebrate Systems In Vitro*, Kurstak, E., Maramorosch, K., and Dübendorfer, A., Eds., Elsevier, Amsterdam, 1980, 337.
473. **Varma, M. G. R.,** Progress in the study of human and animal pathogens in primary and established tick cell lines, in *Invertebrate Cell System Applications*, Vol. 2, Mitsuhashi, J., Ed., CRC Press, Boca Raton, FL, 1989, 119.
474. **Hasegawa, H., Yoshida, M., Shiosaka, T., Fujita, S., and Kobayashi, Y.,** Mutations in the envelope protein of Japanese encephalitis virus affect entry into cultured cells and virulence in mice, *Virology*, 191, 158, 1992.
475. **Kimura, T., Gollins, S. W., and Porterfield, J. S.,** The effect of pH on the early interaction of West Nile virus with P388 D1 cells, *J. Gen. Virol.*, 67, 2423, 1986.
476. **Kimura, T. and Ohyama, A.,** Association between the pH-dependent conformational change of West Nile flavivirus E protein and virus-mediated membrane fusion, *J. Gen. Virol.*, 69, 1247, 1988.
477. **Guirakhoo, F., Heinz, F. X., Mandl, C. W., Holzmann, H., and Kunz, C.,** Fusion activity of Flaviviruses: comparison of mature and immature (prM-containing) tick-borne encephalitis virions, *J. Gen. Virol.*, 72, 1323, 1991.
478. **Summers, P. L., Cohen, W. H., Ruiz, M. M., Hase, T., and Eckels, K. H.,** Flaviviruses can mediate fusion from without in *Aedes albopictus* mosquito cell cultures, *Virus Res.*, 12, 383, 1989.

479. **Hase, T., Summers, P. L., and Eckels, K. H.,** Flavivirus entry into cultured mosquito cells and human peripheral blood monocytes, *Arch. Virol.*, 104, 129, 1989.
480. **Hase, T., Summers, P. L., and Ray, P.,** Entry and replication of Japanese encephalitis virus in cultured neurogenic cells, *J. Virol. Meth.*, 30, 205, 1990.
481. **Ishak, R., Tovey, D. G., and Howard, T.,** Morphogenesis of yellow fever virus 17D in infected cell cultures, *J. Gen. Virol.*, 69, 325, 1988.
482. **Ng, M. L. and Lau, C. L.,** Possible involvement of receptors in the entry of Kunjin virus into Vero cells, *Arch. Virol.*, 100, 199, 1988.
483. **Wengler, G., Wengler, G., and Gross, H. J.,** Studies on virus-specific nucleic acids synthesized in vertebrate and mosquito cells infected with Flaviviruses, *Virology*, 89, 423, 1978.
484. **Cleaves, G. R., Ryan, T. E., and Schlesinger, R. W.,** Identification and characterization of type 2 Dengue virus replication intermediate and replicative form RNAs, *Virology*, 111, 73, 1981.
485. **Ng, M. L., Pedersen, J. S., Toh, B. H., and Westaway, E. G.,** Immunofluorescent sites in Vero cells infected with the Flavivirus Kunjin, *Arch. Virol.*, 78, 177, 1983.
486. **Smith, G. W. and Wright, P. J.,** Synthesis of proteins and glycoproteins in Dengue type 2 virus-infected Vero and *Aedes albopictus* cells, *J. Gen. Virol.*, 66, 559, 1985.
487. **Ng, M. L.,** Ultrastructural studies of Kunjin virus-infected *Aedes albopictus* cells, *J. Gen. Virol.*, 68, 577, 1987.
488. **Hong, S. S. and Ng, M. L.,** Involvement of microtubules in Kunjin virus replication, *Arch. Virol.*, 97, 115, 1987.
489. **Takegami, T. and Hotta, S.,** In vitro synthesis of Japanese encephalitis virus (JEV) RNA: membrane and nuclear fractions of JEV-infected cells possess high levels of virus-specific RNA polymerase activity, *Virus Res.*, 13, 337, 1989.
490. **Mandl, C. W., Kunz, C., and Heinz, F. X.,** Presence of poly (A) in a Flavivirus: significant differences between the 3′ noncoding regions of the genomic RNAs of tick-borne encephalitis virus strains, *J. Virol.*, 65, 4070, 1991.
491. **Mason, P. W.,** Maturation of Japanese encephalitis virus glycoproteins produced by infected mammalian and mosquito cells, *Virology*, 169, 394, 1989.
492. **Ng, M. L. and Westaway, E. G.,** Phenotypic changes in the Flavivirus Kunjin after a single cycle of growth in an *Aedes albopictus* cell line, *J. Gen. Virol.*, 64, 1715, 1983.
493. **Barrett, A. D. T., Pryde, A., Medlen, A. R., Ledger, T. N., Whitby, J. E., Gibson, C. A., De Silva, M., Graves, D. J., Langley, D. J., and Minor, P. D.,** Examination of the envelope glycoprotein of yellow fever vaccine viruses with monoclonal antibodies, *Vaccine*, 7, 333, 1989.
494. **Filshie, B. K. and Rehacek, J.,** Studies of the morphology of Murray Valley encephalitis and Japanese encephalitis viruses growing in cultured mosquito cells, *Virology*, 34, 435, 1968.
495. **Ng, M. L. and Hong, S. S.,** Flavivirus infection: essential ultrastructural changes and association of Kunjin virus NS 3 protein with microtubules, *Arch. Virol.*, 106, 103, 1989.
496. **Hase, T., Summers, P. L., Eckels, K. H., and Baze, W. B.,** Maturation process of Japanese encephalitis virus in cultured mosquito cells in vitro and mouse brain cells in vivo, *Arch. Virol.*, 96, 135, 1987.
497. **Hase, T., Summers, P. L., Eckels, K. H., and Baze, W. B.,** An electron and immunoelectron microscopic study of Dengue-2 virus infection of cultured mosquito cells: maturation events, *Arch. Virol.*, 92, 273, 1987.
498. **Ko, K. K., Igarashi, A., and Fukai, K.,** Electron microscopic observations on *Aedes albopictus* cells infected with Dengue viruses, *Arch. Virol.*, 62, 41, 1979.
499. **Whitfield, S. D., Murphy, F. A., and Sudia, W. D.,** St. Louis encephalitis virus: an ultrastructural study of infection in a mosquito vector, *Virology*, 56, 70, 1973.
500. **Suitor, E. C. and Paul, F. J.,** Syncytia formation of mosquito cell cultures mediated by type 2 Dengue virus, *Virology*, 38, 482, 1969.
501. **Cory, J. and Yunker, C. E,** Arbovirus plaques in mosquito cell monolayers, *Acta Virol.*, 16, 90, 1972.
502. **Varma, M. G. R. and Pudney, M.,** The growth of serial passage of cell lines from *Aedes aegypti* (L) larvae in different media, *J. Med. Entomol.*, 6, 432, 1969.
503. **Deubel, V. and Digoutte, J. P.,** Morphogenesis of yellow fever virus in *Aedes aegypti* cultured cells. I, *Am. J. Trop. Med. Hyg.*, 30, 1060, 1981.
504. **Koblet, H., Kohler, U., and Igarashi, A.,** Fusion of *Aedes albopictus* cells, clone C6/36, by Japanese encephalitis virus is triggered by low pH, *Trop. Med. Nagasaki*, 32, 145, 1990.
505. **Randolph, V. B. and Stollar, V.,** Low pH-induced cell fusion in Flavivirus-infected *Aedes albopictus* cell cultures, *J. Gen. Virol.*, 71, 1845, 1990.
506. **Higgs, S. and Gould, E. A.,** Differences in fusogenicity and mouse neurovirulence of Japanese encephalitis viruses, *Arch. Virol.*, 119, 119, 1991.
507. **Cane, P. A. and Gould, E. A.,** Immunoblotting reveals differences in the accumulation of envelope protein by wild-type and vaccine strains of yellow fever virus, *J. Gen. Virol.*, 70, 557, 1989.
508. **Randolph, V. B., Winkler, G., and Stollar, V.,** Acidotropic amines inhibit proteolytic processing of Flavivirus prM protein, *Virology*, 174, 450, 1990.
509. **Stollar, V. and Thomas, V. L.,** An agent in the *Aedes aegypti* cell line (Peleg) which causes fusion of *Aedes albopictus* cells, *Virology*, 64, 367, 1975.

510. **Igarashi, A., Harrap, K. A., Casals, J., and Stollar, V.**, Morphological, biochemical and serological studies on a viral agent (CFA) which replicates in and causes fusion of *Aedes albopictus* (Singh) cells, *Virology,* 74, 174, 1976.
511. **Řeháček, J.**, Persistent infection of mosquito cells grown in vitro with Murray Valley encephalitis and Japanese encephalitis viruses, *Acta Virol.,* 12, 340, 1968.
512. **Igarashi, A.**, Characteristics of *Aedes albopictus* cells persistently infected with Dengue viruses, *Nature,* 280, 690, 1979.
513. **Kuno, G.**, Persistent infection of a nonvector mosquito cell line (TRA-171) with Dengue viruses, *Intervirology,* 18, 45, 1982.
514. **Sinarachatanant, P. and Olson, L.**, Replication of Dengue virus type 2 in *Aedes albopictus* cell cultures, *J. Virol.,* 12, 275, 1973.
515. **Randolph, V. B. and Hardy, J. L.**, Establishment and characterization of St. Louis encephalitis virus persistent infections in *Aedes* and *Culex* mosquito cell lines, *J. Gen. Virol.,* 69, 2189, 1988.
516. **Randolph, V. B. and Hardy, J. L.**, Phenotypes of St. Louis encephalitis virus mutants produced in persistently infected mosquito cell cultures, *J. Gen. Virol.,* 69, 2199, 1988.
517. **Lee, C.-H. and Schloemer, R. H.**, Mosquito cells infected with Banzi virus secrete an antiviral activity which is of viral origin, *Virology,* 110, 402, 1981.
518. **Lee, C.-H. and Schloemer, R. H.**, Identification of the antiviral factor in culture medium of mosquito cells persistently infected with Banzi virus, *Virology,* 110, 445, 1981.
519. **Hommel, G. A. and Schloemer, R. H.**, Site of suppression of Banzi viral replication by an antiviral factor released from *Aedes albopictus* cells persistently infected with Banzi virus, *Virus Res.,* 4, 37, 1985.
520. **Lewis, J. G., Chang, G.-J., Lanciotti, R. S., and Trent, D. W.**, Direct sequencing of large Flavivirus PCR products for analysis of genome variation and molecular epidemiological investigations, *J. Virol. Methods,* 38, 11, 1992.
521. **Schoepp, R. J., Beaty, B. J., and Eckels, K. H.**, Infection of *Aedes albopictus* and *Aedes aegypti* mosquitoes with Dengue parent and progeny candidate vaccine viruses: a possible marker of human attenuation, *Am. J. Trop. Med. Hyg.,* 45, 202, 1991.
522. **Varelas-Wesley, I. and Calisher, C. H.**, Antigenic relationships of Flaviviruses with undetermined arthropod-borne status, *Am. J. Trop. Med. Hyg.,* 31, 1273, 1982.
523. **Kuno, G. and Oliver, A.**, Maintaining mosquito cell lines at high temperatures: effects on the replication of Flaviviruses, *In Vitro Cell. Dev. Biol.,* 25, 193, 1989.
524. **Okuno, Y., Igarashi, A., Fukunaga, T., Tadano, M., and Fukai, K.**, Electronmicroscopic observation of a newly isolated Flavivirus-like virus from field-caught mosquitoes, *J. Gen. Virol.,* 65, 803, 1984.
525. **Kuno, G.**, A method for isolating continuous cell lines from *Toxorhynchites amboinensis* (Diptera: Culicidae), *J. Med. Entomol.,* 18, 140, 1981.
526. **Oelofsen, M. J., Gericke, A., Smith, M. S., and Van der Linde, T. C. de K.**, Establishment and characterization of a cell line from the mosquito *Culex (Culex) theileri* (Diptera: Culicidae) and its susceptibility to infection with arboviruses, *J. Med. Entomol.,* 27, 939, 1990.
527. **Deubel, V., Kinney, R. M., and Trent, D. W.**, Nucleotide sequence and deduced amino acid sequence of the structural proteins of Dengue type 2 virus, Jamaica genotype, *Virology,* 155, 365, 1986.
528. **Deubel, V., Kinney, R. M., and Trent, D. W.**, Nucleotide sequence and deduced amino acid sequence of the nonstructural proteins of Dengue type 2 virus, Jamaica genotype: comparative analysis of the full-length genome, *Virology,* 165, 234, 1988.
529. **Mackow, E., Makino, Y., Zhao, B., Zhang, Y. M., Markoff, L., Buckler-White, A., Guiler, M., Chanock, R., and Lai, C. J.**, The nucleotide sequence of Dengue type 4 virus: Analysis of genes coding for nonstructural proteins, *Virology,* 159, 217, 1987.
530. **Hahn, Y. S., Galler, R., Hunkapiller, T., Dalrymple, J. M., Strauss, J. H., and Strauss, E. G.**, Nucleotide sequence of Dengue 2 RNA and comparison of the encoded proteins with those of other flaviviruses, *Virology,* 162, 167, 1988.
531. **Sumiyoshi, H., Mori, C., Fuke, I., Morita, K., Kuhara, S., Kondou, J., Kikuchi, Y., Nagamatsu, H., and Igarashi, A.**, Complete nucleotide sequence of the Japanese encephalitis virus genome RNA, *Virology,* 161, 497, 1987.
532. **Dalgarno, L., Trent, D. W., Strauss, J. H., and Rice, C. M.**, Partial nucleotide sequence of the Murray Valley encephalitis virus genome. Comparison of the encoded polypeptides with yellow fever virus structural and nonstructural proteins, *J. Mol. Biol.,* 187, 309, 1985.
533. **Trent, D. W., Kinney, R. M., Johnson, B. J., Vorndam, A. V., Grant, J. A., Deubel, V., Rice, C. M., and Hahn, C.**, Partial nucleotide sequence of St. Louis encephalitis virus RNA: Structural proteins, NS 1, NS 2a, and NS 2b, *Virology,* 156, 293, 1987.
534. **Castle, E., Leidner, U., Nowak, T., Wengler, G., and Wengler, G.**, Primary structure of the West Nile flavivirus genome region coding for all nonstructural proteins, *Virology,* 149, 10, 1986.
535. **Coia, G., Parker, M. D., Speight, G., Byrne, M. E., and Westaway, E. G.**, Nucleotide and complete amino acid sequences of Kunjin virus: definitive gene order and characteristics of the virus-specified proteins, *J. Gen. Virol.,* 69, 1, 1988.

536. **Rice, C. M., Lenches, E. M., Eddy, S. R., Shin, S. J., Sheets, R. L., and Strauss, J. H.,** Nucleotide sequence of yellow fever virus: implications for Flavivirus gene expression and evolution, *Science,* 229, 726, 1985.
537. **Pletnev, A. G., Yamshchikov, V. F., and Blinov, V. M.,** Nucleotide sequence of the genome and complete amino acid sequence of the polyprotein of tick-borne encephalitis virus, *Virology,* 174, 250, 1990.
538. **Mandl, C. W., Iacono-Connors, L., Wallner, G., Holzmann, H., Kunz, C., and Heinz, F. X.,** Sequence of the genes encoding the structural proteins of the low-virulence tick-borne Flaviviruses Langat TP 21 and Jelantsev, *Virology,* 185, 891, 1991.
539. **Parsonson, I. M. and Mc Phee, D. A.,** Bunyavirus pathogenesis, *Adv. Virus Res.,* 30, 279, 1985.
540. **Bishop, D. H. L.,** Ambisense RNA genomes of Arenaviruses and Phleboviruses, *Adv. Virus Res.,* 31, 1, 1986.
541. **Kuno, G.,** Arbovirus replication in cell cultures of nonhematophagous mosquitoes, in *Arboviruses in Arthropod Cells In Vitro,* Vol. 2, Yunker, C. E., Ed., CRC Press, Boca Raton, FL, 1987, 43.
542. **Pettersson, R. F., Gahmberg, N., Kuismanen, E., Kääriäinen, L., Rönnholm, R., and Saraste, J.,** Bunyavirus membrane glycoproteins as models for Golgi specific proteins, *Modern Cell Biology,* Alan R. Liss, New York, 6, 65, 1988.
543. **Nicoletti, L., Marchi, A., Bellisario, L., and Verani, P.,** Toscana virus (Phlebovirus) replication in cultured arthropod cells: establishment of persistent infections, in *Invertebrate Cell System Applications,* Vol. 2, Mitsuhashi, J., Ed., CRC Press, Boca Raton, FL, 1989, 169.
544. **Peters, C. J., Liu, C.-T., Anderson, G. W., Morrill, J. C., and Jahrling, P. B.,** Pathogenesis of viral hemorrhagic fevers: Rift Valley fever and Lassa fever contrasted, *Rev. Infect. Dis.,* 11, Suppl. 4, 743, 1989.
545. **Venter, E. H. and Smith, M. S.,** Mosquito cell cultures provide possible clues as to the natural history of Rift Valley fever virus in southern Africa, in *Invertebrate Cell System Applications,* Vol. 2, Mitsuhashi, J., Ed., CRC Press, Boca Raton, FL, 1989, 133.
546. **Bishop, D. H. L.,** Bunyaviridae and their replication, Part I: Bunyaviridae, in *Virology,* Vol. 1, 2nd ed., Fields, B. N. and Knipe, D. M., Eds., Raven Press, New York, 1990, 1155.
547. **Elliott, R. M.,** Molecular biology of the Bunyaviridae, *J. Gen. Virol.,* 71, 501, 1990.
548. **Gonzalez-Scarano, F. and Nathanson, N.,** Bunyaviruses, in *Virology,* Vol. 1, 2nd ed., Fields, B. N. and Knipe, D. M., Eds., Raven Press, New York, 1990, 1195.
549. **Schmaljohn, C. S. and Patterson, J. L.,** Bunyaviridae and their replication. Part II: Replication of Bunyaviridae, in *Virology,* Vol. 1, 2nd ed., Fields, B. N. and Knipe, D. M., Eds., Raven Press, New York, 1990, 1175.
550. **Wright, K. E., Schmaljohn, C., Schmaljohn, A., and Buchmeier, M. J.,** Arenaviridae and Bunyaviridae, in *Immunochemistry of Viruses II,* van Regenmortel, M. H. V. and Neurath, A. R., Eds., Elsevier, Amsterdam, 1990, 237.
551. **Elliott, R. M., Schmaljohn, C. S., and Collett, M. S.,** Bunyaviridae genome structure and gene expression, *Curr. Top. Microbiol. Immunol.,* 169, 91, 1991.
552. **Gonzalez-Scarano, F., Endres, M. J., and Nathanson, N.,** Pathogenesis, *Curr. Top. Microbiol. Immunol.,* 169, 217, 1991.
553. **Kolakofsky, D., Ed.,** Bunyaviridae, *Curr. Top. Microbiol. Immunol.,* 169, 1991.
554. **Matsuoka, Y., Chen, S. Y., and Compans, R. W.,** Bunyavirus protein transport and assembly, *Curr. Top. Microbiol. Immunol.,* 169, 161, 1991.
555. **Gonzalez-Scarano, F., Jacoby, D., Griot, C., and Nathanson, N.,** Genetics, infectivity and virulence of California serogroup viruses, *Virus Res.,* 24, 123, 1992.
556. **Gonzalez-Scarano, F., Endres, M., Jacoby, D. R., Griot, C., and Nathanson, N.,** Molecular approaches to the study of Bunyavirus encephalitis, in *Molecular Neurovirology,* Roos, R. P., Ed., Humana Press, Totowa, NJ, 1992, 449.
557. **Munz, E., Reimann, M., and Mahnel, H.,** Nairobi sheep disease virus and Reovirus-like particles in the tick cell line TTC-243 from *Rhipicephalus appendiculatus,* in *Arboviruses in Arthropod Cells In Vitro,* Vol. 1, Yunker, C. E., Ed., CRC Press, Boca Raton, FL, 1987, 133.
558. **Sundin, D. R., Beaty, B. J., Nathanson, N., and Gonzalez-Scarano, F.,** A G1 glycoprotein epitope of La Crosse virus: a determinant of infection of *Aedes triseriatus, Science,* 235, 591, 1987.
559. **Beaty, B. J., Holterman, M., Tabachnick, W., Shope, R. E., Rozhon, E. J., and Bishop, D. H. L.,** Molecular basis of Bunyavirus transmission by mosquitoes: role of the middle-sized RNA segment, *Science,* 211, 1433, 1981.
560. **Schopen, S., Labuda, M., and Beaty, B.,** Vertical and venereal transmission of California group viruses by *Aedes triseriatus* and *Culiseta inornata* mosquitoes, *Acta Virol.,* 35, 373, 1991.
561. **Wilson, M. L., Gonzalez, J. P., Cornet, J.-P., and Camicas, J. L.,** Transmission of Crimean-Congo hemorrhagic fever virus from experimentally infected sheep to *Hyalomma truncatum* ticks, *Res. Virol. (Inst. Pasteur),* 142, 395, 1991.
562. **Booth, T. F., Steele, G. M., Marriott, A. C., and Nuttall, P. A.,** Dissemination, replication, and trans-stadial persistence of Dugbe virus (Nairovirus, Bunyaviridae) in the tick vector *Amblyomma variegatum, Am. J. Trop. Med. Hyg.,* 45, 146, 1991.

563. **Grady, L. J. and Kinch, W.,** Two monoclonal antibodies against La Crosse virus show host-dependent neutralizing activity, *J. Gen. Virol.*, 66, 2773, 1985.
564. **Rossier, C., Raju, R., and Kolakofsky, D.,** La Crosse virus gene expression in mammalian and mosquito cells, *Virology*, 165, 539, 1988.
565. **Kolakofsky, D., Bellocq, C., and Raju, R.,** The translational requirement for La Crosse virus S-mRNA synthesis, *Cold Spring Harbor Symp. Quant. Biol.*, 52, 373, 1988.
566. **Hacker, D., Raju, R., and Kolakofsky, D.,** La Crosse virus nucleocapsid protein controls its own synthesis in mosquito cells by encapsidating its mRNA, *J. Virol.*, 63, 5166, 1989.
567. **Raju, R. and Kolakofsky, D.,** The ends of La Crosse virus genome and antigenome RNAs within nucleocapsids are base paired, *J. Virol.*, 63, 122, 1989.
568. **Raju, R., Raju, L., and Kolakofsky, D.,** The translational requirement for complete La Crosse virus mRNA synthesis is cell-type dependent, *J. Virol.*, 63, 5159, 1989.
569. **Carvalho, M. G. C., Frugulhetti, I. C., and Rebello, M. A.,** Marituba (Bunyaviridae) virus replication in cultured *Aedes albopictus* cells and in L-A9 cells, *Arch. Virol.*, 90, 325, 1986.
570. **Kuismanen, E., Bang, B., Hurme, M., and Pettersson, R. F.,** Uukuniemi virus maturation: immunofluorescence microscopy with monoclonal glycoprotein-specific antibodies, *J. Virol.*, 51, 137, 1984.
571. **Frugulhetti, I. C., Silva Filho, F. C., and Rebello, M. A.,** The surface charge of L-A9 cells and *Aedes albopictus* cells infected with Marituba (Bunyaviridae) virus, *Virus Res.*, 7, 219, 1987.
572. **Frugulhetti, I. C. and Rebello, M. A.,** Na^+ and K^+ concentration and regulation of protein synthesis in L-A9 and *Aedes albopictus* cells infected with Marituba virus (Bunyaviridae), *J. Gen. Virol.*, 70, 3493, 1989.
573. **Lyons, M. J. and Heyduk, J.,** Aspects of the developmental morphology of California encephalitis virus in cultured vertebrate and arthropod cells and in mouse brain, *Virology*, 54, 37, 1973.
574. **Newton, S. E., Short, N. J., and Dalgarno, L.,** Bunyamwera virus replication in cultured *Aedes albopictus* (mosquito) cells: establishment of a persistent viral infection, *J. Virol.*, 38, 1015, 1981.
575. **Elliott, R. M. and Wilkie, M. L.,** Persistent infection of *Aedes albopictus* C6/36 cells by Bunyamwera virus, *Virology*, 150, 21, 1986.
576. **Nicoletti, L. and Verani, P.,** Growth of the Phlebovirus Toscana in a mosquito *(Aedes pseudoscutellaris)* cell line (AP-61): establishment of a persistent infection, *Arch. Virol.*, 85, 35, 1985.
577. **Delord, B., Poveda, J.-D., Astier-Gin, T., Gerbaud, S., and Fleury, H. J. A.,** Detection of the Bunyavirus Germiston in Vero and *Aedes albopictus* C6/36 cells by in situ hybridization using cDNA and asymmetric RNA probes, *J. Virol. Meth.*, 24, 253, 1989.
578. **Delord, B., Poveda, J.-D., Astier-Gin, T., Gerbaud, S., Wattiaux, J.-P., and Fleury, H. J. A.,** Quantitative in situ hybridization using strand-specific RNA probes: expression of the Bunyavirus Germiston S segment in mosquito cells, *Mol. Cell. Probes*, 4, 247, 1990.
579. **Raju, R. and Kolakofsky, D.,** La Crosse virus infection of mammalian cells induces mRNA instability, *J. Virol.*, 62, 27, 1988.
580. **Hacker, D., Rochat, S., and Kolakofsky, D.,** Anti-mRNAs in La Crosse virus infected cells, *J. Virol.*, 64, 5051, 1990.
581. **Endres, M. J., Griot, C., Gonzalez-Scarano, F., and Nathanson, N.,** Neuroattenuation of an avirulent Bunyavirus variant maps to the L RNA segment, *J. Virol.*, 65, 5465, 1991.
582. **Scallan, M. F. and Elliott, R. M.,** Defective RNAs in mosquito cells persistently infected with Bunyamwera virus, *J. Gen. Virol.*, 73, 53, 1992.
583. **Patel, A. H. and Elliott, R. M.,** Characterization of Bunyamwera virus defective interfering particles, *J. Gen. Virol.*, 73, 389, 1992.
584. **James, W. S. and Millican, D.,** Host-adaptive antigenic variation in Bunyaviruses, *J. Gen. Virol.*, 67, 2803, 1986.
585. **Ludwig, G. V., Israel, B. A., Christensen, B. M., Yuill, T. M., and Schultz, K. T.,** Monoclonal antibodies directed against the envelope glycoproteins of La Crosse virus, *Microb. Pathogenesis*, 11, 411, 1991.
586. **Rowley, W. A., Dorsey, D. C., and Knowles, M. A.,** The replication of two California serogroup viruses in a cell line from the mosquito *Aedes triseriatus* (Diptera: *Culicidae*), *J. Med. Entomol.*, 21, 501, 1984.
587. **Tesh, R. B.,** Establishment of two cell lines from mosquito *Toxorhynchites amboinensis* (Diptera: *Culicidae*) and their susceptibility to infection with arboviruses, *J. Med. Entomol.*, 17, 338, 1980.
588. **Elliott, R. M.,** Nucleotide sequence analysis of the S RNA segment of Bunyamwera virus, the prototype of the family Bunyaviridae, *J. Gen. Virol.*, 70, 1281, 1989.
589. **Elliott, R. M.,** Nucleotide sequence analysis of the large (L) genomic RNA segment of Bunyamwera virus, the prototype of the family Bunyaviridae, *Virology*, 173, 426, 1989.
590. **Marriott, A. C., Ward, V. K., and Nuttall, P. A.,** The S RNA segment of sandfly fever Sicilian virus: evidence for an antisense genome, *Virology*, 169, 341, 1989.
591. **Müller, R., Argentini, C., Bouloy, M., Prehand, C., and Bishop, D. H. L.,** Completion of the genome sequence of Rift Valley fever Phlebovirus indicates that the L RNA is negative sense or ambisense and codes for a putative transcriptase-replicase, *Nucleic Acids Res.*, 19, 5433, 1991.

592. **Ihara, T., Akashi, H., and Bishop, D. H. L.,** Novel coding strategy (ambisense genomic RNA) revealed by sequence analyses of Punta Toro Phlebovirus S RNA, *Virology,* 136, 293, 1984.
593. **Ihara, T., Smith, J., Dalrymple, J. M., and Bishop, D. H. L.,** Complete sequences of the glycoprotein and M RNA of Punta Toro Phlebovirus compared to those of Rift Valley fever virus, *Virology,* 144, 246, 1985.
594. **Rönnholm, R. and Pettersson, R. F.,** Complete nucleotide sequence of the M RNA segment of Uukuniemi virus encoding the membrane glycoproteins G1 and G2, *Virology,* 160, 191, 1987.
595. **Simons, J. F., Hellman, U., and Pettersson, R. F.,** Uukuniemi virus S RNA segment: ambisense coding strategy, packaging of complementary strands into virions, and homology to members of the genus phlebovirus, *J. Virol.,* 64, 247, 1990.
596. **Marriott, A. C. and Nuttall, P. A.,** Comparison of the S RNA segments and nucleoprotein sequences of Crimean-Congo hemorrhagic fever, Hazara, and Dugbe viruses, *Virology,* 189, 795, 1992.
597. **Ward, V. K., Marriott, A. C., El-Ghorr, A., and Nuttall, P. A.,** Coding strategy of the S RNA segment of Dugbe virus (Nairovirus, Bunyaviridae), *Virology,* 175, 518, 1990.
598. **Gateff, E., Gissmann, L., Shrestha, R., Plus, N., Pfister, H., Schröder, J., and Zur Hausen, H.,** Characterization of two tumorous blood cell lines of *Drosophila melanogaster* and the viruses they contain, in *Invertebrate Systems In Vitro,* Kurstak, E., Maramorosch, K., and Dübendorfer, A., Eds., Elsevier, Amsterdam, 1980, 517.
599. **Ofori-Kwakye, S. K., Lipman, M. B., and Buckley, S. M.,** Immunofluorescence and electron microscopy study of Orungo virus morphology and morphogenesis in *Aedes pseudoscutellaris* cells, in *Invertebrate Systems In Vitro,* Kurstak, E., Maramorosch, K., and Dübendorfer, A., Eds., Elsevier, Amsterdam, 1980, 353.
600. **Joklik, W. K.,** Structure and function of the Reovirus genome, *Microbiol. Rev.,* 45, 483, 1981.
601. **Spence, R. P., Moore, N. F., and Nuttall, P. A.,** The biochemistry of Orbiviruses, *Arch. Virol.,* 82, 1, 1984.
602. **Barber, T. L. and Jochim, M. M., Eds.,** Progress in clinical and biological research, in *Bluetongue and related Orbiviruses,* Vol. 178, Alan R. Liss, New York, 1985.
603. **Joklik, W. K.,** Recent progress in Reovirus research, *Annu. Rev. Genet.,* 19, 537, 1985.
604. **Rehacek, J.,** Arthropod cell cultures in studies of tick-borne Togaviruses and Orbiviruses in central Europe, in *Arboviruses in Arthropod Cells In Vitro,* Vol. 1, Yunker, C. E., Ed., CRC Press, Boca Raton, FL, 1987, 115.
605. **Roy, P.,** Bluetongue virus genetics and genome structure, *Virus Res.,* 13, 179, 1989.
606. **Knudson, D. L. and Monath, T. P.,** Orbiviruses, in *Virology,* Vol. 2, 2nd ed., Fields, B. N. and Knipe, D. M., Eds., Raven Press, New York, 1990, 1405.
607. **Mellor, P. S.,** The replication of Bluetongue virus in *Culicoides* vectors, *Curr. Top. Microbiol. Immunol.,* 162, 143, 1990.
608. **Mackow, E. R., Shaw, R. D., Matsui, S. M., Vo, P. T., Dang, M.-N., and Greenberg, H. B.,** The rhesus Rotavirus gene encoding protein VP3: location of amino acids involved in homologous and heterologous Rotavirus neutralization and identification of a putative fusion region, *Proc. Natl. Acad. Sci. U.S.A.,* 85, 645, 1988.
609. **Van Dijk, A. A. and Huismans, H.,** The in vitro activation and further characterization of the Bluetongue virus-associated transcriptase, *Virology,* 104, 347, 1980.
610. **Huismans, H. and Els, H. J.,** Characterization of the tubules associated with the replication of three different Orbiviruses, *Virology,* 92, 397, 1979.
611. **Urakawa, T. and Roy, P.,** Bluetongue virus tubules made in insect cells by recombinant Baculovirus: expression of the NS 1 gene of Bluetongue virus serotype 10, *J. Virol.,* 62, 3919, 1988.
612. **Thomas, C. P., Booth, T. F., and Roy, P.,** Synthesis of Bluetongue virus-encoded phosphoprotein and formation of inclusion bodies by recombinant Baculovirus in insect cells: it binds the single-stranded RNA species, *J. Gen. Virol.,* 71, 2073, 1990.
613. **French, T. J. and Roy, P.,** Synthesis of Bluetongue virus (BTV) corelike particles by a recombinant Baculovirus expressing the two major structural core proteins of BTV, *J. Virol.,* 64, 1530, 1990.
614. **Chandler, L. J., Ballinger, M. E., Jones, R. H., and Beaty, B. J.,** The virogenesis of Bluetongue virus in *Culicoides variipennis, Progr. Clin. Biol. Res.* (Alan R. Liss), 178, 245, 1985.
615. **Buckley, S. M.,** Propagation of 3 relatively solvent-resistant arboviruses in Singh's *Aedes albopictus* and *Aedes aegypti* cell lines, *J. Med. Entomol.,* 9, 168, 1972.
616. **Wechsler, S. J. and McHolland, L. E.,** Susceptibilities of 14 cell lines to Bluetongue virus infection, *J. Clin. Microbiol.,* 26, 2324, 1988.
617. **Wechsler, S. J., McHolland, L. E., and Tabachnick, W. J.,** Cell lines from *Culicoides variipennis* (Diptera: *Ceratopogonidae*) support replication of Bluetongue virus, *J. Invertebrate Pathol.,* 54, 385, 1989.
618. **Vazeille, M. C., Rosen, L., and Guillon, J.-C.,** An Orbivirus of mosquitoes which induces CO_2 sensitivity in mosquitoes and is lethal for rabbits, *J. Virol.,* 62, 3484, 1988.
619. **Mirchamsy, H., Hazrati, A., Bahrami, S., and Shafji, A.,** Growth and persistent infection of African horse sickness virus in a mosquito cell line, *Am. J. Vet. Res.,* 31, 1755, 1970.
620. **Yunker, C. E. and Cory, J.,** Growth of Colorado tick fever (CTF) virus in primary tissue cultures of its vector, *Dermacentor andersoni Stiles* (Acarina: *Ixodidae*), with notes on tick tissue culture, *Exp. Parasitol.,* 20, 267, 1967.

621. **Yunker, C. E. and Cory, J.,** Colorado tick fever virus: growth in a mosquito cell line, *J. Virol.,* 3, 631, 1969.
622. **Mc Phee, D. A., Parsonson, I. M., and Della-Porta, A. J.,** Comparative studies on the growth of Australian Bluetongue virus serotypes in continuous cell lines and embryonated chicken eggs, *Vet. Microbiol.,* 7, 401, 1982.
623. **Haars, R., Zentgraf, H., Gateff, E., and Bautz, F. A.,** Evidence for endogenous Reovirus-like particles in a tissue culture cell line from *Drosophila melanogaster, Virology,* 101, 124, 1980.
624. **Tesh, R. B., Peleg, J., Samina, I., Margalit, J., Bodkin, D. K., Shope, R. E., and Knudson, D.,** Biological and antigenic characterization of Netivot virus, an unusual new Orbivirus recovered from mosquitoes in Israel, *Am. J. Trop. Med. Hyg.,* 35, 418, 1986.
625. **Řeháček, J. and Sutakova, G.,** Virus-like particles in *Dermacentor reticulatus* ticks, *Acta Virol.,* 33, 577, 1989.
626. **Peterson, A. J. and Nuss, D. L.,** Wound tumor virus polypeptide synthesis in productive noncytopathic infection of cultured insect vector cells, *J. Virol.,* 56, 620, 1985.
627. **Murphy, F. A.,** Rabies pathogenesis: brief review, *Arch. Virol.,* 54, 279, 1977.
628. **Bishop, D. H. L. and Compans, R. W., Eds.,** *Nonsegmented negative strand viruses,* Academic Press, Orlando, 1984.
629. **Wagner, R. R.,** Cytopathic effects of viruses: a general review, in *Comprehensive Virology,* Vol. 19, Fraenkel-Conrat, H. and Wagner, R. R., Eds., Plenum Press, New York, 1984, 1.
630. **Wagner, R. R., Thomas, J. R., and Mc Gowan, J. J.,** Rhabdovirus cytopathology: effects on cellular macromolecular synthesis, in *Comprehensive Virology,* Vol. 19, Fraenkel-Conrat, H. and Wagner, R. R., Eds., Plenum Press, New York, 1984, 223.
631. **Francki, R. I. B., Milne, R. G., and Hatta, T.,** Plant Rhabdoviridae, in *Atlas of Plant Viruses,* Vol. 1, CRC Press, Boca Raton, FL, 1985, 73.
632. **Murphy, F. A.,** The pathogenesis of Rabies virus infection, in *World's Debt to Pasteur,* Koprowski, H. and Plotkin, S., Eds., Alan R. Liss, New York, 1985, 141.
633. **Wunner, W. H.,** Structure of Rabies viruses, in *World's Debt to Pasteur,* Koprowski, H. and Plotkin, S., Eds., Alan R. Liss, New York, 1985, 171.
634. **Banerjee, A. K.,** Transcription and replication of Rhabdoviruses, *Microbiol. Rev.,* 51, 66, 1987.
635. **Emerson, S. U.,** Transcription of vesicular stomatitis virus, in *The Rhabdoviruses,* Wagner, R. R., Ed., Plenum Press, New York, 1987, 245.
636. **Holland, J. J.,** Defective-interfering Rhabdoviruses, in *The Rhabdoviruses,* Wagner, R. R., Ed., Plenum Press, New York, 1987, 297.
637. **Jackson, A. O., Francki, R. I. B., and Zuidema, D.,** Biology, structure and replication of plant Rhabdoviruses, in *The Rhabdoviruses,* Wagner, R. R., Ed., Plenum Press, New York, 1987, 427.
638. **Pal, R. and Wagner, R. R.,** Rhabdovirus membrane and maturation, in *The Rhabdoviruses,* Wagner, R. R., Ed., Plenum Press, New York, 1987, 75.
639. **Pringle, C. R.,** Rhabdovirus genetics, in *The Rhabdoviruses,* Wagner, R. R., Ed., Plenum Press, New York, 1987, 167.
640. **Rose, J. K. and Schubert, M.,** Rhabdovirus genomes and their products, in *The Rhabdoviruses,* Wagner, R. R., Ed., Plenum Press, New York, 1987, 129.
641. **Wertz, G. W., Davis, N. L., and Patton, J.,** The role of proteins in vesicular stomatitis virus RNA replication, in *The Rhabdoviruses,* Wagner, R. R., Ed., Plenum Press, New York, 1987, 271.
642. **Wunner, W. H.,** Rabies viruses — pathogenesis and immunity, in *The Rhabdoviruses,* Wagner, R. R., Ed., Plenum Press, New York, 1987, 361.
643. **Baer, G. M.,** Animal models in the pathogenesis and treatment of rabies, *Rev. Infect. Dis.,* 10, Suppl. 4, 739, 1988.
644. **Calisher, C. H., Karabatsos, N., Zeller, H. G., Digoutte, J. P., Tesh, R. B., Shope, R. E., Travassos da Rosa, A. P. A., and St. George, T. D.,** Antigenic relationships among Rhabdoviruses from vertebrates and hematophagous arthropods, *Intervirology,* 30, 241, 1989.
645. **Frerichs, G. N.,** Rhabdoviruses of fishes, in *Viruses of Lower Vertebrates,* Ahne, W. and Kurstak, E., Eds., Springer Verlag, Berlin, 1989, 316.
646. **Baer, G. M., Bellini, W. J., and Fishbein, D. B.,** Rhabdoviruses, in *Virology,* Vol. 1, 2nd ed., Fields, B. N. and Knipe, D. M., Eds., Raven Press, New York, 1990, 883.
647. **Celis, E.,** Rhabdoviruses, in *Immunochemistry of Viruses,* II. van Regenmortel, M. H. V. and Neurath, A. R., Eds., Elsevier, Amsterdam, 1990, 345.
648. **Wagner, R. R.,** Rhabdoviridae and their replication, in *Virology,* Vol. 1, 2nd ed., Fields, B. N. and Knipe, D. M., Eds., Raven Press, New York, 1990, 867.
649. **Wunner, W. H.,** The chemical composition and molecular structure of rabies viruses, in *Natural History of Rabies,* Baer, L., Ed., CRC Press, Boca Raton, FL, 1991, 31.
650. **Rose, J. K. and Gallione, C. J.,** Nucleotide sequence of the mRNAs encoding the vesicular stomatitis virus G and M proteins determined from cDNA clones containing the complete coding regions, *J. Virol.,* 39, 519, 1981.
651. **Schubert, M., Harmison, G. G., and Meier, E.,** Primary structure of the vesicular stomatitis virus polymerase (L) gene: evidence for a high frequency of mutations, *J. Virol.,* 51, 505, 1984.

652. **Tordo, N., Poch, O., Ermine, A., Keith, G., and Rougeon, F.**, Completion of the rabies virus genome sequence determination: highly conserved domains among the L (polymerase) proteins of unsegmented negative-stranded RNA viruses, *Virology*, 165, 565, 1988.
653. **Comer, J. A., Tesh, R. B., Modi, G. B., Corn, J. L., and Nettles, V. F.**, Vesicular stomatitis virus, New Jersey serotype: replication in and transmission by *Lutzomyia shannoni* (Diptera: Psychodidae), *Am. J. Trop. Med. Hyg.*, 42, 483, 1990.
654. **Weaver, S. C., Tesh, R. B., and Guzman, H.**, Ultrastructural aspects of replication of the New Jersey serotype of vesicular stomatitis virus in a suspected sand fly vector, *Lutzomyia shannoni* (Diptera: Psychodidae), *Am. J. Trop. Med. Hyg.*, 46, 201, 1992.
655. **Ilkal, M. A., Goverdhan, M. K., Shetty, P. S., Tupe, C. D., Mavale, M. S., and Dhanda, V.**, Susceptibility of four species of mosquitoes to Chandipura virus and its detection by immunofluorescence, *Acta Virol.*, 35, 27, 1991.
656. **Schloemer, R. H. and Wagner, R. R.**, Mosquito cells infected with vesicular stomatitis virus yield unsialylated virions of low infectivity, *J. Virol.*, 15, 1029, 1975.
657. **Seganti, L., Superti, F., Girmenia, C., Melucci, L., and Orsi, N.**, Study of receptors for vesicular stomatitis virus in vertebrate and invertebrate cells, *Microbiologica*, 9, 259, 1986.
658. **Superti, F., Seganti, L., Ruggeri, F. M., Tinari, A., Donelli, G., and Orsi, N.**, Entry pathway of vesicular stomatitis virus into different host cells, *J. Gen. Virol.*, 68, 387, 1987.
659. **Tsiang, H.**, Rabies virus infection of myotubes and neurons as elements of the neuromuscular junction, *Rev. Infect. Dis.*, 10, Suppl. 4, 733, 1988.
660. **Seganti, L., Superti, F., Bianchi, S., Orsi, N., Divizia, M., and Pana, A.**, Susceptibility of mammalian, avian, fish, and mosquito cell lines to rabies virus infection, *Acta Virol.*, 34, 155, 1990.
661. **Seganti, L., Superti, F., Sinibaldi, L., Marchetti, M., Bianchi, S., and Orsi, N.**, Rabies virus infection in *Aedes pseudoscutellaris cells*: a study on receptorial structures, *Comp. Immun. Microbiol. Infect. Dis.*, 14, 265, 1991.
662. **Reagan, K. J. and Wunner, W. H.**, Rabies virus interaction with various cell lines is independent of the acetylcholine receptor, *Arch. Virol.*, 84, 277, 1985.
663. **Wyers, F., Richard-Molard, C., Blondel, D., Dezélée, S.**, Vesicular stomatitis virus growth in *Drosophila melanogaster* cells: G protein deficiency, *J. Virol.*, 33, 411, 1980.
664. **Gillies, S. and Stollar, V.**, The production of high yields of infectious vesicular stomatitis virus in *Aedes albopictus* cells and comparisons with replication in BHK-21 cells, *Virology*, 107, 509, 1980.
665. **Gillies, S. and Stollar, V.**, Biochemical characterization of vesicular stomatitis virus-infected *Aedes albopictus* cells deprived of methionine, *Virology*, 112, 318, 1981.
666. **Horikami, S. M., de Ferra, F., and Moyer, S. A.**, Characterization of the infections of permissive and nonpermissive cells by host range mutants of vesicular stomatitis virus defective in RNA methylation, *Virology*, 138, 1, 1984.
667. **Gillies, S. and Stollar, V.**, Conditions necessary for inhibition of protein synthesis and production of cytopathic effect in *Aedes albopictus* cells infected with vesicular stomatitis virus, *Mol. Cell. Biol.*, 2, 66, 1982.
668. **Gillies, S. and Stollar, V.**, Translation of vesicular stomatitis and Sindbis virus mRNAs in cell-free extracts of *Aedes albopictus* cells, *J. Biol. Chem.*, 256, 13188, 1981.
669. **Gillies, S. and Stollar, V.**, Protein synthesis in lysates of *Aedes albopictus* cells infected with vesicular stomatitis virus, *Mol. Cell. Biol.*, 2, 1174, 1982.
670. **Reading, C. L., Penhoet, E. E., and Ballou, C. E.**, Carbohydrate structure of vesicular stomatitis virus glycoprotein, *J. Biol. Chem.*, 253, 5600, 1978.
671. **Rose, J. K., Adams, G. A., and Gallione, C. J.**, The presence of cysteine in the cytoplasmic domain of the vesicular stomatitis virus glycoprotein is required for palmitate addition, *Proc. Natl. Acad. Sci. U.S.A.*, 81, 2050, 1984.
672. **Kotwal, G. J., Buller, R. M. L., Wunner, W. H., Pringle, C. R., and Ghosh, H. P.**, Role of glycosylation in transport of vesicular stomatitis virus envelope glycoprotein, *J. Biol. Chem.*, 261, 8936, 1986.
673. **Shakin-Eshleman, S. H., Remaley, A. T., Eshleman, J. R., Wunner, W. H., and Spitalnik, S. L.**, N-linked glycosylation of rabies virus glycoprotein, *J. Biol. Chem.*, 267, 10690, 1992.
674. **Lodish, H. F. and Porter, M.**, Heterogeneity of vesicular stomatitis virus particles: implications for virion assembly, *J. Virol.*, 33, 52, 1980.
675. **Blondel, D., Dezélée, S., and Wyers, F.**, Vesicular stomatitis virus growth in *Drosophila melanogaster* cells. II. Modifications of viral protein phosphorylation, *J. Gen. Virol.*, 64, 1793, 1983.
676. **Blondel, D., Petitjean, A.-M., Dezélée, S., and Wyers, F.**, Vesicular stomatitis virus in *Drosophila melanogaster* cells: regulation of viral transcription and replication, *J. Virol.*, 62, 277, 1988.
677. **Frey, T. K. and Youngner, J. S.**, Novel phenotype of RNA synthesis expressed by vesicular stomatitis virus isolated from persistent infection, *J. Virol.*, 44, 167, 1982.
678. **Dezélée, S., Blondel, D., Wyers, F., and Petitjean, A.-M.**, Vesicular stomatitis virus in *Drosophila melanogaster* cells: lack of leader RNA transport into the nuclei and frequent abortion of the replication step, *J. Virol.*, 61, 1391, 1987.
679. **Coulon, P. and Contamine, D.**, Role of the *Drosophila* genome in Sigma virus multiplication, *Virology*, 123, 381, 1982.

680. **Brun, G.**, Are the *Drosophila* ref genes for Piry and Sigma virus identical?, in *The Replication of Negative Strand Viruses*, Bishop, D. H. L. and Compans, R. W., Eds., Elsevier, Amsterdam, 1981, 912.
681. **Tesh, R. B., Chaniotis, B. N., and Johnson, K. M.**, Vesicular stomatitis virus (Indiana serotype): transovarial transmission by *phlebotomine* sandflies, *Science*, 175, 1477, 1972.
682. **Lazzarini, R. A., Keene, J. D., and Schubert, M.**, The origin of defective-interfering particles of the negative-strand RNA viruses, *Cell*, 26, 145, 1982.
683. **Emi, N., Friedmann, T., and Yee, J.-K.**, Pseudotype formation of murine leukemia virus with the G protein of vesicular stomatitis virus, *J. Virol.*, 65, 1202, 1991.
684. **Jordan, J. A. and Youngner, J. S.**, Dominance of temperature-sensitive phenotypes, *Virology*, 158, 407, 1987.
685. **Depolo, N. J. and Holland, J. J.**, Very rapid generation/amplification of defective interfering particles by vesicular stomatitis virus variants isolated from persistent infection, *J. Gen. Virol.*, 67, 1195, 1986.
686. **Gillies, S. and Stollar, V.**, Generation of defective interfering particles of vesicular stomatitis virus in *Aedes albopictus* cells, *Virology*, 107, 497, 1980.
687. **Yang, Y. J., Stoltz, D. B., and Prevec, L.**, Growth of vesicular stomatitis virus in a continuous culture line of *Antheraea eucalypti* moth cells, *J. Gen. Virol.*, 5, 473, 1969.
688. **Ferris, D., Hanson, R. P., Dicke, R. J., and Robers, R. H.**, Experimental transmission of vesicular stomatitis virus by diptera, *J. Infect. Dis.*, 96, 184, 1955.
689. **Longworth, J. F.**, Small isometric viruses of invertebrates, *Adv. Virus Res.*, 23, 103, 1978.
690. **Harrap, K. A. and Payne, C. C.**, The structural properties and identification of insect viruses, *Adv. Virus Res.*, 25, 273, 1979.
691. **Podgwaite, J. D. and Mazzone, H. M.**, Latency of insect viruses, *Adv. Virus Res.*, 31, 293, 1986.
692. **Moore, N. F. and Tinsley, T. W.**, The small RNA-viruses of insects, *Arch. Virol.*, 72, 229, 1982.
693. **Moore, N. F., Reavy, B., and King, L. A.**, General characteristics, gene organization and expression of small RNA viruses of insects, *J. Gen. Virol.*, 66, 647, 1985.
694. **Reavy, B. and Moore, N. F.**, The replication of small RNA-containing viruses of insects, *Microbiologica*, 5, 63, 1982.
695. **Gallagher, T. M. and Rueckert, R. R.**, Assembly dependent maturation cleavage in provirions of a small icosahedral insect ribovirus, *J. Virol.*, 62, 3399, 1988.
696. **Schneemann, A., Zhong, W., Gallagher, T. M., and Rueckert, R.**, Maturation cleavage required for infectivity of a Nodavirus, *J. Virol.*, 66, 6728, 1992.
697. **Ball, L. A.**, Cellular expression of a functional Nodavirus RNA replicon from vaccinia virus vectors, *J. Virol.*, 66, 2335, 1992.
698. **Ball, L. A., Amann, J. M., and Garrett, B. K.**, Replication of Nodamura virus after transfection of viral RNA into mammalian cells in culture, *J. Virol.*, 66, 2326, 1992.
699. **Bailey, L., Newman, J. F. E., and Porterfield, J. S.**, The multiplication of Nodamura virus in insect and mammalian cell cultures, *J. Gen. Virol.*, 26, 15, 1975.
700. **Tesh, R. B.**, Infectivity and pathogenicity of Nodamura virus for mosquitoes, *J. Gen. Virol.*, 48, 177, 1980.
701. **Selling, B. H., Allison, R. F., and Kaesberg, P.**, Genomic RNA of an insect virus directs synthesis of infectious virions in plants, *Proc. Natl. Acad. Sci. U.S.A.*, 87, 434, 1990.
702. **Selling, B. H. and Rueckert, R. R.**, Plaque assay for black beetle virus, *J. Virol.*, 51, 251, 1984.
703. **Agrawal, D. K. and Johnson, J. E.**, Sequence and analysis of the capsid protein of Nudaurelia capensis ω virus, an insect virus with T = 4 icosahedral symmetry, *Virology*, 190, 806, 1992.
704. **Moore, N. F. and Pullin, J. S. K.**, The use of invertebrate tissue culture to characterize picorna-like viruses of insects: intracellular proteins induced by cricket paralysis virus in infected *Drosophila* cells, in *Invertebrate Systems In Vitro*, Kurstak, E., Maramorosch, K., and Dübendorfer, A., Eds., Elsevier, Amsterdam, 1980, 403.
705. **Becht, H.**, Infectious bursal disease virus, *Curr. Top. Microbiol. Immunol.*, 90, 107, 1980.
706. **Duncan, R., Nagy, E., Krell, P. J., and Dobos, P.**, Synthesis of the infectious pancreatic necrosis virus polyprotein, detection of a virus-coded protease, and fine structure mapping of genome segment A coding regions, *J. Virol.*, 61, 3655, 1987.
707. **Jones, L. D., Davies, C. R., Steele, G. M., and Nuttall, P. A.**, A novel mode of arbovirus transmission involving a nonviremic host, *Science*, 237, 775, 1987.
708. **Clay, W. C. and Fuller, F. J.**, Nucleotide sequence of the tick-borne Orthomyxo-like Dhori India 1313/61 virus membrane protein gene, *J. Gen. Virol.*, 73, 2609, 1992.
709. **Arif, B. M.**, The entomopoxviruses, *Adv. Virus Res.*, 29, 195, 1984.
710. **Granados, R. R., Dwyer, K. G., and Derksen, A. C. G.**, Production of viral agents in invertebrate cell cultures, in *Biotechnology in Invertebrate Pathology and Cell Culture*, Academic Press, Orlando, 1987, 167.
711. **Dropulic, B. and Masters, C. L.**, Entry of neurotropic arboviruses into the central nervous system: an in vitro study using mouse brain endothelium, *J. Infect Dis.*, 161, 685, 1990.
712. **Pudney, M., Varma, M. G. R., and Leake, C. J.**, Establishment of a cell line (XTC-2) from the South African clawed toad, *Xenopus laevis*, *Experientia*, 29, 466, 1973.

713. **Morier, L., Cantelar, N., and Soler, M.,** Infection of a poikilothermic cell line (XL-2) with Eastern equine encephalitis and Western equine encephalitis viruses, *J. Med. Virol.,* 21, 277, 1987.
714. **Pudney, M. and Varma, M. G. R.,** *Anopheles stephensi* var. *mysorensis.* Establishment of a larval cell line (Mos 43), *Exp. Parasitol.,* 29, 7, 1971.
715. **Hsu, S. H.,** Preliminary observations on growth of arboviruses in a newly established line of mosquito cells (*Culex quinquefasciatus,* Say), *Curr. Top. Microbiol.,* 55, 140, 1971.
716. **Oro, G.,** Establishment of a mosquito cell line from *Haemagogus equinus* larvae, *In Vitro,* 20, 153, 1984.
717. **Tesh, R. B. and Modi, G. B.,** Development of a continuous cell line from the sand fly *Lutzomyia longipalpis* (Diptera: *Psychodidae*), and its susceptibility to infection with arboviruses, *J. Med. Entomol.,* 20, 199, 1983.
718. **Kurtti, T. J., Munderloh, U. G., and Stiller, D.,** The interaction of *Babesia caballi* kinetes with tick cells, *J. Invertebr. Pathol.,* 42, 334, 1983.
719. **Yunker, C. E., Cory, J., and Meibos, H.,** Continuous cell lines from embryonic tissues of ticks *(Acari: Ixoidae), In Vitro Cell. Dev. Biol.,* 17, 139, 1981.
720. **Řeháček, J.,** Cultivation of different viruses in tick tissue cultures, *Acta Virol.,* 9, 332, 1965.
721. **Řeháček, J.,** Use of invertebrate cell culture for study of animal viruses and Rickettsiae, in *Invertebrate Tissue Culture,* Vol. 2, Academic Press, New York, 1972, 279.
722. **Yunker, C. E., Cory, J., and Meibos, H.,** Tick tissue and cell culture: applications to research in medical and veterinary acarology and vector borne diseases, in *Acarology IV,* Vol. 2, Griffiths, D. A. and Bowman, D. E., Eds., Ellis Harwood, Chichester, 1984, 1082.
723. **Grace, T. D.,** The development of clones from lines of *Antheraea* eucalypti cells grown in vitro, *Exp. Cell Res.,* 52, 451, 1968.
724. **Iwata, H., Yamagawa, M., and Roy, P.,** Evolutionary relationships among the gnat-transmitted arboviruses that cause African horse sickness, Bluetongue, and epizootic hemorrhagic disease as evidenced by their capsid protein sequences, *Virology,* 191, 251, 1992.
725. **Weaver, S. C., Scott, T. W., and Lorenz, L. H.,** Patterns of Eastern equine encephalomyelitis virus infection in *Culiseta melanura* (Diptera: *Culicidae*), *J. Med. Entomol.,* 27, 878, 1990.
726. **Vazeille-Falcoz, M., Ohayon, H., Gounon, P., and Rosen, L.,** Unusual morphology of a virus which produces carbon dioxide sensitivity in mosquitoes, *Virus Res.,* 24, 235, 1992.
727. **Moss, S. R., Jones, L. D., and Nuttall, P. A.,** Comparison of the major structural core proteins of tick-borne and *Culicoides*-borne Orbiviruses, *J. Gen. Virol.,* 73, 2585, 1992.
728. **Portela, A., Jones, L. D., and Nuttall, P. A.,** Identification of viral structural polypeptides of Thogoto virus (a tick-borne Orthomyxo-like virus) and functions associated with the glycoprotein, *J. Gen. Virol.,* 73, 2823, 1992.
729. **Maldov, D. G., Karganova, G. G., and Timofeev, A. V.,** Tick-borne encephalitis virus interaction with the target cells, *Arch. Virol.,* 127, 321, 1992.
730. **Gaedigk-Nitschko, K. and Schlesinger, M. J.,** Site-directed mutations in Sindbis virus E_2 glycoprotein's cytoplasmic domain and the 6K protein lead to similar defects in virus assembly and budding, *Virology,* 183, 206, 1991.

Chapter 8

JAPANESE INVERTEBRATE CELL CULTURE PIONEERS

Yukiaki Kuroda

TABLE OF CONTENTS

0-8493-7642-4/94/$0.00+$.50

Yukiaki Kuroda, D.Sc., is Professor of Gene and Cell Biology at Azabu University, Sagamihara and Emeritus Professor at National Institute of Genetics, Mishima, Japan. He received his B.S. at Kyoto University in 1950 and doctorate in Science at Osaka University 1959. NIH International Postdoctoral Research Fellow at University of Chicago 1961–1962. Associate Professor at Osaka University 1963–1966; Head of Laboratory at National Institute of Genetics 1966–1976; Head of Dept. at N.I.G. 1976–1984; Professor at N.I.G. 1984–1990; Professor and Director of Research Institute of Bioscience at Azabu University 1990–present. President of Japan Tissue Culture Assoc.; Chairman of 7th International Conf. of Invertebrate and Fish Tissue Culture at Ohito, Japan; Japan. Soc. of Cell Biol.; Genet. Soc. of Japan; Japan Soc. of Develop. Biol.; Japan Environ. Mutagen Soc. Gene expression and cell biology of cultured insect and mammalian cells during embryonic development, chemical mutagenesis and ageing.

I. INTRODUCTION

Invertebrate tissue culture started later than vertebrate tissue culture. Lower invertebrates, such as sea urchins, were known to survive in filtered sea water for a relatively long time. The higher invertebrates require more complex media for their survival. Tissue culture of arthropods, especially insect tissue culture, has found applications in genetics, developmental biology, physiology, and virology. Goldschmidt[1] in 1915 cultured spermatocytes of silkworm in physiological saline containing silkworm body fluid and observed cell division. Some 20 years later, Trager[2] used physiological saline containing maltose, ovalbumin hydrolyzate, and silkworm blood for his cultures of ovarian tissues of silkworm.

In Japan, in the 1950s several cytogenetics workers cultured testes of grasshoppers in physiological saline for a relatively short period to observe the movement of chromosomes in miotic divisions of spermatocytes by cinemicroscopy. The first attempt of insect tissue culture using synthetic medium was carried out in 1955 by Kuroda and Tamura[3], who cultured melanotic tumors of *Drosophila melanogaster* in K-6 medium of 29 constituents containing 6 inorganic salts, glucose, 4 amino acids, 13 vitamins, casein hydrolysate, sodium acetate, glutathione, ribonucleic acid (RNA), and diphosphopyridine nucleotide (DPN). Two years later, Takami[4] cultured whole embryos from nondiapausing silkworms in nondiapausing egg extract and observed that the embryos developed well beyond the appendage formation stage.

In the 1960s several Japanese scientists collaborated with U.S. scientists. J. Mitsuhashi, H. Hirumi, T. Tokumitsu, and later K. Yamada carried out studies with K. Maramorosch at the Boyce Thompson Institute for Plant Research and at Rutgers University. They used insect tissue culture techniques for cultivation of insect and plant viruses.

Hirumi and Maramorosch[5] who devised a medium consisting of modified Vago's medium No. 22 and Morgan's TC-199[10] medium used this mixture for primary culture of leafhopper embryonic tissues. Mitsuhashi and Maramorosch[6] devised the MM medium originally used for leafhopper embryonic cell cultures. Subsequently this very simple medium, consisting of inorganic salts, glucose, lactalbumin hydrolysate, TC-yeastolate, and fetal bovine serum, became widely used for mosquito, lepidopteran, and dipteran cell lines. Kitamura[7] devised a medium for cultivation of mosquito adult ovarian cells, later used for various mosquito cell cultures.

Japanese scientists have made some unique attempts for cultivation of marine invertebrate cells. Machii[8] cultured mantle tissue fragments from veliger larvae of the pearl oyster in his organ culture medium. On his explant culture from pearl oyster mantle tissue, the depositions of trefoil-shaped material were found. These deposits, absorbing soft X-rays, had a high content of calcium and demonstrated distinct birefringency. He concluded that the crystal must be a prism, a kind of pearl formed *in vitro.* He also developed a medium, Pf35, exclusively for pearly oyster cell culture and established a pearl oyster cell line. Machii and his collaborators also initiated attempts to culture cells from the lug worm (Annelida), the abalone (Mollusca), the sea hare (Mollusca), and the sea urchin (Echinodermata).[9]

Japanese scientists established more than 37 insect cell lines from *Drosophila,* rice stem borers, mosquitos, leafhoppers, swallow tails, cabbage armyworms, fleshflies, silkworms, and beetles, and one cell line from pearl oyster.

In Japan, three international meetings on invertebrate tissue culture were held. The first one was the Seminar on Invertebrate Tissue Culture Application in Fundamental Research, held in Tokyo on December 9 through 12, 1974. This was sponsored by the U.S.-Japan Cooperative Science Program, organized by S. Fukuda and K. Maramorosch. Twenty papers were presented at that time. The second one was a small symposium on Growth and Genetics of Invertebrate Cells *In Vitro,* held during the 3rd International Cell Culture Congress at Sendai, September 10 through 13, 1985, and organized by Maramorosch and Mitsuhashi. The third one was the 7th International Conference on Invertebrate and Fish Tissue Culture, at Ohito, Shizuoka on May 10 through 16, 1987, organized by Y. Kuroda, E. Kurstak, and K. Maramorosch. The Proceedings containing 66 papers presented in this conference were published in 1988. The present chapter describes some pioneering and creative works of Japanese scientists in the field of invertebrate tissue culture.

II. DEVELOPMENT OF CULTURE MEDIA FOR INVERTEBRATE CELLS

A. *DROSOPHILA* MEDIA

Kuroda[3] devised the technique for culturing organs and tissues from 3rd instar larvae of *Drosophila melanogaster.* This technique involves the rearing of mature larvae grown under sterile conditions from surface-sterilized eggs and the preparation of a new culture medium K-6 (Table 1).

Until that time, only physiological salt solution was used for maintenance of isolated insect cells *in vitro.* There were no papers dealing with the cultivation of insect cells and tissues. For the cultivation of chick embryonic muscle tissue, Parker and his collaborators devised synthetic media composed of biologically active chemicals. After many attempts they developed a synthetic medium TC-199.[10]

In Kuroda's medium K-6, inorganic salts, which are concerned with osmotic pressure of the culture medium, and glucose, which is a carbon source of explants, are contained in the same concentration as those in Carlson's physiological solution.[11] Most amino acids in medium K-6 were from casein hydrolyzate, which was lacking L-tryptophan and L-cysteine. Therefore, these two amino acids were added. The amino acid nitrogen of the body fluid of insects is 1.34 g/l in *Dytiscus* adult, 3.27 g/l in the pupa of *Attacus,* 3.22 g/l in the pupa of *Sphinx,* 2.34 g/l in larvae of *Cossus,* 1.7 g/l in larvae of *Deilephila* — about 50 to 100 times higher than that in mammalian plasm.[12]

Since an excess concentration of amino acids beyond this level has a toxic effect on cultured tissues, the concentration of amino acids in medium K-6 was devised to be equivalent to that in the body fluid. Cystein, ascorbic acid, and glutathione were added to protect the SH group of enzymes. Tissues involving melanotic tumors from mature larvae of *D. melanogaster* were cultured in medium K-6, and the effects of some chemicals and insect hormones on melanotic growth of tumors were investigated.

Medium K-6 was slightly modified later and medium K-6′ was devised for cultivation of eye-antennal discs of *D. melanogaster* without any supplementation of mammalian serum or insect hemolymph.[13]

For primary cultures of *Drosophila* embryonic cells, Kuroda[14,15] devised medium K-17 (Table 2).

In this medium, the concentrations of inorganic salts and glucose were the same as those in medium K-6. The concentrations of trehalose and amino acids were the same as in Schneider's medium.[16] Four organic acids, yeastolate (Difco) and peptone (Wako Pure Chem. Ind. pancreatic digest of casein) were added. For primary culture of *Drosophila* embryonic cells, medium K-17 was used with supplementation of 0.1 mg/ml fetuin (Grand Island Biol. Co., Deutsch method) and 15% fetal bovine serum (Microbiological Associates).

B. HOMOPTERAN MEDIA

Hirumi[5] in collaboration with Maramorosch made attempts to culture leafhopper cells for maintaining and proliferating the causative agent of aster yellows disease *in vitro.* They modified Vago's *Bombyx mori* medium No. 22, Morgan's synthetic medium TC-199, and added fetal bovine serum.

Mitsuhashi and Maramorosch[6] devised their MM medium (see Chapter 1) for leafhopper embryonic cell cultures. This medium was composed of 6 inorganic salts, glucose, lactalbumin hydrolysate (amino acid source), yeastolate (vitamin source), and fetal bovine serum. Medium MM was later used for cell cultures of mosquitoes, fleshflies, common armyworms, and cabbage armyworms. At present, the medium MM is one of the most widely used media for lepidopteran and dipteran insects.

TABLE 1
Composition of Medium K-6[3]

Components	mg/100 ml	Components	mg/100 ml
NaCl	700	Niacinamide	0.0125
KCl	20	Ca-pantothenate	0.005
$CaCl_2 \cdot 2H_2O$	2	Ascorbic acid	25
$MgCl_2 \cdot 2H_2O$	10	Biotin	0.005
$NaHCO_3$	5	Folic acid	0.005
$NaH_2PO_4 \cdot 2H_2O$	20	Choline hydrochloride	0.25
Glucose	80	Inosite	0.025
Casein hydrolyzate	5000	*p*-Aminobenzoic acid	0.025
L-Cysteine	26	Sodium acetate	5
L-Cystine	100	Vitamin A	0.05
L-Tryptophan	100	Vitamin B_{12}	0.05
Ribonucleic acid	50	Glutathione	1
Thiamin	0.005	Glutamic acid	10
Pyridoxine hydrochloride	0.0125	DPN	0.1

From Kuroda, Y. and Tamura, S., *Zool. Mag.*, 64, 380, 1955. With permission.

TABLE 2
Composition of Medium K-17

Components	mg/100 ml	Components	mg/100 ml
NaCl	700	L-Isoleucine	15
KCl	20	L-Leucine	15
$CaCl_2 \cdot 2H_2O$	2	L-Lysine·HCl	165
$MgCl_2 \cdot 2H_2O$	10	L-Methionine	80
$NaHCO_3$	5	L-Phenylalanine	15
$NaH_2PO_4 \cdot 2H_2O$	20	L-Proline	170
Glucose	80	L-Serine	25
Trehalose	200	L-Threonine	35
β-Alanine	50	L-Tryptophan	10
L-Alanine	40	L-Tyrosine	50
L-Aspartic acid	40	L-Valine	30
L-Cysteine	6	α-Ketoglutaric acid	20
L-Cystine	10	Fumaric acid	10
L-Glutamic acid	80	Malic acid	10
L-Glutamine	180	Succinic acid	10
Glycine	25	Peptone	500
L-Histidine	40	Yeastolate	200

Note: pH was adjusted to 6.7.

Adapted from References 14 and 15.

Mitsuhashi[17] used medium NCM-2B for primary culture of leafhopper tissues. This medium consisted of two sugars, glucose and sucrose, lactalbumin hydrolysate and bactopeptone for the source of amino acids with supplementation of cysteine and tryptophan, yeastolate for the source of vitamins, and TC-199 medium. He devised media MGM-431 (Mitsuhashi[18]), MGM-443 (Mitsuhashi[19]), MGM-446 (Mitsuhashi[20]), and MGM-448 (Mitsuhashi[37]) for cultivation of lepidopteran insect cells. (See Chapter 1.)

These media were relatively complex, compared to the composition of Grace's medium.[21] Mitsuhashi's media were originally used for leafhopper and cabbage armyworm cell cultures, but later for a wide variety of insect species such as swallow tail, silkworm, and mulberry tiger moth.

C. SIMPLE MEDIA

Mitsuhashi[22] made an attempt to formulate simple and low cost media. Almost all earlier insect culture media contained fetal bovine serum. He attempted to modify the medium MM so as to make it simple and inexpensive. Inorganic salts involved in medium MM were substituted with diluted sea water, and the glucose with table sugar. Since the table sugar was then the sole source of carbohydrates in the medium, the utilization of sucrose by insect cells was suspected, and analyses of culture media proved that insect cells were unchanged. The new medium MTCM-1601 consisted of sea water of 1/4 strength, lactalbumin hydrolysate, TC-yeastolate, and table sugar. This medium could be sterilized by autoclaving. Most of the cell lines tested could be passed continuously in this medium, although growth was somewhat retarded in most cases. As described by Mitsuhashi in Chapter 1, insect cells were able to utilize sucrose in the medium.

D. MOSQUITO MEDIA

Kitamura[7] devised a simple medium for culturing mosquito ovarian tissue. The basal part of culture medium consisted of five inorganic salts, sucrose, and lactalbumin hydrolysate. To the mixture of 80 parts of the basal medium and 20 parts of medium TC-199 of Morgan et al.,[10] calf serum was added at 10 volume percent. This medium was satisfactory for the primary cultures of mosquito ovarian tissues and useful for further study on interaction between arboviruses and carrier insect cells. Kitamura[23] established a cell line of *Culex* mosquito ovarian tissue in this medium. Later he succeeded in adaptation of mosquito cell lines to serum-free medium (Kitamura et al.[24]). In early subcultures, mosquito cells were maintained in medium supplemented with 20% calf serum, up to the 10th passage, and then the amount of serum was gradually reduced to 5% by the 123rd passage. At the 152nd passage, cells were transferred to a new medium containing bovine plasma albumin (BPA; Fraction V, Armour Co.) instead of calf serum. The cell lines adapted to new medium grew well continuously for over 6 months at 28°C. The population doubling time of cells was about 30 to 48 h.

E. MARINE INVERTEBRATE MEDIA

Machii[8] developed medium Pf35 (Table 3) exclusively for pearl oyster cell culture. He had previously examined inorganic ion and free amino acid compositions of body fluid from over 20 species of aquatic mollusks. On the basis of these results, he formulated 15 balanced solutions and 23 fundamental culture media adjusted to the amino acid compositions of body fluids.

The medium Pf35 was formulated from an analysis of inorganic ions and free amino acid concentrations of body fluid from the pearl oyster, *Pinctada fucata.* Vitamins and other minor constituents of the medium were those in medium TC-199. Using medium Pf35, Machii[25] established a cell line from veliger larvae of pearl oyster, *P. fucata.* Machii and Wada[25] cultured fragments of mantle tissues and veliger larvae from the pearl oyster, *P. fucata* and the black tip pearl oyster, *P. margaritifera,* in medium Pf35. They also cultured trypsinized cells from the same tissues and larvae in this medium.

Machii and Wada[25] cultured cells from the midgut glands of the Kuruma prawn, *Penaeus japonicus,* in medium Pf35 and medium Pj1, Pj2, and Pj3, which were developed for the prawn. They detected isozyme bands in esterase, malate dehydrogenase, phosphoglucose isomerase, and phosphoglucomutase, and observed the normal karyotype (2n = 86) in cultured prawn cells.

III. ESTABLISHMENT OF INVERTEBRATE CELL LINES

A. INSECT CELL LINES

For obtaining quantitative and reproducible results using tissue culture cells, the establishment of continuous cell lines is the most important and essential requirement. Several reports on the establishment of insect cell lines from ovaries and gonads were published by French and Chinese workers during the second half of the 1950s. However, the most important report

TABLE 3
Composition of Medium Pf35

Components	mg/100 ml	Components	mg/100 ml
NaCl	2622	Folic acid	0.001
KCl	108	Inositol	0.005
$MgSo_4$	318	Niacin	0.003
$MgCl_2$	220	Niacin amide	0.5
$CaCl_2$	112	Pyridoxal HCl	0.0025
$NaHCO_3$	30	Pyridoxine HCl	0.1
NaH_2PO_4	4.4	Riboflavin	0.1
L-Cystine	0.07	Vitamin B_{12}	0.001
L-Tyrosine	0.36	Thiamine HCl	1
Taurine	12.89	Ascorbic acid	4
L-Aspartic acid	1.2	Vitamin A	0.01
L-Threonine	0.36	α-Tocopherol phosphate	0.001
L-Serine	0.95	Calciferol	0.01
L-Asparagine	0.53	Menadione	0.001
L-Glutamic acid	1.91	Adenine	1
L-Glutamine	1.75	Guanine HCl	0.03
L-Proline	0.35	Hypoxanthine	0.03
L-Glycine	1.5	Thymine	0.03
L-Alanine	1.78	Uracil	0.03
α-amino *n*-butyric acid	0.08	Xanthine	0.03
L-Valine	0.47	Adenylic acid	0.02
L-Methionine	0.3	2-Deoxy-D-ribose	0.05
L-Isoleucine	0.26	D-Ribose	0.05
L-Leucine	0.39	Tween 80	0.5
L-Phenylalanine	0.17	Cholesterol	0.02
L-Ornithine	0.09	Glutathione	0.005
L-Lysine	0.73	Sodium acetate	5
L-Tryptophan	0.1	Glucose	30
L-Histidine	0.47	ATP	1
L-Arginine	1.74	Lactalbumine hydrolysate	10
p-Aminobenzoic acid	0.005	Phenol red	0.5
Phosphoserine	0.1	Penicillin	20
Biotin	0.001	Kanamycin	10
Ca-pantothenate	0.1	Fetal calf serum	10
Choline-chloride	25		

From Machii, A., *Tissue Cult. Res. Commun.*, 4, 18, 1985. With permission.

was that of Grace[21] from Australia. He established four cell strains derived from ovaries of *Antheraea eucalypti* in his medium, which contained heated pupal fluid.

In Japan, the first report on the establishment of an insect cell line was made by Horikawa et al.[26] Horikawa initiated tissue culture of *Drosophila* in the author's laboratory in Osaka University and published the results on the long-term culture of embryonic cells of *Drosophila melanogaster* with his collaborators at the University of Wisconsin. They maintained cells for more than 430 days.

The invertebrate cell lines established by Japanese scientists are listed in Table 4. Mitsuhashi at the National Institute of Agricultural Sciences, Tokyo, established many insect cell lines from the rice stem borer (Lepidoptera),[27] green rice leafhopper (Homoptera),[29] swallow tail butterfly (Lepidoptera),[30] cabbage armyworm (Lepidoptera),[31,33,38] mosquito (Diptera),[34] common armyworm (Lepidoptera),[35] mulberry tiger moth (Lepidoptera),[41] and beetle (Coleoptera).[46] He established cell lines from embryos, larval hemocytes, and fat bodies, 1st larvae and pupal ovaries in media CSM-2F,[27] MGM-401,[24] MGM-431,[30,31,33] MGM-448,[41,46] MGM-450,[46] and MM.[32,34,35,41]

TABLE 4
Invertebrate Cell Lines Established by Japanese Scientists

Cell lines	Species	Original tissue	Culture medium	Ref.
	Fruit fly *(Drosophila)*	Embryos	H-6	26
	Rice stem borer *(Chilo suppressalis)*	Hemocytes	CSM-2F	27
	Mosquito *(Culex molestus)*		Kitamura's medium	28
	Green rice leafhopper *(Nephotettix cincticeps)*	Embryos	MGM-401	29
NIAS-PX-58	Swallow tail butterfly *(Papilio xuthus)*	Pupal ovaries	MGM-431	30
NIAS-MB-19	Cabbage armyworm *(Mamestra brassicae)*	Pupal ovaries	MGM-431	31
NIAS-MB-25	Cabbage armyworm *(Mamestra brassicae)*	Pupal ovaries	MGM-431	31
NIAS-MB-32	Cabbage armyworm *(Mamestra brassicae)*	Pupal ovaries	MGM-431	31
NIH-SaPa-4	Fleshfly *(Sarcophaga peregrina)*	Embryos	MM	32
NIAS-MaBr-85	Cabbage armyworm *(Mamestra brassicae)*	Mature larvae fat bodies	MGM-431	33
NIAS-AeA1-2	Mosquito *(Aedes albopictus)*	1st instar larvae	MM and MTCM-1103	34
NIAS-LeSe-11	Common armyworm *(Leucania separata)*	Larval fat bodies	MM	35
	Silkworm *(Bombyx mori)*	Embryos	MGM-443	36
NIAS-SpSe-1	Arctiid moth *(Spilarctia seriatopunctata)*	Larval fat bodies	MGM-443, MGM-448	37
NIAS-MaBr-92	Cabbage armyworm *(Mamestra brassicae)*	Larval hemocytes	MM, MM-SF	38
NIAS-MaBr-93	Cabbage armyworm *(Mamestra brassicae)*	Laval hemocytes	MM, MM-SF	38
SES-MaBr-1	Cabbage armyworm *(Mamestra brassicae)*	Larval fat bodies	MGM-448, MM, MM-SF	39
SES-MaBr-2	Cabbage armyworm *(Mamestra brassicae)*	Larval fat bodies	MGM-448, MM, MM-SF	39
SES-MaBr-3	Cabbage armyworm *(Mamestra brassicae)*	Laval fat bodies	MGM-448, MM. MM-SF	39
SES-MaBr-4	Cabbage armyworm *(Mamestra brassicae)*	Larval fat bodies	MGM-448, MM, MM-SF	39
SES-MaBr-5	Cabbage armyworm *(Mamestra brassicae)*	Larval fat bodies	MGM-448, MM, MM-SF	39
	Pearloyster *(Pinctada fucata)*	Veliger larvae	Pf35	8
SES-BoMo-15A	Silkworm *(Bombyx mori)*	Embryo	MGM-448, MM	40
FRI-SpIm-1229	Mulberry tiger moth *(Spilosoma imparilis)*	Larval fat bodies	MGM-448, MGM-450, MM	41
SES-Bm-e21	Silkworm *(Bombyx mori)*	Mature embryos	IPL-41	46
SES-Bm-1 30	Silkworm *(Bombyx mori)*	Mature embryos	IPL-41	42
SES-Bma-01	Wild silkworm *(Bombyx mandarina)*	Mature embryos	IPL-41	42
ML-DmD1–20	Fruit fly *(Drosophila melanogaster)*	Imaginal discs		43
SES-BoMo-C129	Silkworm *(Bombyx mori)*	Embryos	MGM-448	44
SES-BoMo-S125	Silkworm *(Bombyx mori)*	Embryos	MGM-448	44
NC-15	Leafhopper *(Nephotettix cincticeps)*	Embryos	Liu and Black's medium	45
NC-19				
NC-20				
NC-21				

TABLE 4 (Continued)
Invertebrate Cell Lines Established by Japanese Scientists

Cell lines	Species	Original tissue	Culture medium	Ref.
NC-24				
NC-25				
NN-1	Leafhopper *(Nephotettix nigropictus)*	Embryos	Liu and Black's medium	45
NN-2				
NN-5				
NN-7				
RD-5	Leafhopper *(Nephotettix dorsalis)*	Embryos	Liu and Black's medium	45
RD-6				
RD-7				
RD-8				
RD-9				
FRI-AnCo-5B	Beetle *(Anomala costata)*	Embryos	MGM-448	46

Kitamura[28] established a cell line from adult ovaries of *Culex molestus* in his basal medium (BCM: 5 inorganic salts, glucose, and lactalbumin hydrolysate) supplemented with 20% medium TC-199 and 10% heat-inactivated calf serum. The cell line was maintained in decreased concentration of calf serum in subsequent subcultures (Kitamura et al.[47]). At the 152nd passage, the cells began to adapt to a new medium containing BPA instead of calf serum. After the 171st passage, the cells were adapted to a medium which consisted of 60% BCM and 40% TC-199 supplemented with 1% BPA.

Mitsuhashi and Shozawa[38] obtained two cell lines, which can grow in media free of serum and lacking sterols, fatty acids, and protein. These cell lines were initiated to culture larval hemocytes from cabbage armyworm, *Mamestra brassicae* in MGM-443[19] and MMG-448.[37] When the first passage was made, there were prohemocyte-like cells, plasmatocyte-like cells, and granular hemocyte-like cells. However, granular-like cells and plasmatocyte-like cells disappeared as the passage advanced, and the populations consisted of mostly prohemocyte-like cells. These cells can grow in MM-SF medium (a serum-free MM medium) and MTCM-1103 medium (free of protein, sterols, and fatty acids).

Mitsuhashi[48] tried to culture several insect cell lines in a simple serum-free medium. When the cells, cultured in medium MM with 3% fetal bovine serum (FBS), were subcultured, a half of the cell suspension was discarded and an equal volume of medium MM-SF (serum-free MM) was added. This way of subculturing was repeated until the cells began to multiply in 100% MM-SF. By the 10th passage, the concentration of FBS carried over from the original culture was about 0.003% and most cells seems to adapt to serum-free conditions. At this stage, the cells were subcultured in the medium completely free of FBS.

By using this way, the cell line NIAS-PX-58 derived from pupal ovaries of the swallow tail butterfly, *Papilio xuthus* (Mitsuhashi[30], NIAS-MB-32 from pupal ovaries of the cabbage armyworm, *Mamestra brassicae* (Mitsuhashi[31]), NIAS-MaBr-85 from larval fat bodies of *M. brassicae* (Mitsuhashi[33]), NIAS-AeA1-2 from larval tissues of the one-striped mosquito, *Aedes albopictus* (Mitsuhashi[34]), NIH-SaPe-4 from embryonic tissues of the flesh fly, *Sarcophaga peregrina* (Takahashi et al.[32]), and NIAS-LeSe-11 from larval fat bodies of the common armyworm, *Leucania separata* (Mitsuhashi[35]) were cultured continuously in the serum-free media. These cells could also grow in the media apparently lacking cholesterol. The cells, which had been cultured in the serum-free medium, could not tolerate low and ultralow temperature storage.

Ninaki et al.[42] established cell lines of the silkworm, *Bombyx mori,* and the wild silkworm, *B. mandarina.* These lines were derived from matured embryos (stage 27 to 29) and designated as SES-Bm-e 21, SES-Bm-1 30, and SES-Bma-01 cell lines. In order to reintroduce characteristics of wild silkworm to domesticated silkworm, cell fusion between two cell lines was attempted, using PEG (polyethylene glycol). In order to determine the condition of the cell fusion, fused hybrid cells between *Bombyx* and *Antheraea* cell lines were subcultured for about 4 months.

Ui et al.[43] established 20 cell lines, designated as ML-DmD1-20, from wing, haltere, antenna, eye-antenna discs, and disc mixtures, involving 4 cell lines from haltere discs, 2 cell lines from antenna discs, and 2 eye-antenna discs. The cell aggregates formed in culture were examined for their differentiation ability by injecting them into metamorphosing larvae. At the early stage of *in vitro* culture, they differentiated into adult cuticular structures containing bristles and/or trichomes which were indicative of disc origin.

B. MARINE INVERTEBRATE CELL LINES

When Machii[8] established a cell line from veliger larva of the pearl oyster, *Pinctada fucaa,* he homogenized 1-day veliger larvae of the pearl oyster, digested them with pronase, and cultured cells thus obtained in medium Pf35 (Table 3) at 25°C. In cultures of marine invertebrates, the most difficult problems are the contaminations by fungi and protozoa. Machii used frequently fungizon (2 μg/ml) and mycostatin (20 g/ml) as antifungous agents, and oxytetracycline (200 g/ml), Fumagillin dodecylamine salt (0.05 μg/ml), and thymerosal (2 μg/ml) as antiprotozoan agents. After cultivation for 20 days, the proliferated monolayer sheet of primary cultured cells was subcultured, and thereafter subcultures were repeated every 15 days. After 20 passages, a trefoil-like deposition, which was 10 μm in diameter, and birefringement crystal was formed on a monolayer cell sheet.

IV. STUDIES IN THE FIELD OF DEVELOPMENTAL BIOLOGY

A. DEVELOPMENT AND DIFFERENTIATION IN ORGAN CULTURES

Kuroda and Tamura[3,51] at Osaka University cultured melanotic tumors of the mid-intestine from 3rd instar larvae of *Drosophila melanogaster* in hanging drops of medium K-6 (Table 1). The cephalic complex (brain) and the ring gland of the 3rd instar larvae, which were cocultured with melanotic tumors in the same medium, inhibited the melanotic growth of tumors. Phenylthiourea (PTC) at a concentration of 1 m*M* inhibited completely the increase of melanotic tumors (Kuroda and Tamura[49]). Some heavy metals, Cu^{2+} or Fe^{3+}, at concentrations of 1 to 5 m*M* promoted markedly the growth of melanotic tumors (Kuroda and Tamura[50,52]).

Kuroda and Yamaguchi[53] cultured the eye-antennal discs from mature larvae (95 h after hatching) of wild-type and *Bar* (1-57.2) eye mutant strains of *D. melanogaster* by the hanging-drop procedure. The cephalic complex of the wild type had an effect on both wild type and *Bar* eye-antennal discs in enhancing the wild-type morphogenesis of the eye-antennal discs. On the other hand, the cephalic complex of the *Bar* mutant promoted the *Bar* type morphogenesis for both wild-type and *Bar* eye-antennal discs.

Many other morphological mutants of the eye in *D. melanogaster* have also been tested by graduate students in Kuroda's laboratory (Horikawa;[54] Fujio[55]) for the activity of the cephalic complexes, to promote the differentiation of eye-antennal discs and for the responsiveness of the eye-antennal discs to cephalic complexes.

The cephalic complex of the *Drop* (3-99.2) mutant had a similar effect as the *Bar* cephalic complex in promoting the differentiation of the eye-antennal discs. The cephalic complex of the *bar-3* (3-79.1) and *eyeless*^{2} (4-2.0) mutants had a slight effect on the wild-type differentiation of eye-antennal discs. The eye-antennal discs of the *Drop* mutant did not respond to the wild-type cephalic complexes. The eye-antennal discs of *bar-3* and *eyeless* responded to wild-type cephalic complexes in differentiation of ommatidium-forming cells.

Ammonium lactate and acetamide at concentrations of 1 and 0.1 m*M*, respectively, had a significant effect in inducing the wild-type differentiation of eye-antennal discs from *Bar, bar-3,* and *eyeless*^{2} in the presence of their own cephalic complexes. Urea at a concentration of 1 m*M* also had a similar effect on the differentiation of these mutant eye-antennal discs.

The conditioned medium from cultures of the wild-type cephalic complexes had effects in promoting the differentiation of the eye-antennal discs of *Bar, bar-3,* and *eyeless*^{2} mutants (Fujio[56]). The conditioned medium from cultures of *Bar* cephalic complexes had both quantitatively and qualitatively different effects on the wild-type eye-antennal discs, as compared to the conditioned medium obtained from cultures of the wild-type cephalic complexes.

The differentiation of the ommatidial cell layer of cultured eye-antennal discs was markedly promoted by the addition of externally isolated ecdysterone analogs to the culture medium (Figure 1). Among ecdysone analogs of plant origin, such as ecdysterone, inokosterone, ponasterone C, and rubrosterone, rubrosterone was most effective in promoting cluster formation of the ommatidium-forming cells. At concentrations as low as 10^{-2} ng/ml, rubrosterone was active in the ommatidium differentiation in cultured eye-antennal discs (Table 5).[57]

Eye-antennal discs cultured in the presence of 0.1 μg/ml rubrosterone and 10^{-5} *M* 5-bromo-2′-deoxyuridine (BUdR) showed a differentiation of ommatidia similar to that in cultures without BUdR. Actinomycin D (1 μg/ml) and puromycin (10 μg/ml) did not inhibit the hormone-induced differentiation of ommatidia (Kuroda[58]). These results indicate that the process of ommatidial cell cluster formation may be conducted by already-existing macromolecules. These macromolecules may be in a masked or inactive state at the beginning of culture and are unmasked or activated by ecdysteroids.

The eye-antennal discs showed differential sensitivity to X-rays in different regions of the ommatidium-forming cell portion and at different stages of cultivation (Kuroda[58]). The ommatidium-forming cells become insensitive to X-rays as their differentiation proceeds. This change in X-ray sensitivity of the cells may begin at the posterior portion and extend to the anterior portion of the eye discs.

In studies on the differentiation of the wing disc of *D. melanogaster,* Kuroda[59] cultured wing discs removed from mature larvae of three *vestigial* alleles, *vestigial (vg), vestigial-notched* (vg^{no}), and *vestigial-nipped* (vg^{np}), to examine the response in the morphogenesis of the wing discs to high temperature. In cultures at 25°C, the posterior circular epithelial cell mass extended inward and formed a pouch in the center of the wing-forming portion after 48 h. The size of the pouch of the wing discs from different *vestigial* alleles had the following sequence:

$$vg^+ > vg^{np} > vg^{no} > vg$$

When these wing discs were cultured at 31°C, the size of the wing pouches formed in the wing discs of each strain had the following different sequence:

$$vg^+ > vg^{no} > vg > vg^{np}$$

The wing pouches of the vp^{np} strain formed at 31°C showed a marked response to become smaller than those cultured at 25°C. The mechanism by which the size of the wing pouch of the different alleles is affected differently by high temperature is not known.

The morphogenetic effects of ecdysterone on the wing discs were examined in the cabbage armyworm, *Mamestra brassicae* L., by Agui and Fukaya.[60] They cultured larval wing discs in Grace's[21] culture medium. Wing discs developed rapidly at high concentration (0.15 to 30 μg/ml) of ecdysterone, but they degenerated early. At low concentration (0.03 μg/ml) of ecdysterone, they developed slowly, and visible morphogenetic changes occurred about 9 days after the onset of the culture. The activities of 5 ecdysterone analogs on morphogenesis of wing discs were compared and found that their activities ranged in the order of ponasterone A = cyasteron = ecdysteron >

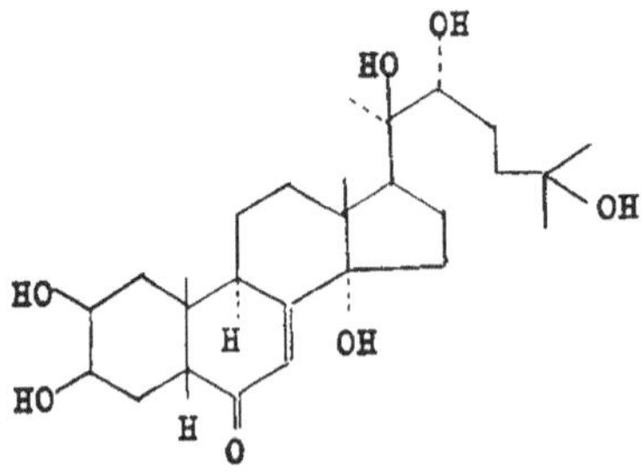

Ecdysterone

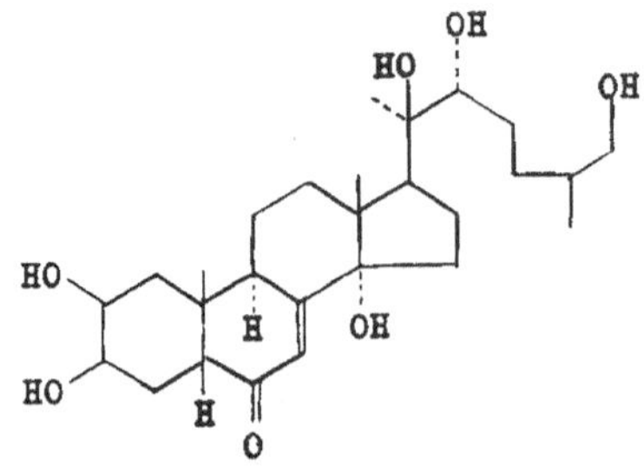

Inokosterone

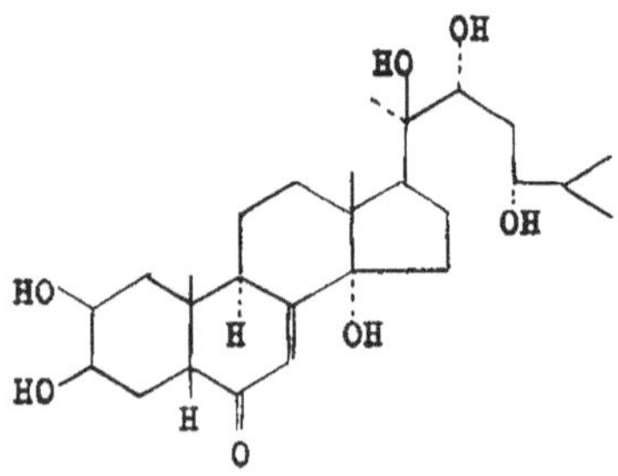

Ponasterone C

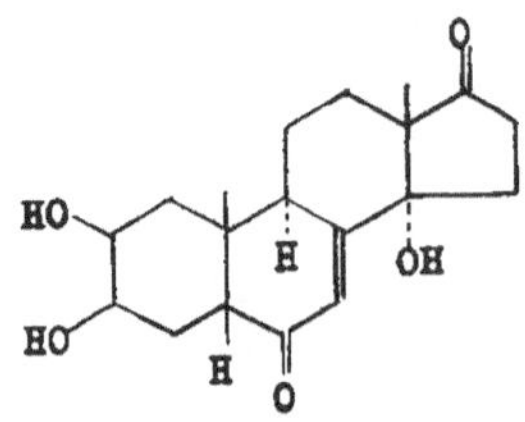

Rubrosterone

FIGURE 1. Chemical structures of ecdysone analogs.

TABLE 5
Effect of Ecdysone Analog on the Differentiation of Eye-Antennal Discs in Organ Culture

Substances	Concentration (mg/ml)	No. of explants tested	No. of explants in which ommatidia differentiated	Percent of differentiation
Control	0	16	0	0
Ecdysterone	10.0	16	15	94
	1.0	8	7	88
	0.1	6	2	33
	0.01	7	2	27
Inokosterone	1.0	3	3	100
	0.1	9	4	44
	0.01	8	3	38
Ponasterone C	1.0	8	5	63
	0.1	8	2	25
Rubrosterone	10.0	8	6	75
	1.0	20	14	70
	0.1	12	11	92
	0.01	11	10	91
	0.001	12	11	92
	0.0001	13	12	92
	0.00001	11	10	91
	0.000001	12	6	50
	0.0000001	12	5	42
	0.00000001	9	2	22

From Kuroda, Y., *Jpn. J. Genet.*, 44, Suppl. 1, 42, 1969. With permission.

ecdysone = rubrosterone. The order of activities of ecdysterone analogs on the wing discs of *Mamestra* was different from that on the eye-antennal discs of the previously described *Drosophila*. In cultures of wing discs of fleshfly, *Sarcophaga peregrina*, Ohmori and Ohtaki[61] reported the ecdysterone and ponasterone A had stimulating effects on the development and the synthesis of DNA, RNA, and protein in the wing discs cultured.

B. DEVELOPMENT IN WHOLE EMBRYO CULTURES

Takami[62,63] cultured whole embryos from the silkworm, *Bombyx mori*, to examine the effects of egg extracts on development of diapausing embryos. Nondiapausing embryos developed well beyond the appendage formation in salt solution containing nondiapausing egg extract. Fully diapausing embryos did not grow in the same medium. The extract from fully diapausing eggs was less effective in promoting the development of whole embryos. The addition of yeast extract, lactalbumin hydrolysate, dextran, tryptophan, cysteine, glutathione, and ascorbic acid to inactive medium promoted the growth of embryos (Takami[64]). It is of interest that fully diapausing embryos developed well, when they were cocultured with nondiapausing embryos side by side in the same hanging drop of culture medium (Takami[65]). When nondiapausing embryos were decapitated before differentiation of neuroblasts, or cultured before differentiation of the prothoracic glands, the brainless embryos detached the cuticle. It was suggested that the "molt" of cultured embryos does not require any hormonal action of the brain in the silkworm, at least for its initiation (Takami[66]).

C. DIFFERENTIATION OF EMBRYONIC CELLS

Mitsuhashi[67] cultured tissues from diapausing larvae of the rice stem borer, *Chilo suppressalis*, in a physiological salt solution supplemented with lactalbumin hydrolyzate, Bacto-peptone, choline chloride, TC-yeastolate, TC-199 medium, and fetal bovine serum. From the nervous systems, three types of cells were obtained. Heart and alary muscles also produced three types of migrating cells. Alimentary canals, Malpighian tubules, fat bodies, and silk glands each liberated one type of cell.

Kuroda[68] cultured cells from lethal embryos of *deep orange (dor)* of *Drosophila melanogaster* in medium K-17 (Table 2) supplemented with 0.1 mg/ml fetuin and 15% fetal bovine serum. The pulsation of muscle cells, the maturation of epithelial cells, the growth of fibroblastic cells, and the extension and branching of nerve fibers of nerve cells were normally observed as those in cultures of wild-type embryonic cells. However, the syncytium formation of muscle cells, growth of epithelial cells, the formation of cellular spheres, and droplet (secreted granules) formation on the nerve fibers of nerve cells were not observed (Figure 2, Table 6) (Kuroda[69,70]).

These results indicate that some type of cells survived and were maintained in a functionally active state for an extended period over the prospective lethal phase of *dor* embryos, although some defects in muscle cells, cellular spheres, and nerve cells were observed. The various defects observed in different cells of the *dor* embryos may be due to some single defect, such as a single alteration in the cell membrane, which may produce some defects in cell fusion of muscle cells, in cell contact to form a cellular sphere, and in secretion of droplets on nerve fibers, resulting in the pleiotrophic effects on different functions of various differentiated cells.

When *dor* embryonic cells were cultured in a medium containing the wild-type egg extract (100 eggs per ml), defects in *dor* embryonic cells were repaired. The effective substances in the wild-type egg extract lost their activity by heat treatment at 80°C for 10 min. This suggests that they are enzymes or heat-labile substances. It was found that in heterozygous embryos *(dor/+)*, the effective substances may be produced by the paternal wild-type gene between 3 and 5 h in embryonic development.

When Kuroda[71,72] cultured tissues and cells from *Drosophila* embryos in medium containing 20-hydroxyecdysone (1 to 10 μg/ml), some characteristic adult tissues and structures developed (Figure 3). Among them the structures which were very similar to the imaginal disc (Figure 3a) and antennal disc (Figure 3b) were remarkable. The eye disc-like structures

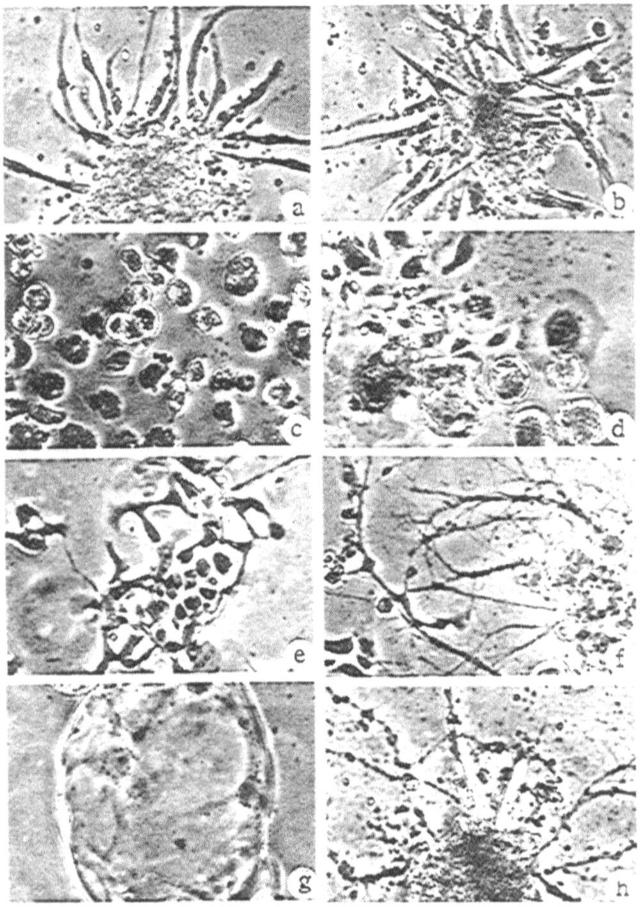

FIGURE 2. Various types of cells from *dor* lethal embryos of *Drosophila melanogaster.* (a) Muscle cells after cultivation for 4 days; (b) muscle cells pulsating synchronously after cultivation for 34 days; (c) epithelial cells with many cytoplasmic granules after cultivation for 3 days; (d) epithelial cells showing chitinous pigmentation after cultivation for 7 days; (e) fibroblastic cells after cultivation for 3 days; (f) nerve cells with network of nerve fibers after cultivation for 8 days; (g) a cellular sphere after cultivation with wild-type egg extract after cultivation for 10 days; (h) nerve fibers with deposits of small droplets on them after cultivation with wild-type egg extract after cultivation for 10 days. (From Kuroda, Y., *Nature,* 252, 40, 1974b. With permission.)

(Figure 3c) also developed with a regular arrangement of ommatidium-forming cells. The bristle (Figure 3d) and tracheal structures (Figure 3e) also developed very well.

The most striking example of the development of adult organs was the leg-like structures (Figure 3f). After 1 day of cultivation, small leg disc-like buds appeared, then they extended gradually, and finally every segment of the adult leg developed. They consisted of segments that corresponded to the adult segments of the leg, such as the coxa, the femur, the tibia, and the tarsus. At the tip of the tarsus, claw-like structures also developed. However, the leg-like structures formed in culture did not contain muscle and nerve cells, but consisted only of cuticle. This may be due to the experimental conditions under which the embryonic tissue fragments separated from other fragments containing muscle and nerve cells.

In the absence of ecdysterone, cells from post-gastrula embryos of *D. melanogaster* differentiated into various embryonic or larval-type cells. Electron microscopy revealed the formation of typical myofibrillar structure in muscle cells (Figure 4a).[73,74]

In epithelial cells, interdigitations, gap junctions, and desmosomes were formed between adjacent cells (Figure 3d). In nerve cells, neurofilaments and neurotubules were formed

TABLE 6
Growth and Differentiation of Various Types of Embryonic Cells from *Drosophila melanogaster* In Vitro

Cell type	Growth and differentiation	Embryonic cells from		
		Wild-type	*deep orange* without WEE[a]	*deep orange* with WEE
Muscle cells	Pulsation	+	+	+
	Syncytium formation	+	–	±
Epithelial cells	Monolayer sheet	+	±	+
	Maturation	+	+	+
Fibroblastic cells	Monolayer sheet	+	+	+
Cellular spheres	Sphere formation	+	–	+
Small cells	Monolayer sheet	+	–	–
	Regular arrangement	+	–	–
Nerve cells	Fiber extension	+	+	+
	Fiber branching	+	+	+
	Droplet formation	+	–	+

[a] Wild-type egg extract.

From Kuroda, Y., *Nature,* 252, 40, 1974b. With permission.

(Figure 3b). In the presence of ecdysterone, cells differentiated into various adult structures. The formation of trachea structure was marked (Figure 5).

D. DEVELOPMENT OF GERM CELLS

Mitsuhashi[67] cultured tissues from diapausing larvae of the rice stem-borer. Male and female gonads gave rise to three or four types of cells, respectively. Spermatocytes developed into spermatids in a medium which contained neither insect hemolymph nor purified insect hormone. Yagi et al.[75] cultured the testes from diapausing rice stem borer larvae in medium CSM-2F (Mitsuhashi[76]), lacking any insect blood. Morphological changes of male germ cells were examined, and the authors found that the spermatogenesis was induced in the testes a long time after the onset of cultivation. The effect of ecdysterone on cultured testes was examined. Ecdysterone promoted the development of testes and spermatocytes from diapausing larvae.

Kuroda[77] cultured testicular fragments obtained from 48-h pupae of *Drosophila melanogaster* in medium K-17 supplemented with 0.1 mg/ml fetuin and 15% fetal calf serum. Under these conditions, without any added ecdysterone, spermatogonia in the anterior fragments showed no, or only slight, changes. In the middle fragments, which contained germ cells at more advanced stages of spermatogenesis, a slight increase in the number of spermatocytes or spermatids was observed. In the posterior fragments, which contained sperm at early stages of spermiogenesis, there were marked elongation of the sperm bundles along their long axis.

The result suggested the possibility that there may be some differences in hormone dependence among germ cells at various stages of spermatogenesis. The process of growth in spermatogenesis to form spermatocytes may strictly require the hormonal stimulation. As the germ cell differentiates, the requirement for hormone may decrease. The process of spermiogenesis may be completely independent of hormones.

The single cyst containing mature spermatid was isolated from the middle part of the testis. It was cultured in medium K-17 with fetuin and fetal calf serum. After cultivation for 16 h, the apical part of the cyst began to coil. The elongation of the tail part continued after cultivation for 20 h. In further cultivation, the tail of spermatids extended considerably and the mature form of sperm was produced after cultivation for 48 h.

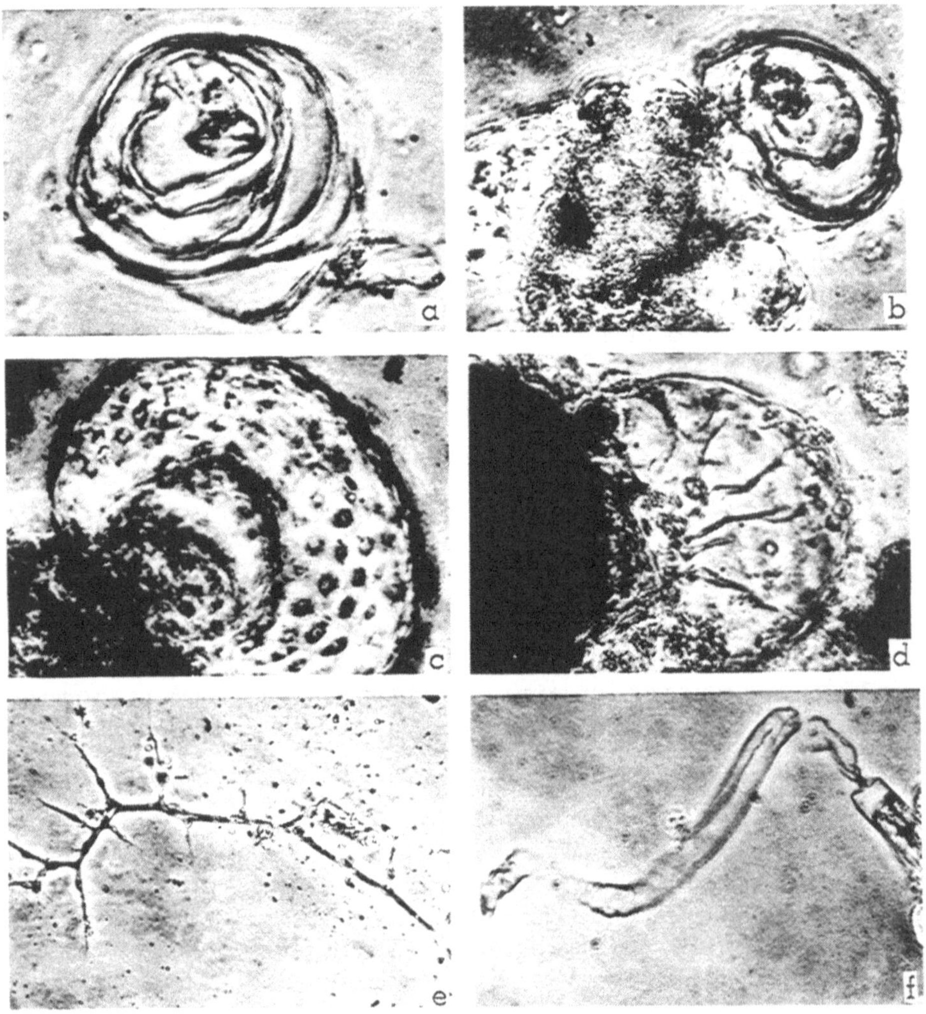

FIGURE 3. Adult structures developed from *Drosophila* embryonic cells in culture. (a) Wing disc-like structure; (b) antenna disc-like structure; (c) eye disc-like structure; (d) bristle-like structure; (e) trachea-like structure; (f) leg-like structure. (From Kuroda, Y., *The Ultrastructure and Functioning of Insect cells*, Akai, H., King, R. C., and Morohoshi, S., Eds., Society of Insect Cells, Tokyo, 91, 1982.)

V. USE OF INSECT CELLS FOR VIROLOGICAL RESEARCH

One of the practical and important applications of cultured insect cells is in virus studies. Initial attempts to obtain the propagation of polyhedrosis virus were carried out in primary cultures of lepidopteran cells. Later when insect cell lines became established, the stable and continuous propagation of virus was obtained in these cell lines.

In Japan, many scientists have succeeded to propagate various types of viruses in primary insect cultures or established cell lines (Table 7). Nuclear polyhedrosis virus (NPV) and cytoplasmic polyhedrosis virus (CPV) were most frequently used in lepidopteran cells in culture.

Yamada and Maramorosch[79] titrated the NPV of *Heliothis zea* in a *Heliothis zea* cell line (IPLB-1075) by using 1% mixed agarose containing a mixture of Seakem (Marine Colloids, Inc.) and Ultra (Bio Rad Laboratory) pure agarose. Visible plaques were observed 8 days post-infection. These plaques ranged in diameter from 0.5 to 2 mm. Dose-response experiments

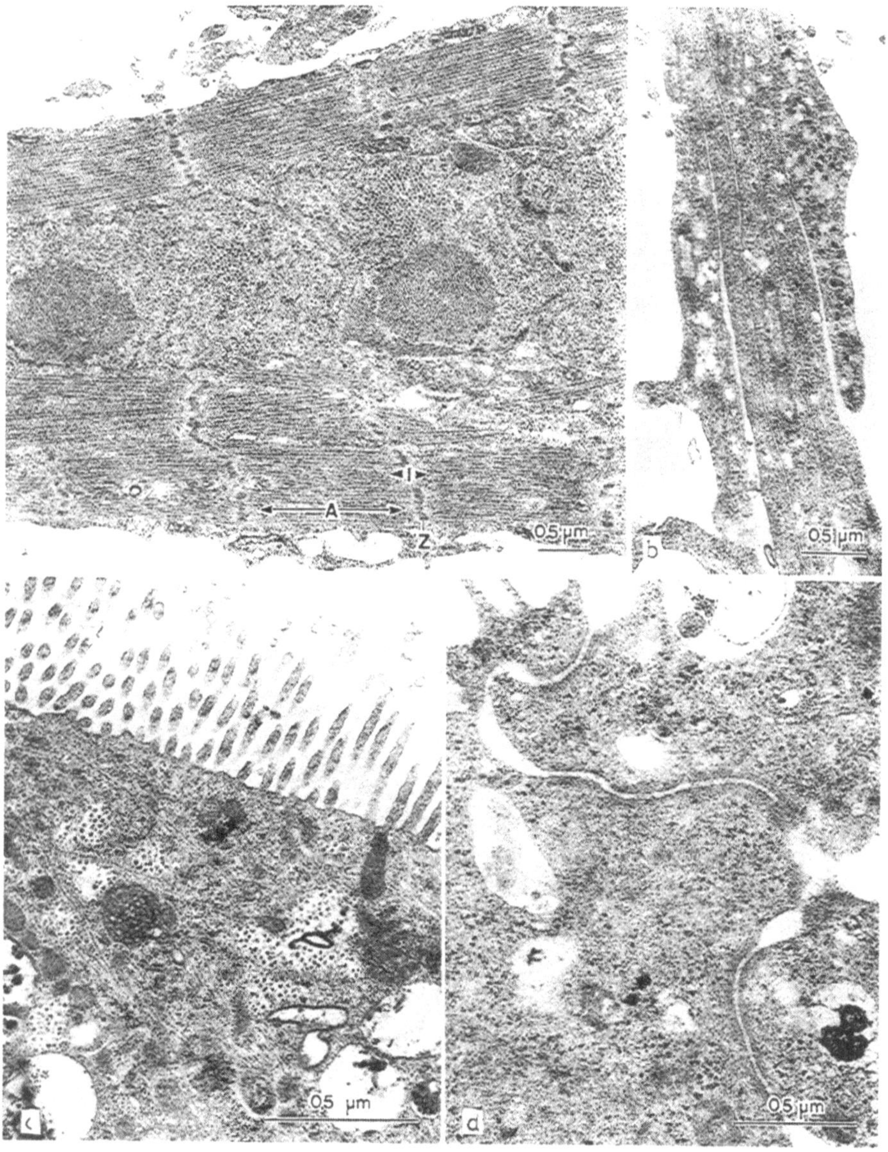

FIGURE 4. Electron micrographs of embryonic or larval-type cells differentiated *in vitro* from undifferentiated embryonic cells of *Drosophila melanogaster* in the absence of ecdysterone. (a) Muscle cells; (b) nerve fibers; (c) epithelial-type cells; (d) epithelial-type cells. (From Kuroda, Y. and Shimada, Y., *Invertebrate and Fish Tissue Culture*, Kuroda, Y., Kurstak, E., and Maramorosch, K., Eds., Japan Scientific Societies Press, Tokyo/Springer-Verlag, Berlin, 1988, 95. With permission.)

indicated that a single particle initiated the formation of a plaque. The titration of *Heliothis zea* baculovirus by this new plaque method provides an accurate technique for the determination of virus concentration.

In serial passages of *Heliothis zea* NPV in a *Heliothis zea* cell line, Yamada et al.[88] demonstrated that long-term serial passages of NPV in homologous host cell culture decreased both the total number of polyhedral inclusion bodies (PIBs) produced and the infectivity of the supernatant as measured by $TCID_{50}$.

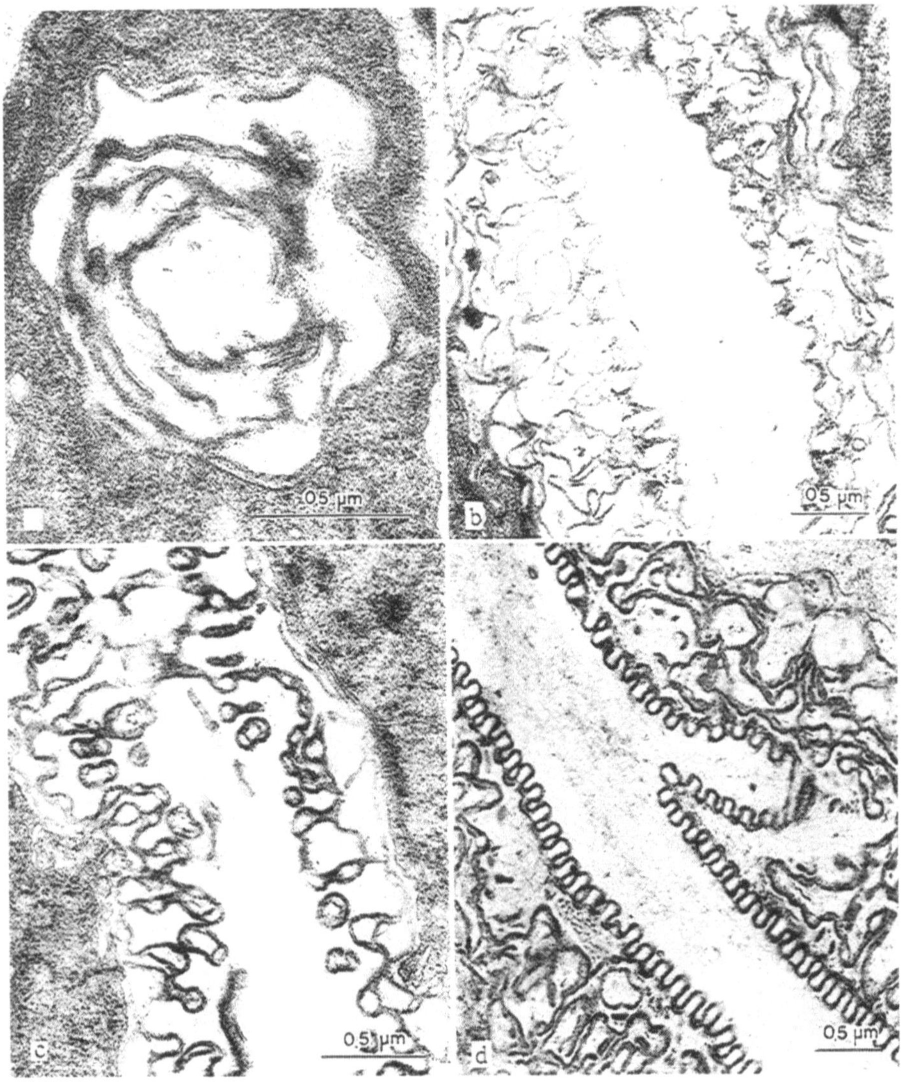

FIGURE 5. Electron micrographs of differentiating trachea in cultures of *Drosophila* embryonic cells in the presence of ecdysterone. (From Kuroda, Y. and Shimada, Y., *Invertebrate and Fish Tissue Culture,* Kuroda, Y., Kurstak, E., and Maramorosch, K., Eds., Japan Scientific Societies Press, Tokyo/Springer-Verlag, Berlin, 1988, 95. With permission.)

Igarashi and Srivastava[89] set up a laboratory-scale mass production system of high virus-yielding clone C6/36 cells of the mosquito *(Aedes albopictus)* cell line for analyzing the genome and structural protein of Japanese encephalitis (JE) virus. The cell suspension of C6/36 clone was made from the stationary culture and cultured in a spinner flask with a rotating blade at 60 rpm at 38°C.

The cells were adapted to grow in Eagle's medium in Earle's saline supplemented with 0.2 m*M* each of nonessential amino acids and 10% heat-inactivated fetal calf serum. The cells were then adapted to attach to a microcarrier. The cells were infected with JE virus and infected fluid was (1) continuously kept, or (2) daily harvested and replaced with fresh maintenance medium. The virus yield in the fluid was assayed by infectivity titration, as well

TABLE 7
Insect Viruses in Cultured Insect Cells

Virus	Culture system	Ref.
Nuclear polyhedrosis virus (NPV)		
bombyx mori virus (No, 20, 22)	Silkworm embryo culture	78
Mamestra brassicae virus	NIAS-MB-19,-25,-32 cell line	31
Heliothis zea virus	IPLB-1075 cell line	79
Bombyx mori virus	Silkworm cell line	36
Cytoplasmic polyhedrosis virus (CPV)		
Chilo iridescent virus (CIV)	Primary cultures of hemocyte cells, circulatory organs, fat bodies, brains, ovaries of rice stem borer	80
Bombyx mori virus	Silkworm embryo culture	81
Chilo iridescent virus (CIV)	Primary cultures of embryos of green rice leafhopper and small brown planthopper	82
Plant virus		
Leafhopper wound	Leafhopper cells	83
tumor virus		84
Rice dwarf virus	Green rice leafhopper embryo cells	85, 86
Aster yellows "virus" (MLO)	Leafhopper embryo cells	87

as the peak height of the purified virion in the sucrose gradient. It was found that the infective virus yield was almost the same in both methods.

Substances and factors for controlling the growth and the infectivity of NPV were investigated. Funakoshi and Aizawa[90] examined the inhibitory effects of metabolic and antimicrobial inhibitors on a NPV of the silkworm, *Bombyx mori,* by plaque assay method using insect cell lines, BM-n and S.P.C. Bm36. Some drugs decreased the infectivity of virus. However, the therapeutic effect of these drugs has not been demonstrated except for tannic acid and Actinomycin A. Several drugs, such as Tunicamycin (0.5 g/ml), phosphonoacetic acid (3.5 m*M*), and trisodium phosphonoformate (3.5 m*M*), remarkably inhibited the replication of *Bombyx mori* NPV.

The NPV infectivity was enhanced by a factor present in the capsule of an armyworm granulosis virus (GV). Hukuhara and Zhu[91] purified a synergistic factor (SyF) from the capsules of the Hawaiian strain of *Pseudaletia unipuncta* granulosis virus. They examined the enhancing effects of the SyF on the infection of *Spodoptera frugiperda* cells with *Trichoplusia ni* multiple nucleocapsid NPV. Specific binding of the SyF on the cell surface and on the envelope of the virus was visualized by immunofluorescent microscopy and electron microscopy using colloidal immuno-gold markers. The SyF had no specific affinity to cell lines from *Bombyx mori, T. ni,* and *Leucania separata* nor to polyhedra-derived virions of *B. mori* NPV. Anti-SyF antibodies and chloroquine inhibited the activity of the SyF.

The transposon *copia* in genomic DNA of *Drosophila melanogaster* could produce the retrovirus-like particles (RLVP), when transfected into the *copia*-free cell line, KUN-DH-33, of *D. hydei.* This suggests the causal relation between the transposon *(copia)* and the retrovirus or RVLP. Miyake et al.[92] isolated the genomic *copia* of *D. simulans* and analyzed the structure and ability to produce RVLP. The *copia* isolated were classified in three types, based on their structure. Among these three types, the longest one (5 kb) could produce RVLP, while the middle one (3 kb) could not. The RVLP produced by 5-kb *simulans copia* in the transfections were very similar to those produced by a *melanogaster copia* in various aspects such as size, morphology, and chemical constitution. They were similar also to RVLP spontaneously appearing in *melanogaster* cell lines and *simulans* cell lines, except for the activity of reverse transcriptase.

VI. CONCLUSION

Many Japanese scientists have contributed to the development of invertebrate tissue culture in various fields of cell biology, developmental biology, physiology, endocrinology, pathology, virology, and microbiology. The devices and development of new technology for culturing new materials were carried out by Japanese scientists by finding procedures for rearing new materials under aseptic conditions, the device of new culture media, the attempts to use new culture vessels, and the specific growth factor(s) for invertebrate cells.

In this chapter, limited numbers of findings and results, published by prominent Japanese scientists, were described. It has to be emphasized that these important achievements by Japanese scientists contributed to international close cooperation with many scientists in foreign countries.

ACKNOWLEDGMENTS

The author gratefully acknowledges the invitation by the editors to contribute to this book. Thanks are also due to Miss Yuko Takada for her assistance in the preparation of the manuscript.

REFERENCES

1. **Goldschmidt, R.** Some experiments on spermatogenesis *in vitro*. *Proc. Natl. Acad. Sci. U.S.A.*, 1, 220, 1915.
2. **Trager, W.** Multiplication of the virus of grasserie in silkworm tissue culture. *J. Exp. Med.*, 61, 501, 1935.
3. **Kuroda, Y. and Tamura, S.** The tissue culture of tumors in *D. melanogaster*. I. The nature of the melanotic tumors and the effect of the metamorphic hormone upon the melanotic growth of tumors. *Zool. Mag.*, 64, 380, 1955.
4. **Takami, T.** *In vitro* culture of embryos in the silkworm, *Bombyx mori* L. I. Culture with silkworm egg extract in special reference to some characteristics of the diapausing egg. *Bull. Seric. Exp. Stn.*, 14, 577, 1957.
5. **Hirumi, H. and Maramorosch, K.** The *in vitro* cultivation of embryonic leafhopper tissues. *Exp. Cell Res.*, 36, 625, 1964.
6. **Mitsuhashi, J. and Maramorosch, K.**, Leafhopper tissue culture: embryonic, nymphal and imaginal tissues from aseptic insects. *Contrib. Boyce Thompson Inst.*, 22, 435, 1964.
7. **Kitamura, S.** The *in vitro* cultivation of tissues from the mosquito, *Culex pipiens* var. *molestus*. II. An improved culture medium useful for ovarian tissue culture. *Kobe J. Med. Sci.*, 11, 23, 1965.
8. **Machii, A.** Establishment of cell line from valiger larva of the pearl oyster, *Pinctada fucata*. *Tissue Cult. Res. Commun.*, 4, 18, 1985.
9. **Machii, A., Wada, K. T., Townsley, S. J., Sasaki, K., and Awaji, M.** Tissue culture of some marine invertebrates. Application of invertebrate cells *in vitro*. Satel, Sym. 3rd Internat. Cell Culture Congrees, Sendai, Japan, Sept. 14, 1985.
10. **Morgan, J. F., Morton, H. J., and Parker, R. C.** Nutrition of animal cells in tissue culture. I. Initial studies on a synthetic medium. *Proc. Soc. Exp. Biol. Med.*, 73, 1, 1950.
11. **Carlson, J. G.** Protoplasmic viscosity changes in different regions of the grasshopper neuroblast during mitosis. *Biol. Bull.*, 90, 109, 1946.
12. **Wigglesworth, V. B.** *The Principles of Insect Physiology*. Metheun & Co., London, 277, 1960.
13. **Kuroda, Y.** Differentiation of ommatidium-forming cells of *Drosophila melanogaster* in organ culture. *Exp. Cell Res.*, 59, 429, 1970.
14. **Kuroda, Y.** Studies on *Drosophila* embryonic cells *in vitro*. I. Characteristics of cell types in culture. *Dev. Growth Differ.*, 16, 55, 1974.
15. **Kuroda, Y.** Spermatogenesis in pharate adult testes of *Drosophila* in tissue cultures without ecdysones. *J. Insect Physiol.*, 20, 637, 1974.
16. **Schneider, I.**, Differentiation of larval *Drosophila* eye-antennal discs *in vitro*. *J. Exp. Zool.*, 156, 91, 1964.
17. **Mitsuhashi, J.** *In vitro* cultivation of the embryonic tissues of the green rice leafhopper, *Nephotettix cincticeps* Uhler (Homoptera: Cicadellidae). *Jpn. J. Appl. Entomol. Zool.*, 9, 107, 1965.
18. **Mitsuhashi, J.** Primary culture of the hemocyte-poietic tissue of *Papilio xuthus* Linne. *Appl. Entomol. Zool.*, 7, 39, 1972.
19. **Mitsuhashi, J.** Requirements of amino acids by insect cell lines. In *Invertebrate Systems In Vitro*, Kurstak, E., Maramorosch, K., and Dübendorfer, A., Elsevier, Amsterdam, 47, 1980.

20. **Mitsuhashi, J.** Establishment and some characteristics of a continuous cell line derived from fat bodies of the cabbage armyworm (Lepidoptera, Noctuidae). *Dev. Growth Differ.*, 23, 63, 1981.
21. **Grace, T. D. C.** Establishment of four strains of cells from insect tissues grown *in vitro*. *Nature*, 195, 788, 1962.
22. **Mitsuhashi, J.** Simplification of media and utilization of sugars by insect cells in cultures. In *Invertebrate and Fish Tissue Culture*, Kuroda, Y., Kurstak, E., and Maramorosch, K., Eds., Japan Sci. Soc. Press and Springer-Verlag, Berlin, 15, 1988.
23. **Kitamura, S.**, Establishment of cell line from *Culex* mosquito. *Kobe J. Med. Sci.*, 16, 41, 1970.
24. **Kitamura, S., Imai, T., and Grace, T. D. C.** Adaptation of two mosquito cell lines to medium free of calf serum. *J. Med. Entomol.*, 10, 488, 1973.
25. **Machii, A. and Wada, T.** Some marine invertebrate tissue culture. In *Invertebrate Cell System Applications*, Vol. 2, Mitsuhashi, J., Ed., CRC Press, Boca Raton, FL, 225, 1989.
26. **Horikawa, M., Ling, L. N., and Fox, A. S.** Long-term culture of embryonic cells of *Drosophila melanogaster*. *Nature*, 210, 183, 1966.
27. **Mitsuhashi, J.** Establishment of an insect cell strain persistently infected with an insect virus. *Nature*, 215, 863, 1967.
28. **Kitamura, S.** Establishment of cell line from *Culex* mosquito. *Kobe J. Med. Sci.*, 16, 41, 1970.
29. **Mitsuhashi, J.** Subculturing of cells from the embryo of the green rice leafhopper, *Nephotettix cincticeps* Uhler (Hemiptera, Deltocephalidae). *Annot. Zool. Jpn.*, 48, 139, 1971.
30. **Mitsuhashi, J.** Establishment of cell lines from pupal ovaries of the swallow tail, *Papilio xuthus* Linne (Lepidoptera: Papilionidae). *Appl. Entomol. Zool.*, 8, 64, 1973.
31. **Mitsuhashi, J.** Establishment and characterization of continuous cell lines from pupal ovaries of the cabbage armyworm, *Mamestra brassicae* (Lepidoptera, Noctuidae). *Dev. Growth Differ.*, 19, 337, 1977.
32. **Takahashi, M., Mitsuhashi, J., and Ohtaki, T.** Establishment of a cell line from embryonic tissues of the fleshfly, *Sarcophaga peregrina* (Insecta: Diptera). *Dev. Growth Differ.*, 22, 11, 1980.
33. **Mitsuhashi, J.** Establishment and some characteristics of continuous cell line derived from fat bodies of the cabbage armyworm (Lepidoptera, Noctuidae). *Dev. Growth Differ.*, 23, 63, 1981a.
34. **Mitsuhashi, J.** A new continuous cell line from larvae of the mosquito *Aedes albopictus* (Diptera, Culicidae). *Biomed. Res.*, 2, 599, 1981b.
35. **Mitsuhashi, J.** A continuous cell line derived from fat bodies of the common armyworm, *Leucania separata* (Lepidoptera: Noctuidae) *Appl. Entomol. Zool.*, 18, 533, 1983.
36. **Inoue, H. and Mitsuhashi, J.** A *Bombyx mori* cell line susceptible to a nuclear polyhedrosis virus. *J. Seric. Sci. Jpn.*, 53, 108, 1984.
37. **Mitsuhashi, J.** Isolation of a continuous cell line from larval fat bodies of an arctiid moth, *Spilarctia seriatopunctata* (Insecta, Lepidoptera, Arctiidae). *Zool. Sci.*, 1, 415, 1984.
38. **Mitsuhashi, J. and Shozawa, A.** Continuous cell lines from larval hemocytes of the cabbage armyworm, *Mamestra brassicae*. *Dev. Growth Differ.*, 27, 599, 1985.
39. **Inoue, H. and Mitsuhashi, J.** Further establishment of continuous cell lines from larval fat bodies of the cabbage armyworm, *Mamestra brassicae* (Lepidoptera: Noctuidae). *Appl. Entomol. Zool.*, 20, 496, 1985.
40. **Inoue, H. and Mitsuhashi, J.** Characterization of a new continuous cell line from silkworm *(Bombyx mori)* embryos. *Appl. Entomol. Zool.*, 23, 8, 1988.
41. **Mitsuhashi, J. and Inoue, H.** Obtainment of a continuous cell line from the larval fat bodies of the mulberry tiger moth, *Spilosoma imparilis* (Lepidoptera: Arctiidae). *Appl. Entomol. Zool.*, 23, 488, 1988.
42. **Ninaki, O., Nagayasu, K., Kameoka, H., Fujiwara, H., Miyajima, N., Takada, N., and Maekawa, H.** Establishment of cell lines from *Bombyx mori* and *B. mandarina*, and attempt of cell fusion. In *Invertebrate and Fish Tissue Culture*, Kuroda, Y., Kurstak, E., and Maramorosch, K., Eds., Japan Scientific Societies Press, Tokyo/Springer-Verlag, Berlin, 243, 1988.
43. **Ui, K., Ueda, R., and Miyake, T.** Continuous cell lines from imaginal discs of *Drosophila melanogaster*. In *Invertebrate and Fish Tissue Culture*, Kuroda, Y., Kurstak, E., and Maramorosch, K., Eds., Japan Scientific Societies Press, Tokyo/Springer-Verlag, Berlin, 251, 1988.
44. **Imanishi, S., Sato, S., and Mori, T.** Characteristics of cell lines derived from embryos in the silkworm, *Bombyx mori*. In *Invertebrate and Fish Tissue Culture*, Kuroda, Y., Kurstak, E., and Maramorosch, K., Eds., Japan Scientific Societies Press, Tokyo/Springer-Verlag, Berlin, 255, 1988.
45. **Kimura, I.** Establishment of cell lines of vector insects, and infection of these cell monolayers with rice dwarf virus (RDV). In *Invertebrate and Fish Tissue Culture*, Kuroda, Y., Kurstak, E., and Maramorosch, K., Eds., Japan Scientific Societies Press, Tokyo/Springer-Verlag, Berlin, 274, 1988.
46. **Mitsuhashi, J.** A continuous cell line from a beetle, *Anomala costata* Hope (Coleoptera, Scarabaeidae). *J. Jpn. Forest Soc.*, 71, 491, 1989.
47. **Kitamura, S., Imai, T., and Grace, T. D. C.** Adaptation of two mosquito cell lines to medium free of calf serum. *J. Med. Entomol.* 10, 488, 1973.

48. **Mitsuhashi, J.** Continuous cultures of insect cell lines in media free of sera. *Appl. Entomol. Zool.*, 17, 575, 1982.
49. **Kuroda, Y. and Tamura, S.** The tissue culture of tumors in *Drosophila melanogaster.* II. The effects of phenylthiocarbamide upon the melanotic growth of tumors. *Zool. Mag.*, 65, 219, 1956a.
50. **Kuroda, Y. and Tamura, S.** The tissue culture of tumors in *Drosophila melanogaster.* III. Effects of Cu-ion upon the melanotic growth of tumors. *Zool. Mag.*, 65, 301, 1956b.
51. **Kuroda, Y. and Tamura, S.** A technique for the tissue culture of melanotic tumors of *Drosophila melanogaster* in the synthetic medium. *Med. J. Osaka Univ.*, 7, 137, 1956c.
52. **Kuroda, Y. and Tamura, S.** The tissue culture of tumors in *Drosophila melanogaster.* IV. The effects of Fe^{+++} upon the melanotic growth of tumors. *Zool. Mag.*, 66, 6, 1957.
53. **Kuroda, Y. and Yamaguchi, K.** The effects of the cephalic complex upon the eye discs of *Drosophila melanogaster. Jpn. J. Genet.*, 31, 97, 1956.
54. **Horikawa, M.** Developmental-genetic studies of tissue-cultured eye-antennal discs of *Drosophila melanogaster.* II. Effects of the metamorphic hormone (cephalic complex) upon growth and differentiation of eye-antennal discs, and strain differences in relation to the metamorphic hormone. *Jpn. J. Genet.*, 35, 76, 1960.
55. **Fujio, Y.** Studies on development of eye-antennal discs of *Drosophila melanogaster* in tissue culture. I. Effects of the facet-increasing substances upon the growth and differentiation of eye antennal discs. *Jpn. J. Genet.*, 35, 361, 1960.
56. **Fujio, Y.** Studies on development of eye-antennal discs of *Drosophila melanogaster* in tissue culture. II. Effects of substances secreted from cephalic complexes upon eye-antennal discs of eye-mutant strains. *Jpn. J. Genet.*, 37, 110, 1962.
57. **Kuroda Y.** Growth and differentiation of embryonic cells of *Drosophila melanogaster in vitro. Jpn. J. Genet.*, 44, Suppl. 1, 42, 1969.
58. **Kuroda, Y.** Differentiation of ommatidium-forming cells of *Drosophila melanogaster* in organ culture. *Exp. Cell Res.*, 59, 429, 1970.
59. **Kuroda, Y.** The tissue culture of the wing discs in the *vestigial* series of *Drosophila melanogaster. Med. J. Osaka Univ.*, 10, 1, 1959.
60. **Agui, N. and Fukaya, M.** Effects of moulting hormones and prothoracic glands on the development of wing discs of the cabbage armyworm (*Memestra brassicae* L.) *in vitro* (Lepidoptera: Noctuidae). *Appl. Entomol. Zool.*, 8, 73, 1973.
61. **Ohmori, K. and Ohtaki, T.** Effects of ecdysone analogues on development and metabolic activity of wing discs of the fleshfly, *Sarcophaga peregraina, in vitro. J. Insect Physiol.*, 19, 1199, 1973.
62. **Takami, T.** *In vitro* culture of embryos in the silkworm, *Bombyx mori* L. I. Culture with silkworm egg extract in special reference to some characteristics of the diapausing egg. *Bull. Seric. Exp. Stn.*, 14, 577, 1957.
63. **Takami, T.** *In vitro* culture of embryos in the silkworm. I. Culture in silkworm egg extract with special reference to some characteristics of the diapausing egg. *J. Exp. Zool.*, 35, 286, 1958.
64. **Takami, T.** *In vitro* culture of embryos in the silkworm, *Bombyx mori* L. II. The effects of extract from inactive eggs on the cultured embryo. *Bull. Sericult. Exp. Stn.*, 15, 477, 1959a.
65. **Takami, T.** Induced growth of diapausing silkworm embryos *in vitro. Science*, 130, 98, 1959b.
66. **Takami, T.** *In vitro* culture of embryos in the silkworm, *Bombyx mori* L. III. Mount of brainless embryos in culture. *J. Exp. Biol.*, 40, 735, 1963.
67. **Mitsuhashi, J.** Tissue culture of the rice stem borer, *Chilo suppressalis* Walker (Lepidoptera: Pyralidae). I. Cell migration from explanted tissues of diapausing larvae. *Jpn. J. Appl. Entomol. Zool.*, 9, 217, 1965a.
68. **Kuroda, Y.** Studies on *Drosophila* embryonic cells *in vitro.* I. Characteristics of cell types in culture. *Dev. Growth Differ.*, 16, 55, 1974a.
69. **Kuroda, Y.** *In vitro* activity of cells from genetically lethal embryos of *Drosophila. Nature*, 252, 40, 1974b.
70. **Kuroda, Y.** Studies on *Drosophila* embryonic cells *in vitro.* II. Tissue- and time-specificity of a lethal gene, *deep orange. Dev. Growth Differ.*, 19, 57, 1977.
71. **Kuroda, Y.** Differentiation of adult structures from *Drosophila* embryonic cells in culture. In *The Ultrastructure and Functioning of Insect Cells*, Akai, H., King, R. C., and Morohoshi, S., Eds., Society of Insect Cells Japan, Tokyo, 91, 1982.
72. **Kuroda, Y.** Differentiation of adult structures in cultures of embryonic tissues from *Drosophila melanogaster.* In *Techniques in In Vitro Invertebrate Hormones and Genes, Techniques in the Life Sciences*, Cell Biology, C2, Elsevier, Ireland, C215/1, 1986.
73. **Kuroda, Y. and Shimada, Y.** Differentiation of embryonic cells of *Drosophila* studied with electron microscope. In *Invertebrate and Fish Tissue Culture*, Kuroda, Y., Kurstak, E., and Maramorosch, K., Eds., Japan Scientific Societies Press, Tokyo/Springer-Verlag, Berlin, 95, 1988.
74. **Kuroda, Y. and Shimada, Y.** Electron microscopic studies on *in vitro* differentiated cells from *Drosophila* embryos. In *Invertebrate Cell System Applications*, Vol. 1, Mitsuhashi, J., Ed., CRC Press, Boca Raton, FL, 77, 1989.

75. **Yagi, S., Kondo, E., and Fukaya, M.** Hormonal effect on cultivated insect tissues. I. Effect of ecdysterone on cultivated tissues of diapausing rice stem borer larvae (Lepidoptera: Pyralidae). *Appl. Entomol. Zool.*, 4, 70, 1969.
76. **Mitsuhashi, J.** Tissue culture of the rice stem borer, *Chilo suppressalis* Walker (Lepidoptera: Pyralidae). III. Effects of temperature and cold-storages on the multiplication of the cell line from larval hemocytes. *Appl. Entomol. Zool.*, 3, 1, 1968.
77. **Kuroda, Y.** Spermatogenesis in pharate adult testes of *Drosophila* in tissue cultures without ecdysones. *J. Insect Physiol.*, 20, 637, 1974c.
78. **Takami, T., Sugiyama, H., Kitazawa, T., and Kanda, T.** Infection of cultured silkworm embryos with nuclear-polyhedrosis virus. *Jpn. J. Appl. Entomol. Zool.*, 10, 197, 1966.
79. **Yamada, K. and Maramorosch, K.** Plaque assay of *Heliothis zea* baculovirus employing a mixed agarose overlay. *Arch. Virol.*, 67, 187, 1981.
80. **Mitsuhashi, J.** Multiplication of *Chilo iridescent* virus in the *Chilo suppressalis* tissues cultivated *in vitro* (Lepidoptera: Pyralidae). *Appl. Entomol. Zool.*, 1, 199, 1966.
81. **Takami, T. Sugiyama, H., Kitazawa, T., and Kanda, T.** Infection of cultured silkworm embryos with the cytoplasmic-polyhedrosis virus. *Jpn. J. Appl. Entomol.*, 11, 182, 1967.
82. **Mitsuhashi, J.** Infection of leafhopper and its tissues cultivated *in vitro* with *Chilo* iridescent virus. *J. Invertebr. Pathol.*, 9,432, 1967.
83. **Shikata, E., Orenski, S. W., Hirumi, H., Mitsuhashi, J., and Maramorosch, K.** Electron micrographs of wound-tumor virus in an animal host and in a plant tumor. *Virology*, 23, 441, 1964.
84. **Hirumi, H. and Maramorosch, K.** Recovery of aster yellows virus from various organs of the insect vector, *Macrosteles fascifrons. Contrib. Boyce Thompson Inst.*, 22, 141, 1963.
85. **Mitsuhashi, J.** Preliminary report on the plant virus multiplication in the leafhopper vector cells grown *in vitro. J. Appl. Entomol. Zool.*, 9, 137, 1965b.
86. **Mitsuhashi, J. and Nasu, S.** An evidence for the multiplication of rice dwarf virus in the vector cell cultures inoculated *in vitro. Appl. Entomol. Zool.*, 2, 113, 1967.
87. **Hirumi, H. and Maramorosch, K.** Electron microscopy of wound tumor virus in cultured embryonic cells of the leafhopper *Macrosteles fascifrons*, Proc. 2nd Int. Coll. Invertebrate Tissue Culture, 203, 1967.
88. **Yamada, K., Sherman, K. E., and Maramorosch, K.** Serial passage of *Heliothis zea* singly embedded nuclear polyhedrosis virus in a homologous cell line. *J. Invertebr. Pathol.*, 39, 185, 1982.
89. **Igarashi, A. and Srivastava, A. K.** Suspension culture of *Aedes albopictus*, clone, C6/36, cells and its application to the production of flaviviruses. In *Invertebrate and Fish Tissue Culture*, Kuroda, Y., Kurstak, E., and Maramorosch, K., Eds., Japan Scientific Societies Press, Tokyo/Springer-Verlag, Berlin, 137, 1988.
90. **Funakoshi, M. and Aizawa, K.** Effect of inhibitors on the replication of a nuclear polyhedrosis virus of the silkworm, *Bombyx mori*. In *Invertebrate and Fish Tissue Culture*, Kuroda, Y., Kurstak, E., and Maramorosch, K., Eds., Japan Scientific Societies Press, Tokyo/Springer-Verlag, Berlin, 155, 1988.
91. **Hukuhara, T. and Zhu, Y.** The effect of synergistic factor on the *in vitro* infectivity on nuclear polyhedrosis viruses. In *Invertebrate and Fish Tissue Culture*, Kuroda, Y., Kurstak, E., and Maramorosch, K., Eds., Japan Scientific Societies Press, Tokyo/Springer-Verlag, Berlin 159, 1988.
92. **Miyake, T., Mae, N., Shiba, T., Kondo, S., Saigo, K., and Hattori, K.** Production of retrovirus-like particles by transposons in genome of *Drosophila simulans*. In *Invertebrate and Fish Tissue Culture*, Kuroda, Y., Kurstak, E., and Maramorosch, K., Eds., Japan Scientific Societies Press, Tokyo/Springer-Verlag, Berlin, 115, 1988.

INDEX

B

C

D

E

F

G

H

Q

R

S

T

U

V

W

X

Y